Building Evidence for Active

Asghar Zaidi • Sarah Harper
Kenneth Howse • Giovanni
Lamura • Jolanta Perek-Białas
Editors

Building Evidence for Active Ageing Policies

Active Ageing Index and its Potential

palgrave
macmillan

Editors
Asghar Zaidi
University of Southampton
Southampton, UK

Sarah Harper
University of Oxford
Oxford, UK

Kenneth Howse
University of Oxford
Oxford, UK

Giovanni Lamura
INRCA National Institute of Health and Science on Aging
Ancona, Italy

Jolanta Perek-Białas
Jagiellonian University in Kraków
Kraków, Poland

ISBN 978-981-10-6016-8 ISBN 978-981-10-6017-5 (eBook)
https://doi.org/10.1007/978-981-10-6017-5

Library of Congress Control Number: 2018930496

Cover illustration: Andrey Bandurenko / Alamy Stock Photo

Printed on acid-free paper

This Palgrave Macmillan imprint is published by Springer Nature
The registered company is Springer Nature Singapore Pte Ltd.
The registered company address is: 152 Beach Road, #21-01/04 Gateway East, Singapore 189721, Singapore

Contents

List of Figures

List of Tables

1

Towards an Evidence-Based Active Ageing Strategy

Asghar Zaidi, Sarah Harper, Kenneth Howse, Giovanni Lamura, and Jolanta Perek-Białas

1.1 Introduction

An active and healthy life remains one of the major aspirations for young and older people. This ambition has become a genuine possibility for people in European countries due to the fact that they have a life expectancy among the highest in the world, and an increasing part of their longer lives is spent in good health. While we rejoice in living longer and in better health, and with more financial security, we also query how

A. Zaidi (✉)
University of Southampton, Southampton, UK

S. Harper • K. Howse
University of Oxford, Oxford, UK

G. Lamura
INRCA National Institute of Health and Science on Aging, Ancona, Italy

J. Perek-Białas
Jagiellonian University in Kraków, Kraków, Poland

A. Zaidi (eds.), *Building Evidence for Active Ageing Policies*,
https://doi.org/10.1007/978-981-10-6017-5_1

these aspirations can be sustained, through our own behavioural responses and through public policy and institutional reforms and innovations.

The challenge for researchers is therefore to identify strategies that are effective in promoting and sustaining activity, independence and health during older ages, with the help of public policies at the national and the local level, by initiatives from civil society organisations as well as by bottom-up behavioural behaviours. One of the policy strategies that has been advocated most strongly over the past two decades is referred to as "active ageing" (WHO 2002; Walker 2002, 2009; Walker and Maltby 2012; Zaidi et al. 2017).

The active ageing strategy is predicated on the insight that, in tackling issues associated with population ageing, successful measures are those which empower older people in increasing their participation in the labour market and in social and family engagement, and recognise that independent, self-reliant, secure and healthy living is an important prerequisite. The multifaceted design of the active ageing policy discourse allows the setting of policy goals to maintain well-being and social cohesion and improve financial sustainability of public welfare systems.

What has also become clear is the necessity for a high-quality and independent evidence base, to demonstrate how ageing experience at the individual level can be combined with higher levels of activity, both paid and unpaid; improved health status, both physical and mental; and a greater degree of autonomy and self-reliance, not just in old age but during earlier phases of life.

In 2012, the EU celebrated the European Year for Active Ageing and Solidarity between Generations (EY2012). The EY2012 renewed the focus on the potential of active ageing as a policy strategy (Council of the European Union 2012). To mark this occasion, the United Nations Economic Commission for Europe (UNECE), the European Commission's Directorate General for Employment, Social Affairs and Inclusion, and European Centre for Social Welfare Policy and Research, Vienna, jointly undertook a major research project to construct a composite measure of active ageing, called the Active Ageing Index (AAI), for 27 EU countries [for further details, see Zaidi et al. (2013) and Zaidi and Stanton (2015)].

The AAI is an analytical tool for policymakers to enable them to devise evidence-informed strategies in promoting active and healthy ageing among older people. The AAI toolkit consists of the overall index, as well as gender and domain-specific indices and their constituting individual indicators. In its design, the AAI draws on the definition offered by the World Health Organisation during the second World Assembly on Ageing (WHO 2002).

The AAI aims to monitor (and compare) active ageing outcomes at different levels: international, national and subnational. It indicates the untapped potential of older people for more active participation in economic and social life and for independent living. It promotes a more active role and greater autonomy of older people in ageing societies. The AAI evidence is used for mutual learning and advocacy of most appropriate policy measures.

The AAI uses a methodology similar to the Human Development Index (HDI) of the UNDP. One of the major benefits is that the AAI enables credible comparisons between EU countries by quantifying the differential extent to which older people have realised and can further realise their potential in distinct domains that determine their active ageing experiences. The AAI also offers a breakdown by gender, thereby highlighting the specific public policy goals of reducing gender disparity in the positive experiences of ageing.

In April 2015, within the programme of the second phase of the AAI project, the United Nations Economic Commission for Europe (UNECE) and the European Commission's Directorate General for Employment, Social Affairs and Inclusion (DG EMPL) organised an International Seminar "Building an evidence base for active ageing policies: Active Ageing Index and its potential" (hereafter referred to as the AAI International Seminar) in Brussels, Belgium. The seminar included papers focusing on how to achieve better active ageing outcomes—measured, for instance, by AAI—through appropriate policies including those that address life course determinants of active and healthy ageing. The seminar brought together close to 150 researchers, civil society representatives, policymakers and other stakeholders. It provided a multidisciplinary forum for those interested in the use of the AAI and other similar research to enhance the knowledge about ageing and older people in Europe and countries across the world.

The AAI International Seminar was structured around five principal topics, which are the same as five parts of this book: (I) Use of Active Ageing Index for Policymaking, (II) Subnational Adaptations of the AAI,

(III) Comparative Analysis of Active Ageing, (IV) Methodological Improvements in Measuring Active Ageing and (V) Use of Active Ageing Index in Non-EU Countries. On each topic a workshop was held, where the authors of papers, submitted under a call for papers launched in June 2014 and selected by an Evaluation Board consisting of the editors of this book, had an opportunity to present and discuss their findings. Fourteen posters were also presented during a poster session covering all of the five above-mentioned topics and beyond.

The Evaluation Board systematically reviewed the 34 papers submitted. In consultation with the UNECE and the DG EMPL, it made the decision that 19 papers suitable for publication would be included in this edited volume, and a selection of six papers more suited for a journal paper would be published in a special AAI volume of the *Journal of Population Ageing*.[1] The papers for the book and for the *Journal* were selected on the basis of the following criteria: (A) intellectual basis of the paper, (B) logic of the analysis, (C) presentation of findings, (D) conclusions and policy implications and (E) whether further work recommended by the editors will improve the paper.

1.2 Use of AAI for Policymaking

The book starts with a set of papers which extend the AAI for the purpose of identifying policies to promote active ageing. The chapter by Dykstra and Fleischmann (Chap. 2) presented at the AAI International Seminar combined Round Four data from the European Social Survey with indicators of the Active Ageing Index and examine conditions favourable to age integration. It uses two specific measures of age integration: the prevalence of cross-age friendships and low levels of ageism. The analyses focus on both "young" (ages 18–30) and "old" (ages 70–90). Interestingly, high levels of independence, health and security in late life, and greater capacity to actively age rather than high levels of working, volunteering, caring and political engagement among the old create the greatest opportunities for age integration. They assert their findings that the phenomenon of active ageing in itself will not lead to greater age integration.

The chapter by Vidovićová (Chap. 3) compares the AAI results with the perceived roles and contributions of older people and older peoples' own preferences and prevalent role sets. The AAI highlights productive roles such as workers and/or volunteers. However, the author argues that older people are socially recognised primarily for their contributions as grandparents and providers of support for families. This is true especially when older people's own views are taken into account as they rate roles of a grandparent, partner, friend and parent of the highest value. They point to the findings that friend, parent, sibling and partner are the most prevalent roles. The discussion calls for strengthening the relative importance of the family roles in the measurement of the AAI.

Kafková starts in her chapter (Chap. 4) by asserting that the goal of active ageing policies is to improve older adults' quality of life. It is therefore important to analyse the relationship between activity and the quality of life. This is achieved by testing the connection between the AAI and subjective well-being, both at a general level and at the level of individual EU countries. The results show a significant correlation but also they point to some problematic indicators. Employment has been identified as the exception among other dimensions having a significant position in the index, but the results cast doubts on its relationship with quality of life. Putting significance on employment leads to overestimation of the position of countries which, despite considerable employment rate, are behind other countries in non-employment indicators.

Wöhrmann, Deller and Pundt (Chap. 5) present the design of the Silver Work Index which complements the AAI centring on work close to the retirement age. It is intended to support evidence-informed strategies for dealing with the challenges of an ageing workforce. The Silver Work Index will allow for a comparison between organisations and promote good practices towards a more active role for older people. This chapter describes the steps of index development, thus laying the groundwork for a new, innovative, meso-level quantitative index that would capture age-friendly employment practices in eight dimensions of good organisational practices. The results contribute to an understanding of how to improve organisational practices involving older employees.

1.3 Subnational Adaptations of the AAI

This part contains papers highlighting challenges faced while constructing the AAI at the subnational level. They present the examples of a particular country (Spain, Italy) and how the AAI can inspire to design the age-friendly environment better. This part emphasises that European countries identify the importance of the subnational analysis of the AAI, not only through the papers presented in this book but also others, such as Breza and Perek-Białas (2014) for Poland and Bauknecht et al. (2016) in the case of Germany. This is despite the challenges in obtaining relevant data at the regional and local level.

Chapter 6 by Bacigalupe, González-Rábago, Martín, Murillo and Unceta presents for the first time main results of the AAI for one of the Southern European regions (Biscay) and how it could be potentially used for policymaking purpose. With an aim to apply the original methodology, apart from the existing and available secondary data, some primary data obtained under a specific survey of respondents aged 55 and over were also used. Thanks to this method, the same questions as the ones used in the original approach were asked. For the Biscay Province in 2012/2014, the AAI reached 39.5 (1.9 points above the EU28 average). Biscay was ranked highly within the 28 EU countries. Thus, Biscay's score places it at a level similar to France, Germany or Luxembourg. This chapter is not only concluded with a discussion on the results and their interpretation in the context of policies but also with limitations and challenges of such analysis. However, it demonstrates that policymakers of this province perceive this AAI tool as useful in a discussion concerning appropriate right policies which constitute their regional responsibility.

The chapter by Ferri, Garcés, Staalduinen, Bond and Hinkema covers evaluation of socio-economic impact of the age-friendly environments (Chap. 7), and the authors show how the idea of the AAI could be conceptually used further at a lower level of governance. The Socio-Economic and Environmental Impact Tool (SEE-IT) also provides a framework for local and regional authorities to evaluate their Age-Friendly initiatives and to better evaluate costs and benefits of their initiatives. A majority of the AAI indicators were included in this framework, next to other types of indicators. In this context, a participatory methodology involving municipalities and regional authorities that have provided their feedback of the tool in an active way could also be perceived as a good example of an activity which engages citizens, stakeholders in discussions on aspects of active and healthy ageing. Now there is a unique opportunity to see the application of this framework in practice in some local communities.

Chapter 8 by Barrio, Rodríguez-Laso, Mayoral, Sancho and Amilibia shows how the Active Ageing Index could be also used as a key source to develop the Strategy of Active Ageing in the Basque Country (Spain). It is another example which presents the AAI results from Spain. In this case, the Basque Country confirms its high position before Spain and the EU28. It is interesting to see how these results could inspire the process of developing the Basque Country's Strategy of Active Ageing and how many measures from the AAI were useful.

The last chapter (9) in this Part II concerns an application of the AAI for all regions in Italy. The chapter by the team of Quattrociocchi, Squillante and Tibaldi covers numerous results which could be useful for policymakers in all Italian regions. This is presented based on time and gender factors. These analyses are particularly interesting as they do not cover only one region, as in the previous chapters. This subnational application provides a unique opportunity to see regional differences in the field of the AAI domains in Italy, also presenting all the indicators over time. In five years, the overall AAI index has increased from 25.9 points to 28.0 points, especially thanks to the higher values in "employment". Generally the Northern regions, the most developed area of the country, have achieved the best records, while the lowest values are found in the Southern area of the country. The variations observed in the AAI at the regional level indicate specific social, economic, political, cultural and demographic differences in all regions. It is worth emphasising that this chapter also includes calculations of the AAI by gender at the subnational level of the country. The gender gap shows a more elevated female disadvantage in the Southern regions, furthermore without any signs of improvement over time of analysis. In most cases, such subnational analysis by gender is not possible, mostly due to the lack of adequate data. However, this application of the AAI for Italy allows to identify the areas where the regions should react by implementing appropriate measures to improve their respective positions. Some regions have managed to improve their position during the analysed period and move it to higher levels; however, there are also regions with lower scores which have lost the position granted in 2007. The current challenge is to identify reasons that have contributed to these changes.

To summarise, one can observe that it is important to address regional and local perspectives in the AAI as well as disparities within the country at the regional and/or local level. Results presented point to whether the policies could be amended and, if yes, in which areas. Such extensive subnational analysis can stimulate a discussion among regional and local policymakers to identify options that would facilitate a better design and implementation of active and healthy ageing strategies.

1.4 Comparative Analysis of Active Ageing

The chapters in this part highlight the importance of comparative analysis of active ageing in drawing policy-relevant insights. The premises for the research reported here is that the AAI as a tool can be applied to contrast active ageing outcomes across countries to evaluate the current situation and identify the most desired directions for the policy changes. The chapter by Karpinska (Chap. 10) examines the policies affecting active ageing outcomes in Poland and the Czech Republic—these two countries sharing the historical legacy of socialist regimes and the transition experience towards the EU structures. Yet, they vary significantly in their active ageing outcomes. While the Czech Republic ranks 13th in the overall AAI for 2012, Poland occupies the bottom positions, in both overall and domain-specific indices. In this chapter, the policies related to the most remarkable differences were pieced together to explain the difference in the active ageing potential in both countries.

The chapter by Cela and Ciommi (Chap. 11) motivates its work by stating that ageing is one of the most important demographic trends in Europe. Migration can be seen as the second major phenomenon shaping European population structure. This introduces a further dimension to the ageing challenge, namely, ageing populations with diverse ethnic and cultural backgrounds. Nonetheless, the phenomenon of ageing migrants is surprisingly understudied. The chapter proposes a modified version of the AAI for the migrant population using data from the first wave (2004/2005) of the SHARE survey in seven European countries. It aims at both verifying whether the AAI is a good index for a population with a migratory background and stimulating the political and academic debate on the necessity to address vulnerabilities and active ageing of older migrants. The work is presented in reference to the framework of an inclusive multicultural Europe committed to social justice.

This chapter by Boehler, Helter, Rohman and Abadie (Chap. 12) outlines how the method of the Discrete Choice Experiments could be meaningfully used to estimate alternative weights for the AAI based on stated preferences. The approach is drawn from the Random Utility Theory and could provide valuable information on marginal substitution rates between the AAI indicators and domains. Authors assert that complementing the current AAI methodology with preference-based weights will allow assessing preference variations across different social, cultural or geographic contexts. This would help define more targeted active and healthy ageing policies and interventions. Furthermore, incorporating stakeholders' preferences in the valuation of policy outcomes will enhance the acceptance of the Index as a tool for policy analysis.

1.5 Methodological Improvements in Measuring Active Ageing

The chapters included in this part covered various aspects of methodological challenges faced in measuring the multifaceted concept of active ageing. The AAI provided the benchmark and the papers went beyond what was covered in the AAI.

The chapter by Olivera (Chap. 13) analyses the differences in active ageing across cohorts for all countries of the EU. It replicates the AAI for cohorts formed by age group, sex and country for 2012. The analysis is performed at the individual level with different regression models at the cohort level and by introducing macro-level variables at the country level among the explanatory factors. The key finding is that there is a gap in active ageing for females and it is larger among older cohorts. Furthermore, the wealth, the income inequality and pension settings of the country are important predictors of active ageing. Finally, in line with the original AAI results, it is found that the countries of social-democratic welfare regime (Nordic countries) offer the most favourable environment for active ageing as measured by the AAI. In contrast, the Post-Communist countries offer the least favourable setting. The standout finding of the paper is that it is important to consider subgroups of individuals when a composite index such as the AAI is computed. This practice will allow identification of high-risk groups and facilitate a more targeted policy response.

Arpino and Bordone (Chap. 14) focus on the social participation domain of the AAI framework and consider participation in different care (e.g., to grandchildren) and non-care (e.g., voluntary work) activities. By using data from the Survey of Health, Ageing and Retirement in Europe and by applying Latent Class Analysis, authors identify three clusters of older people with similar patterns of social participation that vary by type of activities in which they engage and intensity of engagement. The policy-relevant insights are that the oldest persons (aged 76+) comprise a group of people with very low probabilities of engagement (the "inactive" class). Also unemployed and housekeepers are at very high risk of non-engagement. Furthermore, the caregivers are characterised by intensive engagement in care activities and low probability of engagement in other types of activities (e.g., social groups, volunteering). Women are more likely to belong to this group. This study thus identified two target populations on which policies to promote active ageing need to focus: the oldest old—very likely to be excluded from active participation in both kin and non-kin activities—and women, who are more likely to be "trapped" in intensive care activities.

The chapter by Barros and de Almeida (Chap. 15) propose an individual-level measure of active ageing in Portugal following the example of the AAI. It used data from 1021 individuals aged 50 years or more living across 18 regions of Portugal. The analysis covered the domains of socio-demographic characteristics, health care services, activities of daily living (ADL), social participation and access to services (including Ambient Assisted Living), in the pursuit for possible policy instruments to efficiently promote active ageing. One of their key findings is that active ageing is related to higher levels of self-assessed health and other health-related variables. Thus, active ageing can be promoted by an improvement in the general health condition of the older population. Active ageing could in turn benefit health care users and providers by reducing the budgetary burden of health care services. The authors make a strong case for the social investment so as to provide equal opportunities to diverse people of older ages. There is a need to increase the integration of the elderly in society, avoiding feelings of isolation and uselessness in order to assure a sustainable well-being of the population and to adapt to the new demographic realities. They recommend construction of an individual-level measure of active ageing so as to further investigate disparities at regional level, as well as better understand the determinants of active ageing.

Piñeiro Vázquez and co-authors (Chap. 16) calculate the individual-level AAI for the community of Galicia in Spain for deeper subdivision of the results and for the purpose of creating personal profiles. For this, a representative sample of 404 Galician community-dwelling older adults (60+) is used. As for other chapters in this work, the authors argue that the assessment of the AAI at the individual level will improve the adaptation of programmes to the diverse needs of older people and the assessment of its effectiveness. Their results show that the potential of Galician older people should be strengthen in employment and in social participation. It also highlights the good performance in independent living aspects of older Galicians, with the enabling environment for active ageing reflecting an encouraging outlook for this community. The authors promise to continue working on improving the methodologies in measuring active ageing, specifically on the redistribution of the weights assigned to domains. They are aiming to create an observatory to monitor trends in active ageing indicators and offering insights for improvements in public policies.

1.6 Use of Active Ageing Index in Non-EU Countries

The contributions included in this section provide a unique insight into the findings emerging from different attempts to apply the AAI methodology to non-EU contexts. They highlight the various challenges faced by authors in their endeavours, from limitations in the availability of suitable data—which might restrict an in-depth analysis of the investigated topic—to methodological constraints characterising the process of evaluating results, achieved via ad hoc measures developed in the attempt to bypass the mentioned lack of data. Despite—or probably thanks to—these restraints, the chapters presented in this section provide a fascinating overview of how the same tool, like the EU-based AAI or those constructed following a similar approach, might be understood and implemented quite differently in various social, cultural and economic non-EU contexts, thus leading to possibly diverging interpretations also in terms of policy recommendations and/or implications.

In their chapter, *Xiong and Wiśniowski* (Chap. 17) calculate the AAI for China employing data from the China Health and Retirement Longitudinal Study (CHARLS). Their findings are used to argue that, compared to the EU older population, the Chinese one needs more structured policies aiming at maintaining quality of life in older age via initiatives undertaken at the micro-, meso- and macro-levels. Recommendations in this regard include the adoption of more flexible retirement plans to stimulate promotion of a more voluntary participation in the labour market; the strengthening of community-based care services; and the promotion of lifelong learning programmes enabling older Chinese to enjoy a more active and healthy life.

Fanta (Chap. 18) follows the methodological approach used for constructing the AAI to develop an ad hoc aggregated index to analyse the Quality of Life of the Elderly (QLE) in six countries of Latin America (LA). The rapid population ageing experienced by this subcontinent urges the implementation of such tools and methodologies to obtain more reliable information on the needs of the older population and to understand how to best address them politically. The application of the AAI to the LA context seems to suggest that most LA countries are still today unable to ensure a high quality of life for their older population, also due to the lack of data used for planning towards this purpose.

The chapter by *Barysheva and co-authors* (Chap. 19) focuses on the application of the AAI to the Russian context, offering an in-depth analysis of how single AAI domains score at both national and subnational levels (i.e., Russian Federal Districts). Despite the methodological challenges faced in carrying out this exercise, the study allows to point out that, in a country which would rank 22nd among 29 countries (i.e., the 28 EU countries plus Russia), the domain most contributing to Russia's AAI score is "capacity and enabling environment for AA", with "employment" and "societal participation" contributing considerably less. Furthermore, by providing a comparative view of how single Federal Districts score in the different domains, the chapter represents a practical example of how the implementation of the AAI in Russia could be used to advise policies addressing the older population at both national and local levels.

Guntupalli and Chakraborty (Chap. 20) deal with the applicability of the AAI in the Indian context, providing interesting evidence on the need for a "culturally sensitive" interpretation of the AAI approach and results. This is achieved by suggesting, for instance, that the AAI should include a stronger consideration of the social contribution made by older people to society (a domain in which India scores relatively high), besides the economic one. This lies in the fact that, in India, employment in later life cannot univocally be interpreted as active engagement, due to lack of formal support and ageism in the labour market, a situation well reflected by India's poor AAI scoring in the employment domain. Therefore, a more in-depth, cross-culturally validated comparative analysis should be carried out to draw policy conclusions based on the AAI in countries outside the EU context.

Notes

1. The special Active Ageing Index issue of the Journal of Population Ageing came out in March 2017, with the title "Active Ageing Index: Insights into Evidence and Policy". The issue editors were Asghar Zaidi and Kenneth Howse. The papers are available at https://link.springer.com/journal/12062/10/1/page/1

References

Bauknecht, J., Tiemann, E., & Velimsky, J. A. (2016). *Extending the Active Ageing Index to the local level in Germany: Pilot study*. Report prepared by the Institute of Gerontology at the Technical University of Dortmund, under a contract with the United Nations Economic Commission for Europe (Geneva), co-funded by the European Commission's Directorate General for Employment, Social Affairs and Inclusion (Brussels).

Breza, M., & Perek-Białas, J. (2014). *The Active Ageing Index and its extension to the regional level.* Host country paper, Peer Review in Poland: The Active Ageing Index and its extension to the regional level, Kraków. Retrieved from http://ec.europa.eu/social/BlobServlet?docId=12940&langId=en

Council of the European Union. (2012). *Council declaration on the European year for active ageing and solidarity between generations (2012): The way forward*, 17468/12, SOC 992, SAN 322.

Walker, A. (2002). A strategy for active ageing. *International Social Security Review, 55*(1), 121–139. https://doi.org/10.1111/1468-246X.00118.

Walker, A. (2009). Commentary: The emergence and application of active aging in Europe. *Journal of Ageing and Social Policy, 21*(1), 75–93. https://doi.org/10.1080/08959420802529986.

Walker, A., & Maltby, T. (2012). Active ageing: A strategic policy solution to demographic ageing in the European Union. *International Journal of Social Welfare, 21*(Issue Supplement), S117–S130.

WHO. (2002). *Active ageing—A policy framework.* Contribution of the World Health Organisation to the Second United Nations World Assembly on Ageing, Madrid, Spain.

Zaidi, A., Gasior, K., Hofmarcher, M. M., Lelkes, O., Marin, B. Rodrigues, R., Schmidt, A., Vanhuysse, P., & Zolyomi, E. (2013). *Active Ageing Index 2012: Concept, methodology and final results.* Methodology report submitted to European Commission, DG Employment, Social Affairs and Inclusion, and to UNECE, for the project: 'Active Ageing Index (AAI)' (UNECE Grant No: ECE/GC/2012/003). Geneva: European Centre Vienna. Retrieved from http://goo.gl/fDPX6g

Zaidi, A., Gasior, K., Marin, B., Rodrigues, R., Schmidt, A., & Zolyomi, E. (2017). Measuring active and healthy ageing in Europe. *Journal of European Social Policy, 27*(2), 138–157.

Zaidi, A., & Stanton, D. (2015). *Active Ageing Index 2014: Analytical report.* Report produced at the Centre for Research on ageing, University of Southampton, under contract with UNECE (Geneva), co-funded by European Commission, Brussels. Retrieved from http://www.southampton.ac.uk/assets/sharepoint/groupsite/Administration/SitePublisher-document-store/Documents/aai_report.pdf

Asghar Zaidi is Professor in International Social Policy. He is also Visiting Professor at London School of Economics and Political Science. Previously, he worked as Senior Economist at OECD Paris and as Economic Advisor at UK's Department for Work and Pensions. He has been the lead researcher in developing the Active Ageing Index and the Global AgeWatch Index. He has also worked with Age UK to develop the Wellbeing in Later Life (WILL) Index.

Sarah Harper is Professor of Gerontology and co-Director of the Oxford Institute of Population Ageing. She served on the Prime Minister's Council for Science and Technology, and chaired the UK government Foresight Review on Ageing Societies, and the European Active Ageing Index Panel for the UNECE Population Unit. She is the new director of the Royal Institute (RI)—only the second woman to be appointed director of the RI in its 218-year history.

Kenneth Howse is Senior Research Fellow at the Oxford Institute of Population Ageing. His work spans a range of issues, including health policy and the place of religion in later life. Previously, he worked at the Centre for Policy on Ageing and the Institute of Medical Ethics. His current research focus is on ageing in South East Asia where he is leading a study of the role of Older People's Associations in different countries in the region.

Giovanni Lamura is a social gerontologist with an international and interdisciplinary background, working at INRCA (Italy's National Institute of Health and Science on Ageing) since 1992, where he currently leads the Centre for Socio-Economic Research on Ageing. His main research interests focus on long-term care, migrant care work, active ageing, intergenerational relationships and old-age social exclusion. Previously, he was visiting fellow in different research institutions in Germany and Austria.

Jolanta Perek-Białas works at the Institute of Sociology of the Jagiellonian University, Cracow and at the Institute of Statistics and Demography of the Warsaw School of Economics, Poland. She has participated in international projects on active ageing, demographic change, situation of older persons and care. Her current research focus is on ageism in the labour market and application of the active ageing index at the subnational level in Poland.

Part I

Building Evidence for Active Ageing Policies Active Ageing Index and its Potential

2

Are Societies with a High Value on the Active Ageing Index More Age Integrated?

Pearl A. Dykstra and Maria Fleischmann

2.1 Background

In the literature on cleavages in society, age divides have received less attention than separations by gender, race, social class or religion. Notwithstanding heated disputes on the generation gap in the 1970s (e.g., Mead 1970), speculations about impending age wars in the 1980s (e.g., Longman 1986) and conjectures in the most recent decades regarding competition between the young and the old over scarce resources (Emery 2012), researchers have not typically focused on age in analyses of social segmentation. We argue that the separation of age groups in ageing societies merits more systematic attention because it restricts individuals from meeting, interacting and moving beyond stereotyped views.

P.A. Dykstra (✉)
Erasmus University, Rotterdam, Netherlands

M. Fleischmann
University College, London, UK

A. Zaidi (eds.), *Building Evidence for Active Ageing Policies*,
https://doi.org/10.1007/978-981-10-6017-5_2

Hagestad and Uhlenberg (2005, 2006; Uhlenberg 2000) use the term "age segregation" for the separation of children and young adults into schools, the separation of adults into workplaces that exclude the young and the old, and the separation of older people out of the workplace and, in some cases, into institutional housing and care arrangements. Leisure-time activities of young and old tend to occur in different locations because they are often organized by age, as is the case for youth recreation leagues, youth orchestras, senior centre activities and senior tour groups (Hagestad and Uhlenberg 2005). According to Coleman (1982), age segregation deprives the young of a proper view of mid-life and old age, and produces adults who have little experience with and understanding of the young. The author also argues that age segregation can be an antecedent as well as a consequence of *ageism*, defined as prejudice by members of one age group against another age group (Butler 1969). The separation of age groups can contribute to and reinforce ageist attitudes, which, in turn, can create barriers between young and old.

Integration by age occurs when individuals of different ages occupy the same space and hence can engage in face-to-face interactions (Vanderbeck 2007). Lowering age barriers and increasing cross-age interaction may be an effective way of reducing social segmentation and thus promoting more inclusive societies (Hagestad and Uhlenberg 2005, 2006). Uhlenberg (2000) suggests that some societies may be more age integrated than others. In his view, researchers need to consider the extent to which formal and informal barriers restrict opportunities for individuals of different ages to mingle, socialize, collaborate, learn and worship together. We aim to find out whether societies where older adults are visible and active participants [and thus have a high value on the Active Ageing Index (AAI) as developed by Zaidi et al. (2013)] are more age integrated. We use both a behavioural and an attitudinal measure of age integration: the prevalence of cross-age friendships and low levels of ageism. The underlying argument is that high-AAI countries enable *durable interactions* between young and old, thereby promoting a better understanding of people of different ages.

We use data from the fourth round of the European Social Survey (ESS), which had a special module on "Age attitudes and experiences of ageism", and combine them with indicators of Active Ageing.

2.2 Conditions for Cross-age Interaction

2.2.1 Laws and Policies

National policies geared towards services, financial entitlements and amenities for specific age groups are examples of societal forces structuring opportunities and constraints for cross-age interactions. Such policies not only serve to distance specific age groups (e.g., placing older adults in residential facilities; separating children and young adults into schools), but might also shape how age groups perceive one another. Binstock (1983) introduced the term "compassionate ageism" to describe how policies reflect or encourage the view that specific age groups, such as the young and the old, have needs that deserve being cared for. This sense of benevolence might make people feel more sympathetic and open towards people differing in age. In a later publication, Binstock (2010) referred to "double-edged" compassionate ageism: economically assisting older adults is apparently generous, but also reinforces negative stereotypes of frailty, poverty and dependency.

Legislation defining the rights and duties of members of different age groups is a second example of the macro-level organization of age relations. Age discrimination laws, which reflect dominant values about "proper" age relations, are most relevant to the issue at hand. Generally speaking, it is safe to posit that cross-age interactions are more likely in a country that promotes equal treatment of all ages than in a society that does not encourage that kind of equality. Non-discrimination by age became legally enforceable in the European Union in 2006 (Lahey 2010). Some member states have only recently implemented the strand of the Framework Directive prohibiting age discrimination, whilst others have longstanding age discrimination laws. Although similar in many ways, a diverse set of these laws exists across the EU. Each country has unique prohibitions and specific exemptions regarding discrimination, and enforcement mechanisms differ widely.

This brief overview highlights the complexity of delineating policy and legal influences on age integration. Mixed effects are plausible, and as yet, there are few clues as to how to unravel them. For that reason, we decided to focus on the Active Ageing Index rather than legislation and policies. The AAI is a tool that acknowledges that individuals live in a world of multiple jurisdictions and are affected by multiple policies at once (cf. Campbell 2012). Given its focus on the outcomes of and capacities for the participation of older adults in society (Zaidi et al. 2013), the AAI provides an *accumulated appraisal* of the ways in which policies and laws produce and reflect age barriers. We argue that high-AAI countries have created conditions that promote cross-age interaction. The literature on personal relationships provides clues as to what these conditions might be.

2.2.2 Settings

A key principle of the research on personal relationships is that there is "no mating without meeting" (Blau 1977): people find friends and partners among those they encounter in the course of their daily activities. The meeting principle emphasizes that *social settings* offer opportunities to meet particular categories of individuals and thereby influence the kinds of personal relationships that develop. Applied to age integration, this principle suggests that those who can draw upon an age-diverse pool of ties are more likely to engage in cross-age interaction.

What settings provide an age-diverse pool of social personal ties? Families provide a context in which people of different ages continually meet and interact. Hagestad and Uhlenberg (2005) argue in fact that "the family represents the *only* truly age-integrated social institution" (p. 354, emphasis in original). Examples of settings outside the family where old and young can meet and interact are religious communities (Grefe 2011), work (Uhlenberg 2000), volunteer work (Uhlenberg and De Jong Gierveld 2004) and neighbourhoods (Vanderbeck 2007). Few scholars have examined whether age-integrated settings actually facilitate the formation and maintenance of cross-age ties.

Our *first hypothesis* is that high-AAI countries are more likely to provide settings enabling durable cross-age interaction than low-AAI countries. The settings we consider are family, paid work, volunteer work and religious organizations. Next, for both young and old, we test a *second hypothesis*: those whose daily activities are in settings enabling durable cross-age interaction are more likely to report cross-age friendships than those whose daily activities do not bring them to such settings.

2.2.3 Controls

Most research shows that women have larger and more diversely composed networks than men, but patterns are not always clear-cut. Gender differences in personal networks vary by age, life stage and marital history (e.g., Ajrouch et al. 2005). That is why we introduce controls for gender, age and partner status in the analyses. We also control for self-reported health given the role possibly played by selection: health status can contribute to the creation or dissolution of specific network ties or to the formation of networks with particular features (Smith and Christakis 2008).

2.3 Cross-age Interaction and Ageism

When ageism was first defined (Butler 1969), the phenomenon was viewed as something directed at old people only. By now, researchers acknowledge that ageism goes both up and down generational lines (North and Fiske 2012). There is a wealth of evidence showing that building sustained familiarity with individuals across social categories is the safest route for breaking down prejudice, overcoming stigma and preventing discriminatory behaviours (Pettigrew and Tropp 2006). Empirical support for the notion that cross-generational experiences are generalized to age groups as a whole comes from studies of grandparenthood. Higher levels of quality of contact with grandparents are associated with more positive feelings towards older people in general (e.g., Harwood et al. 2005). Acknowledging that durable interactions with people of different ages foster mutual understanding, we test a *third hypothesis*: the young and old who report cross-age friendships are less ageist than those who do not report having cross-age friendships.

2.4 Method

2.4.1 Data

The fourth round of the ESS was collected in 2008–2009 in 31 countries, of which 25 are EU countries (Italy, Malta and Luxembourg do not participate in the ESS). Our analyses are based on these 25 EU countries (see Table 2.1 for an overview). We used the AAI 2010, which is based on 2008 data. We restricted the analyses to two age groups. The "young" are respondents aged 18–30; the "old" are respondents aged 70–90 ($N = 6679$).

2.4.2 Measures

The number of *cross-age friendships* was measured by asking "About how many friends, other than members of your family, do you have who are [younger than 30/aged over 70]?". Answer categories ranged from "none" (1) to "10 or more" (5). The variable was dichotomized, assigning respondents the value 1 when they had two or more cross-age friendship. Sample sizes for the analyses of cross-age friendships are 8716 (young) and 6679 (old).

To assess *ageism*, we used the following question: "Tell me overall how negative or positive you feel towards people in their 20s [people over 70]?". Answer categories ranged from "extremely negative" (0) to "extremely positive" (10). We reverse-coded the answers, so higher scores indicate higher levels of ageism. In the multivariate analyses, we used the log of the ageism score because responses were heavily skewed. Sample sizes for the analyses of ageism are 8612 (young) and 6512 (old).

We had a set of dummy variables representing settings enabling cross-age interaction. The first was whether the younger [older] adult had any *household member* aged 70 and older [aged 18–30] (1 = yes). Given that sustained contact, emotional complexity and perspective-taking are particularly likely to produce a better understanding of people of different ages, we included an assessment of whether the younger adult [the older adult] had any *family member* over 70 [any children or grandchildren] with whom they were able to discuss at least "a few personal issues such as feelings, beliefs or experiences" (1 = yes). We also considered whether

Table 2.1 Prevalence of cross-age friendships and ageism levels, 25 EU countries

	≥2 cross-age friends		Ageism score			
	Age 18–30 ($N = 8716$)	Age 70–90 ($N = 6697$)	Age 18–30 ($N = 8612$)		Age 70–90 ($N = 6512$)	
Country	%	%	M	SD	M	SD
Austria	11.9	21.5	2.82	2.14	3.35	2.26
Belgium	20.5	29.6	3.00	1.66	3.60	1.66
Bulgaria	14.3	22.0	2.17	2.29	2.58	2.41
Croatia	13.1	27.2	2.69	2.28	2.35	2.02
Cyprus	9.3	30.0	1.94	1.80	2.42	2.19
Czech Republic	16.8	30.4	3.19	2.09	3.72	2.08
Denmark	13.2	26.8	2.63	1.80	2.55	1.73
Estonia	15.5	31.7	2.42	1.81	2.88	1.99
Finland	20.8	50.2	1.91	1.36	2.32	1.65
France	19.8	29.7	2.96	1.63	3.75	1.81
Germany	29.0	39.8	2.71	1.78	3.13	1.84
Greece	13.9	27.2	2.19	1.97	2.03	1.89
Hungary	14.1	15.7	2.97	1.95	2.94	1.95
Ireland	35.8	43.5	2.34	1.65	3.12	2.00
Latvia	16.8	21.8	2.30	2.03	2.34	2.05
Lithuania	4.4	5.6	3.10	1.79	3.17	2.08
Netherlands	8.8	18.6	2.58	1.18	3.57	1.62
Poland	16.2	33.3	2.22	1.83	2.73	1.95
Portugal	34.9	40.4	2.93	1.93	3.97	1.92
Romania	19.9	29.6	3.07	2.10	3.89	2.17
Slovakia	17.4	29.1	3.68	2.07	3.72	2.13
Slovenia	20.9	35.2	2.27	1.65	2.23	1.99
Spain	18.7	22.9	2.73	1.80	3.28	1.94
Sweden	18.5	46.2	2.54	1.50	2.53	1.83
United Kingdom	27.7	38.5	2.77	1.71	4.03	2.18
All	18.1	30.6	2.65	1.90	3.13	2.07

the respondent attended *religious services* at least monthly (1 = yes). Another variable indicated whether respondents had done *paid work, volunteer work* or both in the last month (1 = yes). To provide an indication of the age composition of the *work settings*, we added a variable assessing whether respondents spent some, most or (almost) all of that time working with colleagues or volunteers in their 20s [or aged over 70] (1 = yes).

The 2010 values for the overall AAI, as well as those for the separate indicators were taken from the publicly available AAI website.[1] Each of the indicators is expressed in percentage terms, with values ranging from 0 (minimum goalpost) to 100 (maximum goalpost). Note that the upper goalpost of 100 should not be equated with the optimum, as it represents utopian circumstances of fullest possible active ageing (Zaidi et al. 2013). The separate indicators are *employment*; *participation in society*; *independent, healthy and secure living*; and *capacity and enabling environment for active ageing*. All AAI measures were centred on their mean to allow a meaningful model interpretation.

We used four control variables for respondents' *gender* (1 = female), *age* (in years), *partner status* (1 = lives with a partner) and *health status*, respectively. Health status was measured by asking "How is your health in general?" Answer categories ranged from "very good" (1) to "very bad" (5). We reverse-coded the responses, so higher scores indicate better health.

2.4.3 Analytical Strategy

We carried out separate analyses for the two age groups under consideration. The individual-level data on settings enabling cross-age interaction were aggregated to the country level in order to determine whether high-AAI countries are more likely to provide conditions encouraging interactions between young and old. We computed Spearman's correlations between the aggregated measures of settings and the Active Ageing indicators. We used multilevel logistic regression models to analyse the likelihood of having two or more cross-age friends, and multilevel linear regression models to analyse ageism. The same set of country- and individual-level determinants was included in both models—with one exception: having cross-age friends was added to the analysis of ageism.

To better understand the role of the country context, we examined whether indicators of active ageing contribute to an explanation of cross-age friendships and ageism over and above individual-level conditions.

2.5 Results

Across all countries, the young are less likely than are the old to report having two or more *cross-age friends*, 18% and 31%, respectively (see Table 2.2). The proportion of young adults reporting cross-age friendships is lowest in Lithuania (4%) and highest in Ireland (36%). The proportion of older adults reporting cross-age friendships is lowest in Lithuania (6%) and highest in Finland (50%).

Europeans, both young and old, express few negative feelings about age groups different from their own. On a scale from 0 to 10, the average *ageism* score across all countries is 2.7 for the young and 3.1 for the old (see Table 2.2). Older people are generally somewhat more ageist towards the young than the young are towards the old. Ageism levels vary little across countries. Ageism of young towards old is lowest in Finland (1.9) and highest in Slovakia (3.7). Ageism of old towards young is lowest in Greece (2.0) and highest in the UK (4.0).

2.5.1 AAI and Settings Enabling Cross-age Interactions

As predicted (see Table 2.3), we find strong and positive correlations between overall AAI scores and the proportion of people (both young and old) who report engaging in paid or volunteer work and working with colleagues or volunteers who differ in age by several decades. Overall AAI scores are also strongly and positively correlated with the proportion of younger adults who engage in personal disclosure with older family members. Contrary to predictions, strong negative associations with overall AAI scores are observed for intergenerational co-residence (among both young and old), for religious engagement (among both young and old) and for the proportion of older adults engaging in personal disclosure with a (grand)child. An inspection of the findings for the separate Active

Table 2.2 Pearson's correlations between Active Ageing indicators and country-aggregated measures of settings enabling cross-age interaction ($N = 25$)

	Age 18–30 ($N = 8716$)					Age 70–90 ($N = 6697$)				
	Household member >70	Family member >70	Religious organization	Paid or volunteer work	Colleagues or volunteers >70	Household member <30	(Grand) child	Religious organization	Paid or volunteer work	Colleagues or volunteers <30
AAI: Overall	−0.68*	0.44*	−0.42*	0.74*	0.45*	−0.54*	−0.43*	−0.25	0.77*	0.59*
AAI: Employment	−0.29	−0.15	−0.17	0.23	0.13	−0.30	−0.17	−0.10	0.23	0.07
AAI: Participation in society	−0.36	0.41*	−0.44*	0.51*	0.22	−0.38	−0.33	−0.52*	0.70*	0.56*
AAI: Independent, healthy and secure living	−0.63*	0.61*	−0.37	0.73*	0.44*	−0.46*	−0.39	−0.27	0.81*	0.64*
AAI: Capacity and enabling environment	−0.52*	0.62*	−0.38	0.77*	0.52*	−0.38	−0.52*	−0.33	0.78*	0.56*

*$p < 0.05$

Table 2.3 Multilevel logistic regression analysis predicting having at least two cross-age friendships (individual-level predictors)

	Age 18–30 (N = 8716)		Age 70–90 (N = 6697)	
	Baseline model	Individual-level predictors	Baseline model	Individual-level predictors
	OR	OR	OR	OR
Has household member >70 [<30]		1.379*		1.350
Discusses personal issues with family member >70 [with (grand)child]		1.689***		1.127*
Attends religious services at least monthly		1.640***		1.381***
Did paid or volunteer work past month		1.068		1.394**
Worked with colleagues or volunteers in their 70s [20s]		2.720***		2.066***
Gender (female = 1)		0.815***		0.782***
Age		1.005		0.967***
Lives with partner		1.228**		1.200**
Health status		0.932		1.320***
Between-level var. (σ_u^2)	0.239	0.257	0.260	0.264
Within-level var. (σ_e^2)	3.283	3.293	3.288	3.292
Intra-class correlation	0.068	0.072	0.073	0.074

Note: Exponentiated coefficients (odds ratios); *$p < 0.05$, **$p < 0.01$, ***$p < 0.001$

Ageing indicators reveals that the employment rates of older adults show no associations with the aggregated measures of the settings enabling cross-age interaction, whereas social participation generally shows moderate associations. The independent, healthy and secure living indicator and the capacity and enabling environment indicator show the strongest associations with the settings enabling cross-age interaction. Summarizing, our first hypothesis receives mixed support.

2.5.2 Determinants of Cross-age Friendships

Table 2.4 shows, in accordance with hypothesis 2, that the young whose daily activities bring them to settings enabling cross-age interaction have

Table 2.4 Multilevel logistic regression analysis predicting having at least two cross-age friendships (macro-level predictors)

	Age 18–30 ($N = 8716$)				Age 70–90 ($N = 6697$)			
	OR	Between-level variance (σ_u^2)	Within-level variance (σ_e^2)	Intra-class correlation	OR	Between-level variance (σ_u^2)	Within-level variance (σ_e^2)	Intra-class correlation
Macro only								
Baseline model		0.239	3.283	0.068		0.260	3.288	0.073
AAI: Overall	1.018	0.220	3.288	0.063	1.031*	0.203	3.291	0.058
AAI: Employment	1.010	0.233	3.291	0.066	1.021	0.235	3.291	0.067
AAI: Participation in society	1.003	0.239	3.293	0.068	1.010	0.255	3.292	0.072
AAI: Independent, healthy and secure living	1.018	0.218	3.288	0.062	1.029*	0.204	3.294	0.058
AAI: Capacity to actively age	1.014	0.227	3.292	0.064	1.024*	0.222	3.294	0.063
Macro, controlling for individual level								
AAI: Overall	1.017	0.239	3.298	0.068	1.021	0.239	3.293	0.068
AAI: Employment	1.013	0.247	3.292	0.070	1.020	0.240	3.286	0.068
AAI: Participation in society	1.002	0.257	3.293	0.072	1.004	0.263	3.284	0.074
AAI: Independent, healthy and secure living	1.015	0.241	3.283	0.068	1.019	0.242	3.292	0.069
AAI: Capacity to actively age	1.011	0.249	3.293	0.070	1.013	0.254	3.289	0.072

Note: Exponentiated coefficients (odds ratios); $*p < 0.05$

a higher probability of reporting at least two cross-age friendships. There is one exception: having performed paid and/or volunteer work in the past month is not associated with a higher likelihood of having cross-age friends. The young who share a household with someone over the age of 70 have a factor 1.4 higher odds of reporting cross-age friendships, whereas those who have a family member over the age of 70 with whom personal issues can be discussed have a factor 1.7 higher odds of reporting cross-age friendships. Although having done paid or volunteer work per se is not associated with having cross-age friends, having worked with colleagues or volunteers in their 70s is a strong predictor: respondents' odds of having cross-age friendships are a factor 2.7 higher. For the young, the odds of reporting cross-age friendships are a factor 1.6 higher when they attend religious services at least monthly.

Among the older respondents, the individual-level determinants of having cross-age friends are quite similar to those for the young. Again, having worked with colleagues or volunteers quite different in age than oneself strongly increases the likelihood of reporting cross-age friendships (a factor 2.1 higher). Older respondents living with people under the age of 30 and those who attend religious services at least monthly have a factor 1.4 higher odds of reporting at least two friends under 30, keeping other variables constant. Contrary to what was observed for the younger respondents, having done (volunteer) work is associated with having cross-age friendships (the odds are a factor 1.4 higher). Having a younger family member with whom personal issues are discussed also increases the odds of having cross-age friendships by a factor of 1.1.

As the intra-class correlations in Tables 2.4 and 2.5 show, over 90% of the variance in the prevalence of cross-age friendships is at the level of individual respondents, whereas 7% of the variance is attributable to cross-country differences. Do the AAI indicators account for any of the cross-country differences? As the top right-hand part of Table 2.5 shows, the overall AAI is a significant and positive predictor of the likelihood of having cross-age friends among the old, accounting for approximately 21% of the cross-country differences. Further inspection of the table shows that the effect of the overall AAI is actually attributable to two separate indicators: independent, healthy and secure living, and capacity to actively age. As the bottom right-hand part of Table 2.5 shows, after

Table 2.5 Multilevel linear regression analysis predicting ageism (individual-level predictors)

	Age 18–30 (N = 8612)		Age 70–90 (N = 6512)	
	Baseline model	Individual-level predictors	Baseline model	Individual-level predictors
	B	B	B	B
Has household member >70 [<30]		0.018		0.020
Discusses personal issues with family member >70 [with (grand)child]		−0.052**		−0.084**
Attends religious services at least monthly		−0.072**		0.002
Did paid or volunteer work past month		−0.027		−0.025
Worked with colleagues or volunteers in their 70s [20s]		−0.014		−0.038
≥2 cross-age friendships		−0.089**		−0.068**
Gender (female = 1)		−0.109**		−0.100**
Age		−0.005*		0.000
Lives with partner		−0.018		−0.011
Health status		−0.054**		−0.055**
Constant	1.139**	1.575**	1.237**	1.466**
Between-level var. (σ_u^2)	0.018	0.012	0.036	0.019
Within-level var. (σ_e^2)	0.333	0.326	0.335	0.327
Intra-class correlation	0.052	0.035	0.098	0.056

Note: *$p < 0.01$, **$p < 0.001$

controlling for the individual-level predictors, the coefficients for the AAI predictors lose their significance. None of the Active Ageing indicators is a significant predictor of the likelihood of having cross-age friendships among the young. Tables 2.3 and 2.5, taken together, suggest that the high-AAI countries have settings outside the household enabling cross-age interaction, which in turn facilitate the formation and maintenance of cross-age friendships by older adults in particular.

2.5.3 Cross-age Friendship and Ageism

As predicted, those who report cross-age friendships tend to be less ageist (Table 2.5). This finding holds for both young and old. The intra-class correlations show that approximately 5% of the variance in ageism of young towards old and approximately 10% of the variance in ageism of old towards young are attributable to cross-country differences. Nevertheless, none of the Active Ageing indicators contribute to an explanation of ageism over and above the individual-level indicators (table not shown).

2.6 Conclusion

The starting point of our study was to find out whether societies where older adults are visible and active can be characterized as more age integrated in the sense that close ties between the young and the old are not limited to the family and that both the young and the old tend not to be prejudiced towards one another. The answer to the question of whether societies with a high value on the AAI are more age integrated is "yes" and "no". A complex pattern emerges from our findings.

First, our study underscores the importance of *anchoring*, that is, of distinguishing the age group under consideration. A strength of our study is that we focused on both young and old. Whereas none of the Active Ageing indicators significantly predicted whether or not younger adults had cross-age friendships, they did make a difference regarding the likelihood that older adults had cross-age friendships. Older adults were more likely to engage in friendly relations with young individuals in societies with the following characteristics: high levels of independence, health and security in late life and greater capacity to actively age. These findings suggest that quality of life, well-being and autonomy contribute to a general atmosphere facilitating sustained familiarity of the old with the young.

Second, the AAI indicators did not account for any between-country variance in *ageist* tendencies among either young or old. Conceivably, the cultural climate of a country is more pertinent to the explanation of ageism. The importance of emotional and cultural dispositions is evident at the individual level. Those who have a close relationship with a family member who strongly differs in age tend to be less ageist. Among the young, religiosity is inversely associated with ageism.

Third, high AAI levels are generally favourable to work-related (both paid and unpaid) settings enabling durable cross-age interaction, but tend to have low levels of *household-related* (intergenerational co-residence) and *religious* settings that encourage meaningful contacts between young and old. The individual-level data show that sharing a household with a person who strongly differs in age and frequently attending religious services are conducive to the formation and maintenance of cross-age friendship. Such opportunities are greater in low-AAI countries than in high-AAI countries.

We cannot rule out the possibility that the low degree of variance at the country level in the prevalence of friendships and ageist tendencies is responsible for the dearth of significant findings for the Active Ageing indicators. Our individual-level hypotheses received strong support, however. We convincingly demonstrated that people whose daily activities are in settings encouraging cross-age interaction are more likely to have cross-age friends. Previously, few studies have addressed the issue of where, when and how cross-age ties are formed and maintained. This issue has received little attention (Riley and Riley 2000), perhaps because homophily, the tendency to form relationships with similar others, is a strong theme in relationship research. We also received credible evidence supporting the hypothesis that people with cross-age friendships tend to be less ageist.

Interestingly, the Active Ageing indicators representing older adults' activities, that is, employment and social participation, showed no associations with their likelihood of having cross-age friendships. Apparently, high levels of working, volunteering, caring and political engagement among the old in a country are not necessarily conducive to the formation and maintenance of friendships with young people. Corroboration of this observation is provided by our finding that no or only moderate associations were found between the employment and social participation indicators and the aggregate measures of settings enabling cross-age interaction. It seems that we should be cautious to equate high levels of active engagement among the elderly with ample opportunities to encounter young people and to become connected to them. Older adults seem to be primarily active in age-homogeneous ghettos. Our findings suggest that "productive ageing", which is emphasized time and time again (e.g., Gonzales et al. 2015), will in and of itself not lead to greater age integration.

Following the adage that societal ageing is not only about older adults, we included both young and old in our study. Interesting contrasts between the two groups emerged. Lower proportions of the young reported cross-age friendships. Moreover, they were generally less prejudiced towards older people than the old were towards younger people. At first sight, the findings seem contradictory: fewer cross-age friends, but less ageist. More careful scrutiny suggests that the findings are not as puzzling as they initially seem to be. To start, it is important not to confuse macro-level and micro-level associations. At the individual level, our findings show the expected pattern for both young and old: those who report cross-age friendships tend to be less ageist. One should also note that different target groups are being compared: the young are evaluating the old, whereas the old are evaluating the young. We have no information about how the young feel about their age group and neither do we know how the old feel about their peers. So our findings do not allow assessments about which group is most ageist. Finally, contemporary circumstances of the young require consideration. In their study on changes in the homophily of confidants between 1985 and 2004 in the USA, Smith et al. (2014) argue that the young increasingly face institutional barriers to engage in cross-age ties. They contend that younger adults seem to be "somewhat more cloistered in their generational institutions" in recent years. They attribute the growing isolation of the young from older cohorts outside of the family to the growing importance of delayed life course transitions. The authors touch upon an issue that is also critical in Europe given the current economic crisis: the inability of younger adults to start a life of their own and to enter environments that are less age-homogeneous than the worlds of education and recreation.

Notes

1. http://www1.unece.org/stat/platform/pages/viewpage.action?pageId=76287845

References

Ajrouch, K. J., Blandon, A., & Antonucci, T. C. (2005). Social networks among men and women: The effects of age and socioeconomic status. *Journal of Gerontology: Social Sciences, 60*, 311–317.

Binstock, R. H. (1983). The aged as a scapegoat. *The Gerontologist, 23*, 136–143.

Binstock, R. H. (2010). 50th anniversary feature article. From compassionate ageism to intergenerational conflict? *The Gerontologist, 50, 574–585.*

Blau, P. M. (1977). *Inequality and heterogeneity: A primitive theory of social structure*. New York: Free Press.

Butler, R. N. (1969). Age-ism: Another form of bigotry. *The Gerontologist, 9*, 243–246.

Campbell, A. L. (2012). Policy makes mass politics. *Annual Review of Political Science, 15*, 333–351.

Coleman, J. S. (1982). *The asymmetric society*. Syracuse, NY: Syracuse University Press.

Emery, T. (2012). Intergenerational conflict: Evidence from Europe. *Journal of Population Ageing, 5*, 7–22.

Gonzales, E., Matz-Costa, C., & Morrow-Howel, N. (2015). Increasing opportunities for the productive engagement of older adults: A response to population aging. *The Gerontologist, 55*, 252–261.

Grefe, D. (2011). Combating ageism with narrative and intergroup contact: Possibilities of intergenerational connections. *Pastoral Psychology, 60*, 99–105.

Hagestad, G. O., & Uhlenberg, P. (2005). The social separation of old and young: A root of ageism. *Journal of Social Issues, 61*, 343–360.

Hagestad, G. O., & Uhlenberg, P. (2006). Should we be concerned about age segregation? Some theoretical and empirical explorations. *Research on Aging, 28*, 638–653.

Harwood, J., Hewstone, M., Paolini, S., & Voci, A. (2005). Grandparent-grandchild contact and attitudes toward older adults: Moderator and mediator effects. *Personality and Social Psychology Bulletin, 31*, 393–406.

Lahey, J. N. (2010). International comparison of age discrimination laws. *Research on Aging, 32*, 679–697.

Longman, P. (1986). Age wars: The coming battle between young and old. *The Futurist, 20*, 8–11.

Mead, M. (1970). *Culture and commitment: A study of the generation gap*. New York: Wiley.

North, M. S., & Fiske, S. T. (2012). An inconvenienced youth? Ageism and its potential intergenerational roots. *Psychological Bulletin, 138*, 982–997.

Pettigrew, T. F., & Tropp, L. R. (2006). A metaanalytic test of intergroup contact theory. *Journal of Personality and Social Psychology, 90*, 751–783.
Riley, M. W., & Riley, J. W., Jr. (2000). Age integration: Conceptual and historical background. *The Gerontologist, 40*, 266–270.
Smith, K. P., & Christakis, N. A. (2008). Social networks and health. *Annual Review of Sociology, 34*, 405–429.
Smith, J. A., McPherson, M., & Smith-Lovin, L. (2014). Social distance in the United States: Sex, race, religion, age, and education homophily among confidants, 1985 to 2004. *American Sociological Review, 79*, 432–456.
Uhlenberg, P. (2000). Why study age integration? *The Gerontologist, 40*, 276–281.
Uhlenberg, P., & De Jong Gierveld, J. (2004). Age-segregation in later life: An examination of personal networks. *Ageing and Society, 24*, 5–28.
Vanderbeck, R. M. (2007). Intergenerational geographies: Age relations, segregation and re-engagements. *Geography Compass, 1/2*, 200–221.
Zaidi, A., Gasior, K., Hofmarcher, M. M., Lelkes, O., Marin, B., Rodrigues, R., Schmidt, A., Vanhuysse, P., & Zolyomi, E. (2013). *Active Ageing Index 2012: Concept, methodology and final results*. EC/UNECE, Active Ageing Index Project, UNECE Grant ECE/GC/2012/003. Vienna: European Centre for Social Welfare Policy and Research. Retrieved from http://www.euro.centre.org/data/aai/1253897823_70974.pdf

Dr. Pearl A. Dykstra is Professor of Sociology at the Erasmus University Rotterdam. Her research focuses on ageing and intergenerational relations.

Dr. Maria Fleischmann works as Research Associate at University College London. Her research interests are work, retirement and health.

3

The Expected, Evaluated, Perceived, Valued and Prevalent Social Roles of Older People: Are They by Consent?

Lucie Vidovićová

3.1 Introduction

Europe is traditionally called the "old continent", but the ageing of its population is a rather modern phenomenon with a profound impact on all aspects of society. The longer and somewhat healthier lives of both the young-old and old-old have prompted calls for action in order for societies to address the challenges and opportunities brought by these demographic changes (Timonen 2008). Both national- and EU-level policies attempt to address this issue by various means, the call for active ageing, attempts to raise employment rates among older workers and, rather delayed in time, support for informal caregivers of the elderly being the most typical examples of national policies (cf. Avramov and Mašková 2003; Gilbert and Powell 2005). Yet how are these top-down actions

Acknowledgement: This work was supported by the Czech Scientific Agency under Grant number P404/13-34958S. "Role overload: Grandparenting in the Era of Active Ageing".

L. Vidovićová (✉)
Faculty of Social Studies, Masaryk University, Brno, Czech Republic

A. Zaidi (eds.), *Building Evidence for Active Ageing Policies*,
https://doi.org/10.1007/978-981-10-6017-5_3

resonating in the lives of ageing people themselves? Are they in line with the values, wishes and expectations which they themselves hold for the latter phases of their lives? One way to find out the answer to this question is to compare both the socially and politically expected and evaluated roles of older people on the one hand, and the perceived, valued and prevalent roles of older people on the other.

In this chapter, we use the Active Ageing Index (AAI) as an "operationalisation" or representation of the dominant policy paradigm and compare it with selected results of a secondary analysis of the Eurobarometer Active Ageing Survey (2012), Life Roles Survey (2014) and the European Value Study (2008) as sources of information on (a) the perceived roles and contributions of older people in today's Europe and (b) older peoples' own preferences and prevalent role sets. By this exercise, we want to draw attention to the risk of possible discrepancies between these two perspectives.

3.2 The Demography-Related Context and Roles Older People Are Expected to Play in Actively Ageing Societies

Any social development is determined or at least profoundly influenced by the demographic structures of societies. The proportion of older people is increasing rapidly.[1] Within Europe, such a development causes considerable unease. In the Eurobarometer Active Ageing Survey (2012), when asked how much they were concerned about the rising number of people aged 65 and above, respondents from nine countries—Bulgaria, Greece, Poland, Portugal, Latvia, Lithuania, Croatia, Spain and Italy—expressed significant concerns, while only those from Iceland, Sweden and Norway, on average, tended to be less concerned.[2] In other words, going from north-west to south-east, the level of concern about the greying of the population increases, being most pronounced in the new member states.

The EU Green Paper on demographic change (European Commission 2005) comprehensively addresses the nature of the challenge Europe is facing and the urgency to take action. These demographic trends and related concerns, both publicly and privately expressed, call for a response from public policymakers, who need to prepare "today" for the challenges of "tomorrow". As a result, many attempts to make adjustments to the pension systems and to enhance the labour market participation of older workers up to the age of 70 were, have been and will continue to be made (cf. Kryńska and Szukalski 2013), as these are seen as one of the most important ways of ensuring economic stability and sustainability.

The concept of "active ageing" is a general and overreaching umbrella, philosophy or resource for many of these policies. An often-quoted definition given by the WHO (2002) states that active ageing is the process of optimising opportunities for health, participation and security in order to enhance the quality of life as people age. However, many critical voices have been raised in the last decade against the economic reductionism of what was originally (i.e. in the WHO definition) meant to have been a general statement rather than a policy agenda (Walker 2002). In European Commission documents, "active ageing means helping people stay in charge of their own lives for as long as possible as they age and, where possible, to contribute to the economy and society".[3] Active ageing, as presented in various national policy documents Europe-wide, now refers to the situation in which older people continue to have an opportunity to participate in the formal labour market, as well as engage in other unpaid productive activities, which may range from care provision for family members to volunteering.[4] Along these lines, the broad ideas of active ageing as "a quality of life issue" are often distilled into three major policies relating to only three major roles: productive workers, volunteers and carers for the frail elderly. As such, regardless of various criticisms (Katz 2000; Calasanti and King 2005; Gilleard and Higgs 2000; Buys and Miller 2012), "active ageing" has topped the policy agenda in terms of preparation for demographic change in Europe (Foster and Walker 2015; Perek-Bialas et al. 2006; Walker 2006).

3.3 Social Roles in Higher Age

3.3.1 The Active Ageing Index as an Evaluation of Role Performance on the National Level

The Active Ageing Index (hereinafter AAI) is a tool that offers a framework that can be used by policymakers and stakeholders to identify challenges and opportunities linked with population ageing and also enables them to identify priorities and set targets for improvements based on international comparison (Zaidi et al. 2013). The index score for individual countries expresses the extent to which their older people's potential is used, and the extent to which older people are enabled and encouraged to participate in the economy and society. In other words, it may be used to measure how well older people of different nationalities are doing in meeting the abovementioned expectations drawn from active ageing discourses. As one would expect, this tool closely reflects the ideas of active ageing by the selection of indicators and by the weights ascribed to them. If we deconstructed the AAI indicators into nine major topics and multiplied them by the weights ascribed to each of them, including the weight for the domain, we would see that, for example, employment indicators account for 36%, health indicators 18% and volunteering 16% of the overall index evaluation.[5] This leads to the possible conclusion that the three roles of worker, (non)patient and volunteer are those most highly *evaluated*, leaving activities such as grandparenting, caring or leisure somewhat behind. As a result, the countries with the highest performance levels in the employment dimension also tend to be the overall winners of the AAI statistical exercise.

3.3.2 Roles to Play: As "Society" Sees It

Basic analyses of active ageing discourses and policy agendas as well as the analysis of the AAI structure above indicate that there is a demographically driven need for the greater inclusion of older people in society, and for an increase in the use of their potential. The plan to achieve this and to measure its success focuses mainly on productive roles. Let us now have a look at the "vox populi" to assess what society sees as the main roles of older people. Here, the Eurobarometer 378 Active Ageing survey shows, partially in opposition to what the AAI says about the underdeveloped potential of older people, that Europeans see seniors as playing a large part in public life through politics, through economic participation and through local community engagement and engagement in families (Table 3.1), and would welcome them (or even require them?) to play at least the same, or an even bigger role, in these domains in the future (Table 3.2), which is indeed in line with the AAI rationale.

We may notice, however, that in the realm of politics about 22% of respondents and in the field of the economy (employment, consumption?) 18% of respondents would prefer older people to play a somewhat weaker role. In the private sphere of families, such opposition is displayed by only 9% of Eurobarometer respondents. Table 3.3 confirms this picture by showing the perceived contribution of older people, which is again strongly evident in families, where older people contribute with grandparenting and by providing care and financial support.

The presentation of this kind of empirical data in separate lines and columns may sometimes mislead us into perceiving these contributions as if they were the contributions of several distinct people: "the carer", "the grandparent" and "the worker". Yet, in reality, these are very often different *roles* played by a single person in a composite role set. Therefore, we may try to sum these different contributions to arrive at the indicator of total perceived contribution of older people (and this is shown in Table 3.4 in the penultimate seventh column) for different European societies.

Table 3.1 Perceived role of people aged 55+ (%)

	Participation in politics	Participation in the economy	Participation in the local community	Participation in families
Major role	72	66	68	82
Minor role	23	30	28	16
No role at all	5	4	4	2

Source: Eurobarometer 378 Active Ageing (2012); $N = 31{,}280$, own calculation

Table 3.2 Preferred roles for older people according to public opinion (%)

	In politics	In the economy	In local community	In families
More of a role	39	41	43	43
The same role as now	39	41	42	48
Less of a role	22	18	15	9

Source: Eurobarometer 378 Active Ageing (2012); $N = 31{,}280$, own calculation

Table 3.3 Perceived contribution of people 55+ to society—selected roles (%)

Contribution as	Contribute greatly	Contribute a little	Do not contribute at all
Workers	64	32	3
Consumers	69	29	2
Volunteers	54	37	9
Financial support to family	73	24	3
Family care providers	72	25	3
Grandchild care providers	81	17	2

Source: Eurobarometer Active Ageing (2012); *N* = 31,280, own calculation

Table 3.4 Perceived contribution of older people 55+ (in %)—selected countries, ranked by the sum

Rank		Grandparent	Financial support	Worker	Volunteer	Sum	AAI 2014 ranking
1	Cyprus	95	94	73	58	320	13
6	Germany E	89	81	64	85	320	8
7	Germany W	82	83	72	81	318	
8	Italy	83	85	80	68	317	14
10	Austria	80	81	70	73	305	10
11	United Kingdom	82	76	66	71	295	5
15	France	85	77	59	73	294	9
16	Sweden	78	73	80	57	288	1
18	Netherlands	77	61	63	87	287	3
19	Slovakia	82	72	65	57	276	26
21	Poland	88	72	58	37	255	27
23	Croatia	89	67	60	29	245	21
24	Slovenia	82	75	46	42	245	23
31	Czech Rep.	66	61	62	34	222	12

Source: Eurobarometer Active Ageing (2012); *N* = 31,280, own calculation

We should not be surprised that these summed contributions[6] differ according to the particular nation state, which reflects structural and policy differences as well as cultural differences, including objective and subjective opportunities and their take ups. We may also notice that the ranking of the countries by total contribution is not equal to the AAI ranking (eighth column), and we elaborate this relation more in detail in a later section.

Although we suggested above that policies rooted in the active ageing concept expect older citizens to be active, working and productive,[7] we can now see from the above data that public opinion appears to demand slightly different forms of engagement from ageing persons: older people are expected to help, to assist and to provide resources both at home and in the community, with the roles of grandparent, family support provider and carer *perceived* to be the most significant roles of older age.

3.3.3 Preferred Roles: From the Perspective of Older People Themselves

Now let us move on to another piece of the puzzle: the values and preferences of older people themselves. Here, we will use the European Value Study (EVS 2008), in which Europeans were asked about the extent to which different life areas were important to them. The general results are shown in Fig. 3.1.

As Fig. 3.1 shows, the family was the most important realm of life for respondents of the EVS, followed by work and friends, which achieved almost equal scores. From the cohort point of view displayed in Fig. 3.1, the importance of all areas declines with age, (again) except family and

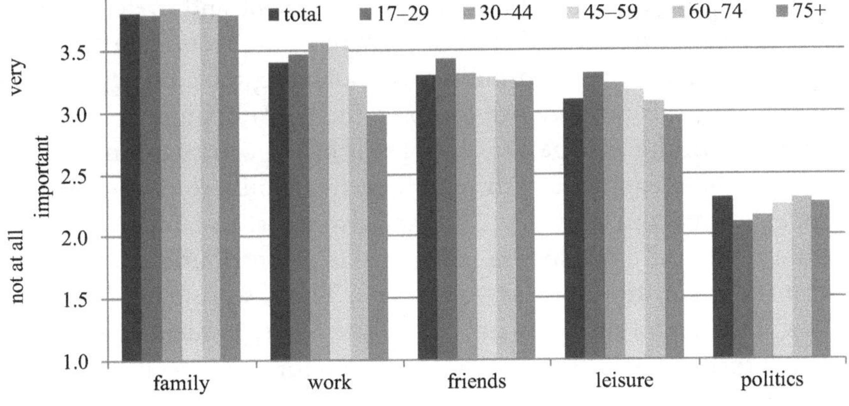

Fig. 3.1 Level of importance of different life domains by age of respondent (mean value). Source: European Value Study (2008), $N = 56{,}190$; own calculations

politics, the first being stable and the second increasing. This age-cohort dynamic of the importance of work also varies by national context (data omitted), which gives a hint about the possible influences of cultural and/or welfare state-policy contexts.

In general, the individually *valued* social roles seem to be family (grand/parent, partner), and leisure roles (e.g. friend). Having said this, it could be interesting, in the next section of this chapter, to examine how the AAI ranking relates to these abovementioned soft measures of the perceived contribution of older people within society and their own preferences.

3.4 Expected, Evaluated, Perceived and Valued Roles: Are They by Consent?

We began this chapter with the suspicion that active ageing policies may be overemphasising (economically) productive activities, and underestimating, not so much in rhetoric as in praxis, other important social roles played by older people. One way to test this is to look at the proxy measures of active ageing success expressed in AAI analytics in order to see whether they are in compliance with socially visible and individually preferred roles on the level of national states. In other words, if the AAI ranking truly mirrors the broad understanding of active ageing as an overreaching concept including the most prevalent and preferred social roles, statistical tests should show a high and unidirectional correlation between all of these selected indicators, indicating the closeness of these concepts. For this purpose, we combine the AAI rankings of individual countries with the average perceived levels of the contributions of older people in those countries with respect to two family roles (providers of grandchild care and providers of financial support to families) and two socio-economically productive roles (making contributions as workers and making contributions as volunteers). Then, we include the mean values of the importance of work and leisure for respondents aged 50–70 years, as these constitute the top target group for active ageing policies.

The logic of the AAI rankings awards the highest values to those EU countries which have the widest gap in potential to fill (better performing countries are in better positions on the list). The contribution of older people was measured by the Eurobarometer Active Ageing Survey (2012) and the higher shares of respondent's agreement show bigger perceived contributions of older people. The importance of life domains was measured on a scale from 1 "of no importance" to 4 "extremely important". Now, all coefficients related to the AAI rankings displayed in Table 3.5 are of solid statistical value, but are a mixture of direct and inverse proportions. First, we see that better AAI rankings were achieved by those EU countries in which the contributions of older people as financial supporters of families, as workers and as volunteers were perceived as the biggest. However, it is not the worker which plays the major role here, but rather the volunteer. In other words, those countries recognised by the AAI as using older people's potential the most are also those whose residents highly recognise and value the role of older people as volunteers, whether it is because of the fact that they volunteer more or because their work is highly recognisable within the community and society, or both. Secondly, even when the AAI strives to include the evaluation of unpaid work and familial contributions (Zaidi et al. 2013), the widely recognised contribution of grandparents devoted to providing care for grandchildren goes in the opposite direction; or, with a little exaggeration, we may say that the AAI "punishes" the countries where the perceived level of work and contribution in what is labelled as the most important life domain—the family—is the highest. Finally, and quite surprisingly, considering the importance of life domains for older respondents, better individual AAI country rankings are correlated with the lower importance of work and the higher importance of leisure time. As time passes, some of the correlations strengthen; for example, the perception of the contribution of older people is more tightly related to the AAI ranking of 2014 than that of 2012. Meanwhile, others weaken, as, for example, the relation of AAI 2014 to the perceived level of financial support to the family and consequently to the total perceived contribution. These changes are a good illustration of the dynamics of background structural influences, and, therefore, also of the viability of their further alteration and cultivation.

3.5 Roles Played

So far, we have looked at the expected, evaluated, perceived and valued roles of older people, taking into account various societal levels and different angles. What seem to be missing are the roles which are actually played by older people, as they may differ from those either expected or valued. To show the most *prevalent* roles of older people, we can present the results of a recent representative survey, "Life roles—2014" ($N = 730$), carried out in the Czech Republic among respondents aged 50–70. Here, the young-old were asked "What are your roles in your current stage of life?" and presented with 14 different roles as choices. The roles of friend (chosen by 91%), parent (90%), partner (79%), sibling (79%) and grandparent (62%) were the most prevalent, followed by worker (56%), learner (54%), enjoyer of leisure time (42%), child (41%), patient (41%), child-in-law (31%), believer (25%), carer (24%), volunteer (7%) and other (1%). Further, the respondents were asked to choose the most important role from their list of choices. Here, the survey once again confirmed the roles of parent, partner and grandparent as the most important (chosen by 35%, 30% and 14%, respectively). The grandparental role, that is, the role covered by no social policies in many national contexts, was the one that brings the greatest satisfaction (level of happiness) to both sexes; in contrast, the role of worker, being the most important for only 6%, was considered the most time consuming and the key reason for perceived role overload.

3.6 Conclusions and Discussion

In this chapter, we considered the question of whether the expected, evaluated, perceived, valued and prevalent social roles of older people are in agreement in contemporary Europe. We have argued that, in line with concerns about future demographic development, active ageing policies and measures expect older people to be active and contributing productively, working, volunteering, learning, providing intergenerational care, consuming and, in many ways, resisting "old age". Different social actors tend to exercise various forms of, oftentimes conflicting, pressure on

Table 3.5 Correlation coefficients (Pearson's) for AAI 2012 and 2014 rankings, the perceived contribution of older people in various areas, and the importance of work and leisure for older people

	AAI ranking 2012	AAI ranking 2014	As grandparents	Financial	As workers	AS volunteers	Total contribution	Work importance
AAI ranking 2014	0.968							
Contribution as grandparents	0.301	0.347						
Financial support to family	−0.227	−0.127	0.735					
Contribution as workers	−0.431	−0.410	0.377	0.605				
Contribution as volunteers	−0.684	−0.691	0.192	0.481	0.441			
Total contribution	−0.511	−0.471	0.644	0.862	0.778	0.784		
Work importance for 50–70	0.333	0.409	0.530	0.346	−0.116	−0.226	0.062	
Leisure importance for 50–70	−0.380	−0.292	0.134	0.399	0.515	0.321	0.470	0.319

older people to keep healthy; exercise; socialise; have fun; keep up to date; stay in the labour market and yet leave it to make space for younger people; act one's age and yet beat it; and value one's work and yet be there for the family. For example, "The senior of the future: studies hard, engages in business, and is gregarious" (Jiřička 2013) is how the key policy document the National Action Plan for Positive Ageing in the Czech Republic is presented to the public in the mass media, clearly illustrating for us the discursive pressure.

Alongside the public and policy pressures, pressures from the compositions of individual role sets may arise. Among the most pronounced social roles an adult individual plays are family roles, such as partner, parent or adult child, which are often understood as the most important and central roles, but, at the same time, may often be filled with duties relating to care and (therefore) be time and energy consuming (Reitzes and Mutran 2002). As there are multiple role responsibilities within each individual role set, these create demands which compete for time and energy, leading to inter-role conflicts. Goode (1960 in Parris Stephens et al. 2001) even holds that role obligations in total make inter-role conflicts the norm. According to the already above-quoted survey on the social roles of the "young-old", people aged 50–70 have, on average, seven different social roles and about half of such people talk about experiencing role overload giving rise to a considerable level of stress (Vidovićová et al. 2015). When asked which role could be blamed the most for this situation, 42% answered "the employee/worker" role, followed significantly less frequently by the role of "the patient" (13%) or some of the family roles. The roles of worker, partner and grandparent were labelled as those which are the hardest to combine. At the same time, the role of the grandparent was recognised not as the most important—that position was taken by the parental (35%) and spousal (30%) roles—but as the one which brought the most joy and happiness to the young-old. If respondents had no grandchildren, the most joyful role was the one of friend.

If policies are ideas guiding actions towards those that are most likely to achieve a desired outcome, there should be an overlap between what people want, need and desire, and what the respective policies are aiming to achieve, and what outcomes are evaluated (cf. Ranzijn 2004). So far, with respect to the active ageing concept, the overlap appears to be rather limited, as Fig. 3.2, recapitulating the different emphases of the different perspectives discussed above, illustrates. Evaluative measures like those considered in the AAI could play an important role in bringing the roles that matter to ageing people more into the foreground, highlighting, even more than today, the weight of non-labour market roles.

Given the limited space here, it is not possible to discuss in detail the different cultural meanings attached to such key concepts as age, values, or work and its importance, and so on. For example, concerns about demographic change may be deeply rooted in the meanings and representations of chronological age and the cultural constructions of old age. Similarly, the importance of work (role) may have quite a different meaning and content for an older Polish steel worker compared to an older Swedish university professor. Therefore, the evaluation of socioeconomic and cultural determinants is open to further examination. Another obvious limitation of our analysis is the fact that the time frames of the analysed datasets are non-parallel. Both values and policies and their outcomes do change over time and the information collected in the 2008 EVS may not correspond fully with surveys from 2012, which were used for the construction of the AAI, and so on.

The importance of the AAI as a tool for examining the progress of active ageing policies—one which covers employment rates, voluntary activities, family care, political participation, lifelong learning, physical fitness and health, living conditions, financial and physical security and indicators of the capacity to age actively—is demonstrable and growing. In the Czech Republic, for example, it has been adopted to evaluate the achievements of the National Action Plan for the Support of Positive Ageing for 2013–2017. By such usage, it is becoming a key point of interest for policymakers, politicians and public opinion. For this reason, the continual refining of this index should remain a high priority. Clear support for this argument lies in the strong correlation ($r^2 = 0.502$) between the AAI and the level of concern about the rising numbers of older people mentioned at the very beginning of this chapter. Countries using the potential of their seniors the most are also the ones that are least afraid of the greying of their populations. Or, if we wonder which came first—the chicken or the egg—maybe the countries that are not so concerned about

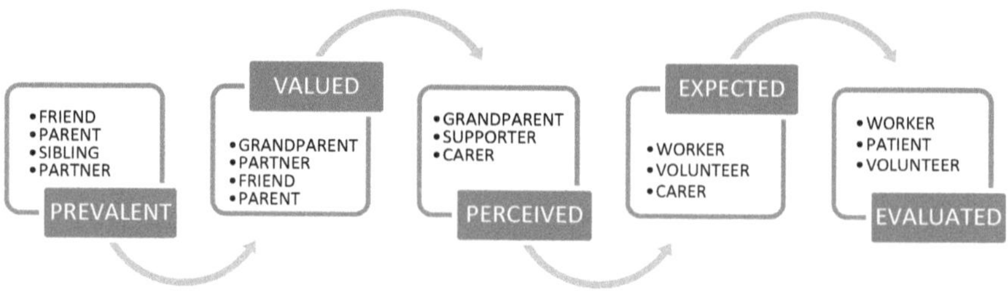

Fig. 3.2 Overview of different approaches to the social roles of older people

(or afraid of) the greying of their populations are therefore able to use the potential of their seniors better.

Notes

1. Source: http://ec.europa.eu/social/main.jsp?catId=1062&langId=en
2. That is, the average answer on a 1- to 4-point scale, where 1 represented "very concerned" and 4 "not at all", obtained in these countries was above 3 points; however, it was still less than 3.5 points; the overall average was 2.7 points; the average value of the indicator in the "very concerned" countries was around 2.3 points (author's own calculations).
3. Source: http://ec.europa.eu/social/main.jsp?catId=1062&langId=en
4. Cf. National Strategic Policy for Active Ageing in Malta (online). Available from: https://activeageing.gov.mt/en/Pages/NSPAA.aspx (cit. 23 January 2015); or National Action Plan for Positive Ageing in the Czech Republic. Available from: http://www.mpsv.cz/files/clanky/22733/NAP_2013_2017.pdf
5. For example, lifelong learning = (10 p. indicator weight × 10 p. domain weight) = 100 points/(9825 sum of all points for all domains/100) = 1%. Other topics include grandparenting 9%; caring 9%; leisure 5%; environment 3%; finances 3%; and learning 1%.
6. These are, of course, perceived contributions as evaluated by the respondents in general, not actual amount of care and work provided by older people. We use this approach to proxy this information.
7. In some sense, caring roles are, today, often considered only as alternatives to paid services, and this rationale is used also in recent discussions on

introducing carers' leave and carers' benefits (i.e. care as work) in the Czech Rep.

References

Avramov, D., & Mašková, M. (2003). *Active ageing in Europe. Population studies*. Strasbourg: Council of Europe Publishing.

Buys, L., & Miller, E. (2012). Active ageing: Developing a quantitative multidimensional measure. In G. Boulton-Lewis & M. Tam (Eds.), *Active ageing, active learning, issues and challenges*. Dordrecht: Springer.

Calasanti, T., & King, N. (2005). Firming the floppy penis. *Men and Masculinities, 8*(1), 3–23. https://doi.org/10.1177/1097184X04268799.

Eurobarometer No. 378 Active Ageing. (2012). [Online] Retrieved from http://ec.europa.eu/public_opinion/archives/ebs/ebs_378_en.pdf

European Commission. (2005). *Green Paper Confronting demographic change: A new solidarity between the generations*. [Online]. Retrieved January 24, 2015, from http://europa.eu/legislation_summaries/employment_and_social_policy/situation_in_europe/c10128_en.htm

European Value Study. (2008). [Online]. Accessed January 24, 2015, from http://www.gesis.org/en/services/data-analysis/survey-data/european-values-study/

Foster, L., & Walker, A. (2015). Active and successful aging: A European policy perspective. *The Gerontologist, 55*(1), 83–90. Retrieved January 7, 2016. doi:https://doi.org/10.1093/geront/gnu028

Gilbert, T., & Powell, J. (2005). Family, caring and ageing in the United Kingdom. *Scandinavian Journal of Caring Sciences, 19*(1), 53–57.

Gilleard, C., & Higgs, P. (2000). *Cultures of ageing. Self, citizenship and the body*. Harlow: Pearson Education.

Goode, W. J. (1960). A theory of role strain. *American Sociological Review, 25*, 483–496.

Jiřička, J. (2013, February 13). Senior budoucnosti: pilně se vzdělává, podniká a nestraní se lidí. *Idnes* [Online]. Retrieved January 7, 2016, from http://zpravy.idnes.cz/strategie-starnuti-seniori-dp4-/domaci.aspx?c=A130213_135510_domaci_jj

Katz, S. (2000). Busy bodies: Activity, aging, and the management of everyday life. *Journal of Aging Studies, 14*(2), 135–152.

Kryńska, E., & Szukalski, P. (2013). *Active ageing measures in selected European Union countries*. Final Report. Lodz: University of Lodz.
Life Roles Survey, Technical Report. (2014). *Technická zpráva ze sběru dat – výzkum na téma "Přetížená role"*. Brno: Focus.
Parris Stephens, M. A., Towsend, A. L., Martire, L. M., & Druley, J. A. (2001). Balancing parent care with other roles: Interrole conflict of adult daughter caregivers. *Journal of Gerontology, 56B*(1), 24–34.
Perek-Bialas, J., Ruzik, A., & Vidovićová, L. (2006). Active ageing policies in the Czech Republic and Poland. *International Social Science Journal, 58*(19), 559–570.
Ranzijn, R. (2004). Role ambiguity: Older workers in the demographic transition. *Ageing International, 29*(3), 281–308.
Reitzes, D. C., & Mutran, E. J. (2002). Self-concept as the organization of roles: Importance, centrality, and balance. *Sociological Quarterly, 43*(4), 647–667.
Timonen, V. (2008). *Ageing societies: A comparative introduction*. Maidenhead: Open University Press.
Vidovićová, L., Galčanová, L., & Petrová Kafková, M. (2015). Význam a obsah prarodičovské role u mladých českých seniorů a seniorek [The Meaning and Performance of the Grandparenting Role among Young Old Czechs]. *Sociologický časopis/Czech Sociological Review, 51*(5), 761–782.
Walker, A. (2002). A strategy for active ageing. *International Social Security Review, 55*(1), 121–139.
Walker, A. (2006). Active ageing in employment: Its meaning and potential. *Asia-Pacific Review, 13*(1), 78–93.
WHO. (2002). *Active ageing: A policy framework*. World Health Organisation. [Online]. Retrieved January 7, 2016, from www.who.int/ageing/publications/active_ageing/en/
Zaidi, A., Gasior, K., Hofmarcher, M. M., Lelkes, O., Marin, B., Rodrigues, R., Schmidt, A., Vanhuysse, P., & Zolyomi, E. (2013). *Active ageing index 2012. Concept, methodology, and final results*. Research Memorandum/Methodology Report. Vienna: European Centre. [Online]. Retrieved January 7, 2016, from http://www.euro.centre.org/data/aai/1253897823_70974.pdf

Lucie Vidovićová, PhD, is a sociologist at the Faculty of Social Studies, Masaryk University, and at the Research Institute for Labour and Social Affairs. Her long-term research interests include sociology of ageing, environmental gerontology, age discrimination and active ageing. She is also involved in research projects in the field of family and social policy. She conducts research for national as well as European bodies and works as a consultant on implementation projects.

4

The Active Ageing Index (AAI) and its Relation to the Quality of Life of Older Adults

Marcela Petrová Kafková

4.1 Introduction

The current form of ageing has been fundamentally shaped by the discourse of activity and primarily by the concept of active ageing (Petrová Kafková 2013) that has successfully influenced both academic and public areas (Hasmanová Marhánková 2014). The main goal of active ageing as defined by the World Health Organisation (Active Ageing 2002) is to enhance the quality of life as people age. The means of enhancing quality of life should entail, according to the WHO definition, continuing participation in social, economic, cultural, spiritual and civic affairs and the optimisation of opportunities to maximise health, participation and security (Active Ageing 2002: 12). This definition is original and the most inclusive. Activity in this concept means participation in society and all its spheres even of those individuals who are frail, disabled or in need of care, because 'older people…, who are ill or live with disabilities can remain

This work was supported by the Czech Scientific Agency under Grant number P404/13-34958 F.

M. Petrová Kafková (✉)
Faculty of Social Studies, Masaryk University, Brno, Czech Republic

A. Zaidi (eds.), *Building Evidence for Active Ageing Policies*,
https://doi.org/10.1007/978-981-10-6017-5_4

active contributors to their families, peers, communities and nations' (Active Ageing 2002: 12). The key component of this concept is to support individual responsibility, autonomy and independence. Other used definitions emphasise various aspects of older adults' participation in society and are mainly rooted in economics and participation in the labour market (Petrová Kafková 2013). The Active Ageing Index (AAI) (Zaidi et al. 2013) measures active ageing in accordance with the broad definition proposed by the WHO. Therefore, a crucial question arises: what is the relation between the Active Ageing Index and the quality of life of older adults? In other words, is the quality of life in older age in countries with a high AAI higher than in countries with a lower AAI? Does a better use of older adults' potential bring the improvement of their quality of life?

'The Active Ageing Index aims to take a comprehensive look at the activity and independence of older people' (Active Ageing Index 2015) and it actually shows this way what is (at the national level) older adults' quality of life. It describes older adults' position in society and provides an analytical tool to measure the potential of ageing society. It tries to picture how much older adults' potential is made use of in different countries and to what extent older adults are able to participate in society (Vidovićová and Petrova Kafková 2016). The unstated assumption is greater involvement of older adults whose greater participation contributes to greater prosperity of the whole society and will bring sustainable prosperity of ageing societies despite the threat of increasing expenditure on health and social security. For these reasons, it is a very attractive tool for policymakers. Considering the form and possibilities of the index, it is necessary to discuss the issue in two main directions. First, it is the specific form of the index, that is, the results that the index shows. The second direction is conceptual and relates to the framing of the index by active ageing. Both of these discussions are very closely related. This text aims to contribute to these discussions focusing on the relation of active ageing and 'happy' ageing or quality of life in old age.

The current focus on activity and participation brings a lot of unintended consequences and casts doubts on benefits of both the concept of active ageing and particularly its reduced concepts. Activity is typically associated with the third age (Laslett 1989) that is seen as young, active old age whereas the oldest-old age is accompanied by loss and decline in this concept (Stathi and Simey 2007). Its fulfilment often comes from a biomedical perspective on physical activity and healthy lifestyle, and it basically hierarchises activities, because demanding physical activities are more valuable than other ones (Lassen 2014). Activities that do not follow the logic of the work ethic are systematically made invisible (Biggs 1993). Therefore, many authors (Clarke and Warren 2007; Lassen and Moreira 2014) express the need for a broader concept of activity—a concept that goes beyond structural factors (money, jobs and retirement) or physical functions and offers other ways in which individuals can 'age actively' (Clarke and Warren 2007). Fulfilling the concept should take various forms due to the different degrees of physical and mental abilities of older adults. The need to include the cultural specificity of the active ageing concept is described also by Walker (2002).

It is necessary to mention that various translations of active ageing can be used by different actors and institutions for various purposes (Hasmanová Marhánková 2014). A number of problematic aspects does not stem from the original idea of the World Health Organization, but from different reductions of the inclusive concept. As a result, they have a strongly normative character and tend to create a new 'appropriate' form of ageing, an ideal form of a very energetic life, the fulfilment of which is in fact strongly conditioned by social class (Gilleard and Higgs 2005).

A whole range of activities not fitting into the reduced definitions of active ageing fulfil its purpose. An example might be playing billiards, as Lassen (2014) showed in the example of Danish men playing regularly in the centre for older adults. Participants considered the cultural significance of this activity and the involvement in the community very important. Even playing billiards supports older adults' social participation and fulfils the vision of active ageing, even though it is not an activity improving physical condition. Moreover, as the author shows, the above-mentioned billiards enables the involvement of fragile individuals, by linking activity and passivity in a single, integral unit. Therefore, it can be understood as an inclusive activity. And the inclusion of older adults is the original intention in the concept of active ageing, as defined by the World Health Organization (Active Ageing 2002).

The composite index obviously cannot cover all possible activities and it must always be somehow reducing [closer to the methodological limits of the index see Vidovićová and Petrová Kafková (2016)]. But it is necessary to know the limits and articulate them clearly. The index of active ageing containing the popular phrase 'active ageing' and protected by the major European institutions (United Nations Economic Commission for Europe and European Commission) can become an influential tool to influence both policy settings and to create and strengthen the notion of 'proper' ageing so it carries considerable responsibility. By its nature, summing up the frequency of activities, it contributes to the belief that the more activity the better and it helps to create an excluding dichotomy between activity and passivity (Lassen and Moreira 2014). This dichotomy brings considerable risk that the inclusive strategy becomes a normative, excluding ideology. Therefore, it is important to present the idea of active ageing with its goals and impacts responsively (Hasmanová Marhánková 2014: 26). The ability of sensitive approach is, of course, conditioned by a thorough and deep knowledge. And this text aims at its deepening by studying the relation between the Active Ageing Index and quality of life.

4.2 Methods, Data and Variables

Two sources of data are used. These are the new AAI results for 2012[1] and the third wave of the Europen Quality of Life Survey (EQLS) from the year 2012. Since the AAI is focused primarily on the population aged 55 years and older, I have excluded from the EQLS analysis respondents less than 55 years of age. There are only the 28 member states of the European Union in my sample, so that the selection of countries matches the AAI data. This gives me a sample of 15,971 respondents. I do not consider individual variation, however, and work only with aggregated data for individual countries, that is, $n = 28$. Because of the small sample size, the regression is not applicable but the principal component analysis (PCA) is used to test the proximity of individual indicators both at a general level and at the level of individual countries. PCA associates the used variables into a smaller number of factors where each of the principal components is the linear combination of all variables, which explains the highest proportion of total variance of initial data (OECD 2006).

Quality of life is operationalised by means of subjective well-being concept. Subjective well-being is as one part of quality of life next to quality of living environment and quality of performance (Veenhoven 2001) and treated here as a multidimensional concept measured through three different indicators. The first of these is a measure of happiness, which is captured by the following question: 'Taking all things together on a scale of 1 to 10, how happy would you say you are?' A score of 1 indicates a respondent is very unhappy and 10 means he or she is very happy. The average value in the file is 7.15 ± 2.03 (median 8.00).

The second indicator is life optimism. Here, a higher value indicates a higher degree of life optimism. I have created the indicator from nine items (each scored from 1 to 5): (a) 'I am optimistic about the future'; (b) 'I generally feel that what I do in life is worthwhile'; (c) 'I feel I am free to decide how to live my life'; (d) 'In my daily life, I seldom have time to do the things I really enjoy'; (e) 'I feel left out of society'; (f) 'Life has become so complicated today that I almost can't find my way'; (g) 'I feel that the value of what I do is not recognised by others'; (h) 'Some people look down on me because of my job situation or income'; and (i) 'I feel close to people in the area where I live'. The reliability of the resulting index has been verified (Cronbach's α = 0.741). The average value of the resulting index file is 3.75 ± 0.69 (median 3.78).

Whereas happiness and optimism capture the long-term, general condition of individuals, the third of the selected indexes refers to their current situation. It is an indicator of psychological discomfort, which asks respondents about the previous two weeks and combines three feelings: (A) 'felt particularly tense'; (B) 'felt lonely'; and (C) 'felt downhearted and depressed'. Respondents determine on a six-point scale how often they were affected by these feelings. I have verified the reliability of the index by Cronbach's α, which was 0.815. The items are scored from 1–6; a higher value indicates less frequent feelings of psychological discomfort. The average value of the resulting index is 4.83 ± 1.17 (median 5.00).

All three selected indicators of quality of life are quite strongly related, but they are not completely interchangeable. For the exact values of mutual correlations, see Table 4.1.

The last indicator referring to quality of life is gross domestic product (GDP) per capita. Although GDP refers to the development and wealth

Table 4.1 Mutual correlation (Spearman's rho) of quality of life indicators

	Happiness	Index of discomfort	Index of life optimism
Index of discomfort	0.852**	–	
Index of life optimism	0.810**	0.799**	–
GDP	0.837**	0.720**	0.752**

Source: EQLS (2012), own calculation
Note: $^{**}p < 0.01$

of countries rather than to quality of life, economic indicators in addition to social and subjective well-being represent one of the three common ways for measuring quality of life (Diener and Suh 1997), since quality of life is a complex, multifaceted construct that requires multiple approaches from different theoretical angles (ibid: 214). And although the relation between GDP and well-being is very complex, Hagerty and Veenhoven (2003) show that higher GDP contributes to greater happiness which has been confirmed also by strong correlations with indicators of subjective well-being (see Table 4.1). For the exact values of selected indicators at the level of countries, see Table 4.2.

4.3 Results

First, I will focus on the relationship between overall AAI scores and the indicators of quality of life. The feeling of happiness as a basic indicator of quality of life correlates well with the overall AAI (Spearman's rho = 0.796, $p > 0.01$), and in the countries with higher levels of AAI, people are, on average, happier. There is also a similarly strong correlation with GDP (Spearman's rho = 0.781, $p > 0.01$). What is slightly weaker is the relationship of the total AAI and the discomfort index (Spearman's rho = 0.692, $p > 0.01$) as same as the life optimism index (Spearman's rho = 0.707, $p > 0.01$). These results suggest that the AAI is relatively successful in capturing the quality of life of respondents.

In the next step, I have looked at the relationship between individual dimensions and items in the AAI and the three selected indicators for quality of life. I also consider the links between the individual items with the total value of the AAI. As can be seen from Table 4.3, out of the four

Table 4.2 Average values of quality of life indicators according to countries

	Happiness	Index of discomfort	Index of life optimism	GDP per capita
Austria	7.88	5.29	4.16	32,400
Belgium	7.77	5.05	3.69	29,900
Bulgaria	5.46	4.16	3.30	11,600
Croatia	7.01	4.81	3.73	15,200
Cyprus	7.31	4.44	3.43	23,700
Czech Republic	6.81	4.68	3.51	20,100
Denmark	8.28	5.56	4.34	31,500
Estonia	6.44	4.74	3.58	16,900
Finland	8.20	5.36	3.92	28,800
France	7.35	4.77	3.81	27,300
Germany	7.34	4.99	3.99	30,300
Greece	6.24	3.92	3.38	24,497
Hungary	6.61	4.52	3.65	16,500
Ireland	8.07	5.37	3.88	32,299
Italy	6.90	4.40	3.70	25,100
Latvia	6.10	4.61	3.60	14,700
Lithuania	6.33	4.62	3.55	16,600
Luxembourg	7.86	4.93	4.03	68,100
Malta	7.14	4.89	3.70	21,800
Netherlands	7.77	5.22	3.96	32,900
Poland	6.93	4.75	3.51	16,200
Portugal	6.78	4.54	3.65	19,500
Romania	6.25	4.41	3.58	11,800
Slovakia	6.55	4.92	3.53	18,400
Slovenia	6.82	4.96	3.72	21,000
Spain	7.57	4.85	3.91	24,700
Sweden	8.16	5.37	4.27	31,800
United Kingdom	7.98	5.14	3.71	27,300
Total	7.14	4.83	3.74	24,675

dimensions of the AAI, independent living and capacity for AA have the closest correlation with the quality of life indicators. Social participation correlates with quality of life indicators in a moderate way. Concerning the fourth dimension, employment, the values of the correlation coefficients are very low.

A closer look at each item within the dimension of employment reveals that quality of life correlates only with the rate of employment for those 55–59 years old and only weakly with the employment rate of those 60–64 years old (although the value is not statistically significant). In these

Table 4.3 Correlation coefficients (Spearman's rho) for indicators of quality of life and individual items in the AAI

	Happiness	Index of discomfort	Index of life optimism	GDP per capita	AAI total
1. Employment	0.288	0.298	0.248	0.228	0.682**
1.1 Employment rate 55–59	0.372	0.411*	0.317	0.375*	0.752**
1.2 Employment rate 60–64	0.250	0.259	0.217	0.216	0.612**
1.3 Employment rate 65–69	0.065	0.085	0.038	−0.026	0.383*
1.4 Employment rate 70–74	0.079	0.125	0.078	−0.028	0.321
2. Social participation	0.789**	0.568**	0.672**	0.820**	0.739**
2.1 Voluntary activities	0.898**	0.824**	0.862**	0.879**	0.871**
2.2 Care to children, grandchildren	−0.119	−0.296	−0.217	−0.047	−0.243
2.3 Care to older adults	0.194	0.053	0.028	0.048	0.092
2.4 Political participation	0.808**	0.620**	0.675**	0.746**	0.814**
3. Independent living	0.884**	0.820**	0.839**	0.840**	0.824**
3.1 Physical exercise	0.774**	0.756**	0.714**	0.728**	0.768**
3.2 No unmet needs of health and dental care	0.569**	0.652**	0.513**	0.668**	0.442*
3.3 Independent living arrangements	0.695**	0.652**	0.633**	0.731**	0.866**
3.4 Relative median income	−0.045	−0.193	0.129	0.107	−0.223
3.5 No poverty risk	0.086	0.143	0.134	0.187	0.240
3.6 No material deprivation	0.796**	0.751**	0.801**	0.771**	0.698**
3.7 Physical safety	0.730**	0.673**	0.584**	0.461*	0.453*

(continued)

Table 4.3 (continued)

	Happiness	Index of discomfort	Index of life optimism	GDP per capita	AAI total
3.8 Lifelong learning	0.633**	0.643**	0.566**	0.591**	0.736**
4. Capacity for AA	0.887**	0.767**	0.770**	0.812**	0.844**
4.1 RLE achievement of 50 years at age 55	0.537**	0.245	0.404*	0.641**	0.439*
4.2 Share of healthy life years in the RLE at age 55	0.523**	0.379*	0.310	0.393*	0.408*
4.3 Mental well-being	0.886**	0.742**	0.840**	0.832**	0.818**
4.4 Use of ICT	0.750**	0.812**	0.721**	0.755**	0.794**
4.5 Social connectedness	0.649**	0.532**	0.606**	0.587**	0.598**
4.6 Educational attainment	−0.160	0.166	−0.079	−0.175	0.041

Source: AAI (2002) from Zaidi and Stanton (2015) and EQLS (2012); own calculation
Note: $^{**}p < 0.01$, $^{*}p < 0.05$

cases, a higher employment rate brings also a higher quality of life. Somewhat troubling, however, are the results in the employment rate of 65–69-year-olds and 70–74-year-olds. Although the negative correlation of some coefficients is not statistically significant, the negative sign does suggest that higher levels of employment may be (weakly) associated with lower quality of life. The highest employment rates for 70–74-year-olds, according to the 2012 AAI, are found in Romania (20.4), Portugal (16.6) and Estonia (12), countries whose GDP is among the lowest in the European Union. The same applies to the level of happiness. The European average is 7.14 (see above), and the average values for Romania, Portugal and Estonia are 6.25, 6.78 and 6.44, respectively. These two items are also significantly less correlated with the total value of the AAI. As Zaidi et al. (2013) suggest, it is possible that high employment rates in higher age groups are not only a reflection of the relative lack of barriers in the labour market, but also a result of low pensions, forcing older people to work at an advanced age.

The second dimension—social participation—correlates with quality of life for two of the four indicators. The correlation between voluntary activities and the quality of life indicators is statistically significant, a finding supported by a number of studies (McMunn et al. 2009; Morrow-Howell et al. 2003; Wilson 2000; Warburton et al. 2001). The different proportion of volunteers in a country is subject in particular to its socio-cultural environment (Erlinghagen and Hank 2006); for example, Southern and Eastern European countries have lower rates of volunteering than Nordic countries. However, the form of assistance varies. While in Nordic countries, it is more formalised, in the South, it takes the form of regular help in the community (Manual 2011). Higher rates of political participation in the country also appear to contribute to quality of life. In contrast, the results suggest that help to others, such as caring for children or grandchildren, or older adults in need, does not benefit the quality of life of individuals. The indication that the increasing amount of care for grandchildren or children decreases the quality of life for individuals is a somewhat surprising result. Due to the low value of the correlation coefficient, however, we only can talk about the sign of the relationship.

The negative impact of care is supported by numerous studies (Minkler and Fuller-Thomson 1999; Park 2006) in which grandparents care for grandchildren, representing missing parental care. This indicator also negatively correlates with the total value of the AAI, even if the value of the correlation coefficient is low. The relationship between care for the elderly and quality of life, according to the correlation coefficients, is only slightly stronger. Given the fact that intensive care of the elderly is limited to a relatively short period of time, the strongest correlation is found with the index of discomfort, which unlike the other indicators of quality of life, captures better the current situation rather than a general, long-term condition. The indicator of care for grandchildren or children, in addition to the overall AAI, does not correlate with quality of life. The reason may be that intensive care for older adults is a highly time-consuming activity, which severely limits the available time for other activities that are in public discourse often recognised as more valuable and more relevant to the concept of active ageing.

Independent living, as a third dimension of the AAI, correlates very closely with quality of life. For the majority of its indicators, increasing their rate increases quality of life. These items are also quite closely connected with the AAI total value. The exceptions are relative median income and no poverty risk. Relative median income is not correlated with quality of life, and with the total value of the AAI, it correlates negatively. In other words, increasing relative median income decreases the total value of the AAI. No poverty risk indicator does not correlate with quality of life indicators at all, with the AAI total value only weakly.

What correlates strongly with quality of life and with the total AAI is the last dimension, that is, capacity for active ageing, but the correlation is not applicable for its individual items. Somewhat weaker correlations are evident in the two first indicators based on healthy life years. The indicator of educational attainment correlates neither with quality of life nor with the total score of AAI. The social connectedness indicator does not reach very strong correlations either.

To summarise, I can state that the correlation coefficients for individual items in the AAI and quality of life show substantial differences in the closeness of the relationship. The majority of indicators relate to quality of life, albeit by varying degrees, but a small proportion of the coefficients are equal to zero or even negative.

As the next step, the relation between quality of life and the Active Ageing Index was tested using principal component analysis. First, individual indicators were included in the analysis, but the result was not statistically significant. However, the inclusion of results for individual dimensions was successful. Four dimensions of the Active Ageing Index and four quality of life indicators disintegrate into two factors[2] based on PCA.

The first factor, 'quality of life', includes most of the indicators and accounts for 69.3% of the variance. The second factor accounts for 13.4% of the variance and is loaded by the variable 'employment'; for this reason, it has been labelled as 'economic participation'. Taken together, these two components capture most of the variability in the data (the two principal components together account for 82.7% of the total variance). The result is clearly visible through component plots (Fig. 4.1). The figure shows clearly that all variables correlate positively with quality of life and the values of these correlations are quite strong. The variable 'employment' is basically a kind of an outlier and its relation with the factor of quality of life is close to zero. PCA confirms the results of the correlations

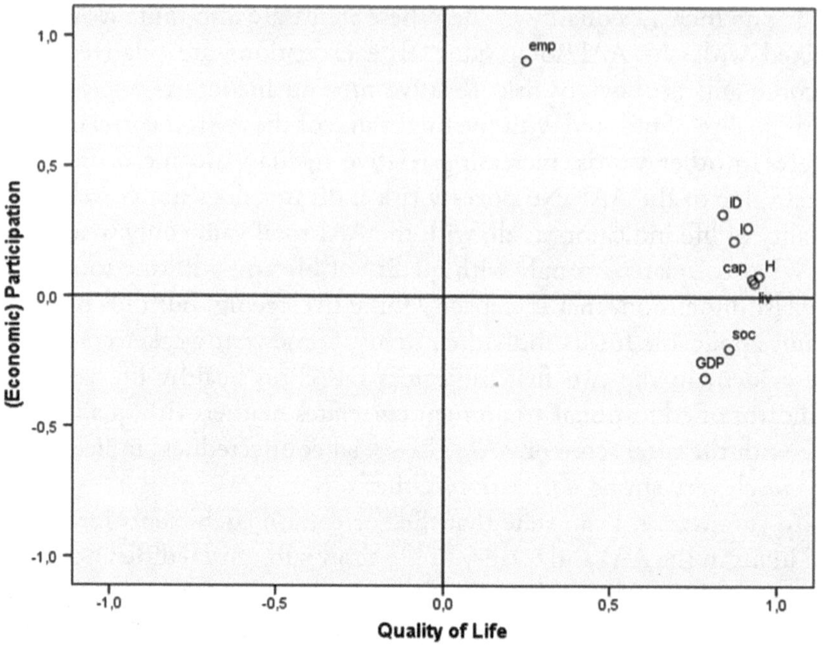

Fig. 4.1 Quality of life and (economic) participation (principal component analysis)

showing that the index of active ageing describes quality of life well except the employment dimension, which is neither beneficial nor reductive for quality of life. The chart also shows that the dimensions 'independent living' and 'capacity for AA' are most closely related with happiness. The dimension 'participation in society' is close to GDP.

The results on the vertical axis are substantial as well. The overall resulting correlations are very weak in this case, in the case of GDP and participation in society even negative. The explanation in the case of GDP might be the fact that economic participation especially in older age is often conditioned by financial necessity. It is necessary to mention in the context of the negative correlation between participation in society and the second factor that the time spent at work reduces the time available for other activities especially leisure time activities, including meeting friends, but also the time for voluntary work or care for others (closer to the issue of complementarity of older adults' roles; Vidovićová et al. 2015).

We may conclude that the results of both analytical procedures are consistent and confirm the low benefit of economic participation to older adults' quality of life. The next step locates countries within the space defined by both principal components (see Fig. 4.2), giving a comprehensive picture of the positions of individual countries. The dimension of economic participation divides the European Union countries into almost halves (Precisely 12: 16) according to the level of older adults' economic participation. The position on the horizontal axis shows that high quality of life exists in the countries with both low and high economic participation. There are even more countries with low economic participation of older adults correlating positively with quality of life than those with high economic participation of older adults. The figure also shows a significant difference between the original member states and the new EU members of Central and Eastern Europe, namely

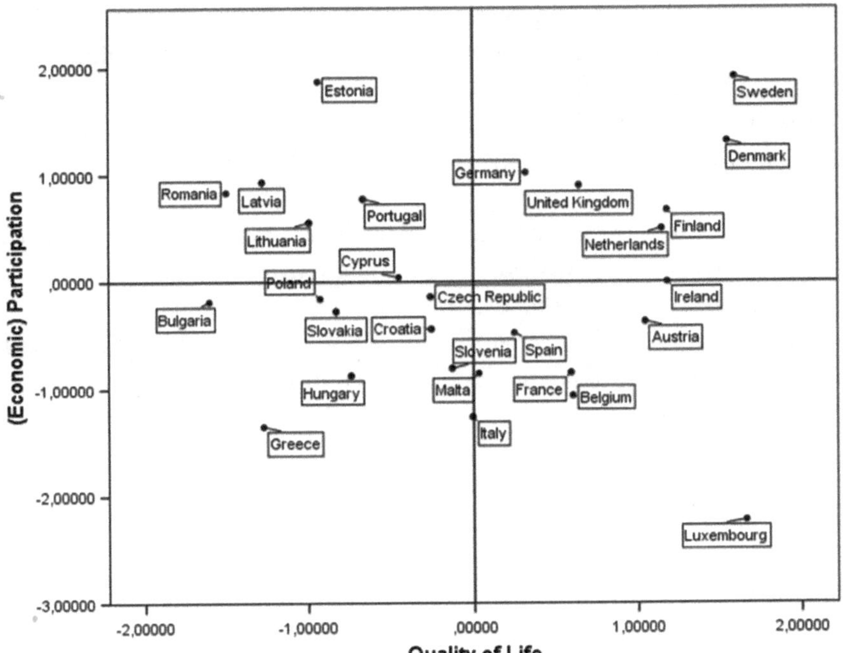

Fig. 4.2 Relation between quality of life and the dimension of the AAI at the level of EU countries

in the distribution along the horizontal axis. The vertical axis reminds us of the fact that there are different types of social states in Europe and it divides the countries into Northern vs. Southern Europe.

An important result is the location of some new EU member countries. Overall AAI scores showed that the top countries are Estonia (10th position) and the Czech Republic (11th position), which are ranked approximately in the middle of EU member countries. The results, presented above, show an important position of Slovenia (23rd position in the total AAI) and Croatia (18th position in the total AAI) on the horizontal axis of quality of life. Estonia is placed more to the left, that is, among countries with lower quality of life, but it is high on the vertical axis (ranked 2nd in the dimension of economic participation after Sweden, which guaranteed a high position in the overall ranking of countries according to the AAI). Also the other two Baltic States dropped in our analyses among the countries with the lowest quality of life despite considerable economic participation. Conversely, Austria (13th in the total AAI) ranked among the countries with high quality of life, and Belgium (15th in total AAI) reached the level of France (7th in the total AAI).

Summing up the above-presented results, we can conclude that although the Active Ageing Index indicates older adults' quality of life in the European Union satisfyingly, putting significance on economic participation leads to inconsistencies. Especially the position of the countries below the average, but with high economic participation of older adults, has been overrated, as the example can serve Estonia. The presented results indicate repeatedly that the importance placed on employment in the index brings this analytical tool further away from the original inclusive definition of active ageing.

4.4 Discussion

The Active Ageing Index is a comprehensive and elaborate instrument, but as with all analytical summarisation indexes, the need for concessions in the interpretation of results is essential. For example, we can discuss the appropriateness of the chosen indicators and weights (cf. Vidovićová and Petrová Kafková 2016). The AAI is a response to the efforts to capture this very broad and multidimensional concept, under which almost anything can be subsumed (Walker 2002), in hopes of creating a lucid tool suitable for the formulation of specific policies. Primarily, it does not apply to individual comparisons, but rather countries or population categories such as men and women. While the WHO (Active Ageing 2002) focuses primarily within its definition on individuals' quality of life and the quality of the environment in which they live, Zaidi et al. (2013: 9) emphasise creating a financially and socially sustainable public welfare system in Europe that aims to uncover the untapped potential of older adults (Active Ageing Index 2015). This emphasis differs from that of the WHO (Active Ageing 2002). In addition to fostering a suitable, safe environment highlighting individual responsibility for each aspect of life in old age, it promotes a lifelong form of active ageing rather than targeting older adults.

Active ageing, as defined by the creators of the AAI (Zaidi et al. 2013), places the main emphasis on the participation of older people in society. Many studies have noted the crucial importance of activity and participation in society for quality of life among older adults, as well as their ability to preserve mental and physical abilities (Galloway and Jokl 2000; Lindenberger 2009; Rowe and Kahn 1997). According to Zaidi et al. (2013: 20), the selection of indicators had been set primarily as an attempt to capture activities beneficial to society, thus downplaying the possible negative impact on the quality of life for individuals. This is evident especially for indicators of care, which may be a beneficial activity for society, but at the same time, it may reduce quality of life (Baumgarten et al. 1992; Meyer 2012).

Although dominated by those that are somehow productive for society, a wide range of activities in the AAI fall under the auspices of participation, including social participation, further education or physical activity. Although the range of activities seems quite broad, labour market participation is dominant in this index, with the weight of the employment domain at 35% of the total results. But my results have indicated that the employment domain has no connection with quality of life for older adults. This emphasis on employment links the AAI to the concept of productive ageing (Moody 2001) rather than to the inclusive WHO definition. The great emphasis put on economic participation in the index may be conditioned by the perception of the European Union, whose interest is concerned with the sustainability of the ongoing socio-political transformations (Lassen and Moreira 2014). The application of active ageing in Europe has been predominantly in the productivist mould (Walker 2009). And although the documents of the European Union have conceptualised active ageing through participation and well-being across the life course, the policy instruments still focus primarily on employment (ibid). The increase of retirement age and the promotion of older workers' employment still offer a seemingly simple solution on how to solve increasing longevity. In a similar manner, the active ageing concept is typically downplayed in the real policies of national governments, not only in the European Union, which is criticised by many authors (Katz 2000; Perek-Białas et al. 2006; Petrova Kafková 2013; Walker and Maltby 2012).

The definition of 'appropriate' activities is problematic in any operationalisation of active ageing, because it may implicitly refer to all other activities as 'unsuitable for proper aging' (Holstein Minkler 2003; Katz 2009). Activities that do not follow the work ethic are systematically made invisible (Biggs 1993). Consequently, such an interpretation of active ageing results in blindness to activities that do not fulfil the ideal of the work ethos or a healthy lifestyle. This perception tends to overshadow ordinary activities as baking a cake or walking the dog (Hasmanová Marhánková 2013) or leisure activities like billiard playing (Lassen 2014). It is culturally contingent, preferring activities typical for Western societies that marginalise minorities with different lifestyles (Ranzijn 2010).

The contemporary emphasis on activities for older adults, along with demographic changes, causes the accumulation of large numbers of socially significant roles for the young old. Young old are economically active, caring for grandchildren and elderly family members, and pursuing their self-development and hobbies (Petrová Kafková 2015). All of these roles are socially strongly supported, but as was shown in the Czech environment, the accumulation of any of these roles brings a significant increase of stressors in everyday life (Vidovićová et al. 2015). Promoting active ageing without careful consideration of all the impacts seems problematic and threatens to decrease the quality of life for ageing people. And the aim of increasing participation in one field could decrease the participation in another field.

Notwithstanding these serious considerations, the Active Ageing Index represents a useful analytical tool enabling the comparison of situation of older adults in different countries or regions and the enhancement of knowledge crucial to setting better ageing politics and improving the quality of life for older adults.

Notes

1. The file 'AAI_2012_Results_5th_December_EU28only' is available at: http://www1.unece.org/stat/platform/pages/viewpage.action?pageId=76287845
2. The resulting PCA has the following parametres: the value of Kayser-Meyer-Olkin Measure of Sampling Adeqacy = 0.892. The significance of Bartlett's test of Sphericity 0.000. Quadrimax rotation with Kaiser Normalization.

References

Active Ageing: A policy framework. (2002). Geneva: World Health Organisation.

Active Ageing Index. (2015, September). (on-line). Presentation available on-line on project's websites [cit. 29.12.2015]. Retrieved from http://www1.unece.org/stat/platform/display/AAI/I.+AAI+in+brief

Baumgarten, M., et al. (1992). The psychological and physical health of family members caring for an elderly person with dementia. *Journal of Clinical Epidemiology, 45*(1), 61–70.

Biggs, S. (1993). *Understanding ageing: Images, attitudes and professional practice.* Buckingham: Open University Press.

Clarke, A., & Warren, L. (2007). Hopes, fears and expectations about the future: What do older people's stories tell us about active ageing? *Ageing and Society, 27*(4), 465–488.

Diener, E., & Suh, E. (1997). Measuring quality of life: Economic, social, and subjective indicators. *Social Indicators Research, 40*(1–2), 189–216.

Erlinghagen, M., & Hank, K. (2006). The participation of older Europeans in volunteer work. *Ageing and Society, 26*(4), 567–584.

EQLS. (2012). *European Quality of Life Survey.* Dublin: Eurofound.

Galloway, M. T., & Jokl, P. (2000). Aging successfully: The importance of physical activity in maintaining health and function. *Journal of the American Academy of Orthopaedic Surgeons, 8*(1), 37–44.

Gilleard, C., & Higgs, P. (2005). *Contexts of ageing. Class, cohort and community*. Cambridge: Policy Press.

Hagerty, M. R., & Veenhoven, R. (2003). Wealth and happiness revisited—Growing national income does go with greater happiness. *Social Indicators Research, 64*(1), 1–27.

Hasmanová Marhánková, J. (2013). *Aktivita jako projekt. Diskurz aktivního stárnutí a jeho odezvy v životech českých seniorů a seniorek. Studie*. Praha: Sociologické nakladatelství.

Hasmanová Marhánková, J. (2014). Aktivní stárnutí jako idea, nástroj a kapitál. Kde hledat kořeny úspěchu koncepty aktivního stárnutí? *Sociální studia, 11*(3), 13–29.

Holstein, M. B., & Minkler, M. (2003). Self, society, and the "new gerontology". *The Gerontologist, 43*(6), 787–796.

Katz, S. (2000). Busy bodies: Activity, aging, and the management of everyday life. *Journal of Aging Studies, 14*(2), 135–152.

Katz, R. (2009). Intergenerational family relations and subjective well-being in old age: A cross-national study. *European Journal of Ageing, 6*(2), 79–90.

Laslett, P. (1989). *A fresh map of life: The emergence of the third age*. Weidenfeld and Nicolson.

Lassen, A. J. (2014). Billiards, rhytms, collectives. Billiards at a Danish Activity Centre as a culturally specific form of active ageing. *Ethnologia Europaea. Journal of European Ethnology, 44*(1), 57–74.

Lassen, A. J., & Moreira, T. (2014). Unmaking old age: Political and cognitive formats of active ageing. *Journal of Aging Studies, 30*, 33–46.

Lindenberger, U. (2009). Technologie ve stáří; šance z pohledu výzkumu chování. In P. Gruss (Ed.), *Perspektivy stárnutí: z pohledu psychologie celoživotního vývoje* (pp. 137–151). Praha: Portál.

Manual on the measurement of volunteer work. Final approved pre-publication version. (2011). IWO [cit. 25.9.2011]. Retrieved from http://www.ilo.org/wcmsp5/groups/public/greports/documents/meetingdocument/wcms_100574.pdf

McMunn, A., Nazroo, J., Wahrendorf, M., Breeze, E., & Zaninotto, P. (2009). Participation in socially-productive activities, reciprocity and wellbeing in later life: Baseline results in England. *Ageing and Society, 29*(5), 765–782. https://doi.org/10.1017/S0144686X08008350.

Meyer, M. H. (2012). Grandmothers juggling work and grandchildren in the United States. In S. Arber & V. Timonen (Eds.), *Contemporary grandparenting: Changing family relationships in global contexts* (pp. 71–90). The Policy Press.

Minkler, M., & Fuller-Thomson, E. (1999). The health of grandparents raising grandchildren: Results of a national study. *American Journal of Public Health, 89*(9), 1384–1389.

Moody, H. (2001). Productive aging and the ideology of old age. In N. Morrow-Howell, J. Hinterlong, & M. W. Sherraden (Eds.), *Productive aging: Concepts and challenges* (pp. 175–196). JHU Press.

Morrow-Howell, N., Hinterlong, J., Rozario, P. A., & Tang, F. (2003). Effects of volunteering on the well-being of older adults. *The Journals of Gerontology Series B: Psychological Sciences and Social Sciences, 58*(3), S137–S145.

OECD. Employment outlook 2006 boosting jobs and incomes: Boosting jobs and incomes. (2006). OECD.

Park, H. H. (2006). The economic well-being of households headed by a grandmother as caregiver. *Social Service Review, 80*(2), 264–296.

Perek-Białas, J., Ruzik, A., & Vidovićová, L. (2006). Active ageing policies in the Czech Republic and Poland. *International Social Science Journal, 58*(190), 559–570.

Petrová Kafková, M. (2013). *Šedivějící hodnoty? Aktivita jako dominantní způsob stárnutí*. Brno: Munipress.

Petrová Kafková, M. (2015). Older people as care givers and their roles in family in the era of active ageing: Case of the Czech Republic. *Studie Socjologiczne, 217*(2), 49–73.

Ranzijn, R. (2010). Active ageing—Another way to oppress marginalized and disadvantaged elders? Aboriginal elders as a case study. *Journal of Health Psychology, 15*(5), 716–723.

Rowe, J. W., & Kahn, R. L. (1997). Successful aging. *The Gerontologist, 37*(4), 433–440.

Stathi, A., & Simey, P. (2007). Quality of life in the fourth age: Exercise experiences of nursing home residents. *Journal of Aging and Physical Activity, 15*(3), 272–286.

Veenhoven, R. (2001). Quality-of-life and happiness: Not quite the same [cit. 6.10.2016]. Retrieved from http://repub.eur.nl/pub/8753/

Vidovićová, L., Galčanová, L., & Petrová Kafková, M. (2015). Význam a obsah prarodičovské role u mladých českých seniorů a seniorek. *Sociologický časopis/Czech Sociological Review, 51*(5), 761–782.

Vidovićová, L., & Petrová Kafková, M. (2016). Index aktivního stárnutí (AAI) v regionální aplikaci. *Demografie: Revue pro výzkum populačního vývoje, 58*(1), 49–66.

Walker, A. (2002). A strategy for active ageing. *International Social Security Review, 55*(1), 121–139.

Walker, A. (2009). Commentary: The emergence and application of active aging in Europe. *Journal of Aging & Social Policy, 21*(1), 75–93.

Walker, A., & Maltby, T. (2012). Active ageing: A strategic policy solution to demographic ageing in the European Union. *International Journal of Social Welfare, 21*(3), S117–S130.

Warburton, J., Terry, D. J., Rosenman, L. S., & Shapiro, M. (2001). Differences between older volunteers and nonvolunteers. *Research on Aging, 23*(5), 586–605. https://doi.org/10.1177/0164027501235004.

Wilson, J. (2000). Volunteering. *Annual Review of Sociology, 26*, 215–240.

Zaidi, A., Gasior, K., Hofmarcher, M. M., Lelkes, O., Marin, B., Rodrigues, R., Schmidt, A., Vanhuysse, P., & Zolyomi, E. (2013). *Active Ageing Index 2012. Concept, methodology and final results.* Methodology report submitted to European Commission's DG Employment, Social Affairs and Inclusion, and to Population Unit, UNECE [cit. 6.3.2015]. Vienna: European Centre Vienna. [online]. Retrieved from http://www.euro.centre.org/data/1364466765_60390.pdf

Zaidi, A., & Stanton, D. (2015). *Active Ageing Index 2014: Analytical report.* Report produced at the Centre for Research on Ageing, University of Southampton, under contract with UNECE (Geneva), co-funded by European Commission, Brussels. Retrieved form http://www.southampton.ac.uk/assets/sharepoint/groupsite/Administration/SitePublisher-documentstore/Documents/aai_report.pdf

Marcela Petrová Kafková is a researcher at Office for Population Studies, Faculty of Social Studies, Masaryk University in Brno. Her research interest comprises a range of issues of social gerontology and environmental gerontology. She devotes to active ageing concept and follow-up policies in the long term. Her current research focuses on ageing with disability.

5

Complementing AAI at the Meso Level: The Silver Work Index

Anne Marit Wöhrmann, Jürgen Deller, and Leena Pundt

5.1 Introduction

The Active Ageing Index (AAI) has been developed to identify and promote active responses to the challenges of a population that is ageing. Its aim is to provide "a new tool for policy makers to enable them to devise evidence-informed strategies in dealing with the challenges of population ageing and its impacts on society" (Zaidi et al. 2013, preface). The AAI focuses on the need for a sound evidence base that allows for comparisons and policy learnings across different national contexts. While AAI monitors (and compares) active ageing outcomes at international, national and subnational levels to indicate the untapped potential of older people for more active participation, it also offers the potential for developing indi-

A.M. Wöhrmann
Federal Institute for Occupational Safety and Health, Dortmund, Germany

J. Deller (✉)
Leuphana University, Lüneburg, Germany

L. Pundt
SiB School of International Business, Bremen, Germany

A. Zaidi (eds.), *Building Evidence for Active Ageing Policies*,
https://doi.org/10.1007/978-981-10-6017-5_5

ces that focus on the organizational level—to identify and promote good, age-friendly practices at the organizational level. The approach presented here for constructing the Silver Work Index (SWI) was indeed inspired by the AAI, although it uses a different method of identifying and validating indicators, and it offers evidence at the meso level. More specifically, it can be seen to complement the AAI with an instrument that moves the focus to good organizational practices in the field of the AAI's first domain—employment. To reach this goal, the SWI allows for a comparison of activities and processes between organizations in order to promote a more active role for older people in those organizations, below the international, national and subnational levels that the AAI is constructed to examine.

For several years now, the European Commission and its member states have emphasized the importance of promoting healthy and active ageing. To reach this objective, a number of policy changes intended to support longer working lives have been adopted across Europe, that is, working conditions have been adapted in ways meant to better suit older workers. At the same time, incentives are geared towards extending working lives and penalizing early withdrawal from the labour market. Later retirement ages have already been legislated in many EU member states, or legislation is currently being planned. It is in the face of such developments that we are constructing the SWI as an index for good, age-friendly employment practices at the organizational level. We aim at identifying attributes of good organizational practices for older workers who are about to enter retirement or who are continuing to work while having reached retirement age. This approach will allow for comparisons to be made between organizations based on a range of good organizational practices.

Demographic changes and a longer working life resulting from later retirement have resulted in a shift of age structures in the workforces of several countries. Placed in this situation, societies and organizations face the challenge of retaining and developing the potential of older employees (Lehr and Kruse 2006). These circumstances form the backdrop for the development of the SWI. At the same time, the employment rates of individuals at or beyond retirement age are on the rise (for Germany, see Deller and Pundt 2014). Such late career activity can have advantages on several levels. On the individual level, working people can profit in several respects, such as in the areas of health and psychological well-being. On the organizational level, using the potential of older workers beyond retirement age could help organizations prevent shortages in availability of skilled employees. Finally, on a societal level, later retirement could help relieve strained social security systems (Büsch et al. 2010; Deller et al. 2009; Deller and Maxin 2009).

In this context, it will be important to find specific ways to use the potential of employees as they reach and enter retirement age. However, the specific measures needed depend on the current personnel practices regarding older employees already in place in each organization. In order to determine the quality of current workforce conditions, a diagnostic tool is needed. Therefore, the aim of this study is to develop such an instrument—the SWI.

5.2 Theoretical Background

Success in using the potential of older employees nearing retirement age or beyond is based on thoughtful managerial approaches towards these individuals. Many opportunities for proactive management have already been identified in the areas of age and ageing, demographic development, and generational aspects. Examples of areas where action could be taken are organizational culture, leadership, recruiting, work design, health promotion, and development of qualifications and competencies, as well as knowledge management (cf. Deller et al. 2008; Klaffke 2014; Naegele and Walker 2006; Schuett 2014). The SWI is developed to enable the assessment, comparison, and evaluation of employment and workplace conditions of individuals nearing retirement age or beyond. SWI's dimensions cover these areas but also address the topic of the transition to retirement as well as work in retirement.

In order to fully cover relevant perspectives, the SWI has been developed in cooperation with scientific experts and field practitioners. Figure 5.1 displays the development process.

The overarching construct of the SWI is "good organisational management of employees nearing retirement age or beyond." Since SWI is a multidimensional construct, relevant dimensions and indicators were

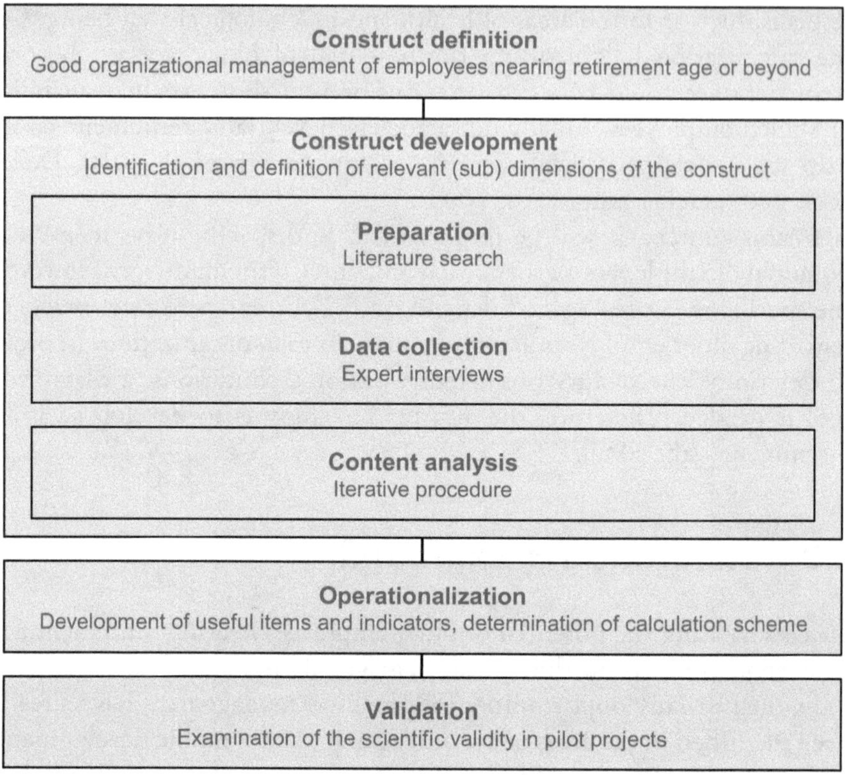

Fig. 5.1 Process of SWI development

defined during construct development. First, what personnel policies and measures should be available in which fields of action was determined in order to assess whether a specific type of management of employees just before or after retirement age is effective (vs. ineffective).

To this end, qualitative techniques were used to shed light on the construct from different perspectives so as to reach a comprehensive understanding of its facets. This article focuses on the SWI's construct development; subsequent steps, as shown in Fig. 5.1, will be realized in future project phases. These steps will include the operationalization of management and employment condition indicators as well as the validation of the index, including its indicators.

5.3 Method

5.3.1 Procedure and Participants

In order to determine the aspects of good organizational management of employees nearing retirement age or beyond, an empirical qualitative research design was applied. This design used semi-structured expert interviews. All subject matter experts had had personal experience with employment nearing retirement age or beyond, or had dealt with such questions in their jobs, either in practice or in research.

Twenty-seven expert interviews were conducted from June to November 2013. The interviewees (P01 to P27) were researchers from various disciplines (demographics, economics, gerontology, Human Resources (HR) management, and psychology) as well as employees of retirement age, HR executives, HR managers, management consultants, executives of placement agencies for paid and voluntarily active working retirees, and representatives of strategic and operational management in various industries.

The interviewees were aged 35–83 years ($M = 52.7$; $SD = 10.6$). Most were male (74%). Experts had, on average, 28 years of professional experience ($M = 28.4$; $SD = 11.4$; range 7–59 years). The experts worked in various industries: 29% professional, scientific, and technical; 26% finance and insurance; 15% manufacturing; 15% administrative, support and other services; 11% human health and social work activities; and 4% information and communication. The industry heterogeneity of the sample was intended to cover a variety of viewpoints on the research topic.

During the first part of the interview, the interviewees were asked to elaborate on aspects that they thought were characteristic of good organizational management practice involving employees aged 60 and older. In addition, in view of subsequent operationalization, the experts were asked to determine which indicators and methods would be useful for describing and measuring the characteristics mentioned. In the second part of the interview, the experts were interviewed on additional aspects related to good organizational management practices concerning employees nearing retirement age or beyond that had earlier been identified through an analysis of the relevant body of literature and had not been mentioned by the experts in the first part of the interview. These aspects were: perception of age/ageing; ways of structuring and designing the work/workplace; procedures in place for retiring or resigning; individual financial situation; methods of guiding older employees; available health management and promotion resources; and the information on the range of possibilities for continued employment after retirement age. Here, too, possibilities for operationalizations were requested. The interviews took 30–60 minutes. They were audio-recorded and transcribed using the software f4 (www.audiotranskription.de).

5.3.2 Qualitative Data Analysis

The aim of the data analysis was to develop a category system that would represent aspects of good organizational management of employees nearing retirement age or beyond. The process of analysis was iterative as phases of formation of categories and their validation alternated, resulting in an increasing precision of the identified category system. Figure 5.2 displays the iterative proceedings in content analysis.

The first step of the data analysis was a qualitative content analysis following Mayring (2010). Dimensions and indicators were derived inductively from the data. To this end, text passages relevant to the research question were paraphrased and either assigned to an already existing category or allocated to a new category. The category system was repeatedly revised in the process. Data analysis was conducted with the software MAXQDA (www.maxqda.de).

Iterative proceedings in the content analysis
Data analysis (qualitative content analysis)
Expert workshop
Revision
Examination of scientific criteria
Final revision
Final category system

Fig. 5.2 Iterative proceedings in the content analysis

The results of the content analysis provided the basis for an expert workshop with six representatives from research and practice. The aim of this workshop was to discuss the adequacy and practicability of the categories and to consolidate the dimensions and indicators identified through data analysis. As a result of the workshop, the complexity of the category system was demonstrably reduced.

In the next step, the category system was defined more precisely by formulating definitions and examples closely linked to the data for the different dimensions and indicators. Then, interrater-reliabilities (cf. Krippendorff 2013) were determined in order to validate the category system. To this end, nine individuals who were not involved in the project independently coded the interview material. Krippendorff's alpha ranged from a dimension average of 0.30–0.76. This provided the authors with indications of necessary adaptations of the categories. The reliability test of the resulting final category system is underway.

5.4 Results

The iterative process of analysis resulted in eight dimensions of good organizational management of employees nearing retirement age or beyond. Additionally, for every dimension, two to four indicators were identified. Dimensions and indicators were organized into the "house of SWI" as depicted in Fig. 5.3. *Organizational* culture and leadership represent the roof of the house, as both are overarching dimensions that affect the design of all of the other dimensions.

The organizational culture dimension includes the prevailing collective patterns of thought and action, as well as the standards and values, within an organization. According to the experts' statements, an organizational culture that fosters good organizational management of employees nearing retirement age or beyond should convey a marked appreciation towards all age groups. This is reflected in the derived indicators equality of opportunity, positive image of age, and target group-oriented communication.

The equality of opportunity indicator specifies that initial conditions should be the same for every employee regardless of age. Furthermore, no discrimination or stigmatization due to age should occur. However, the

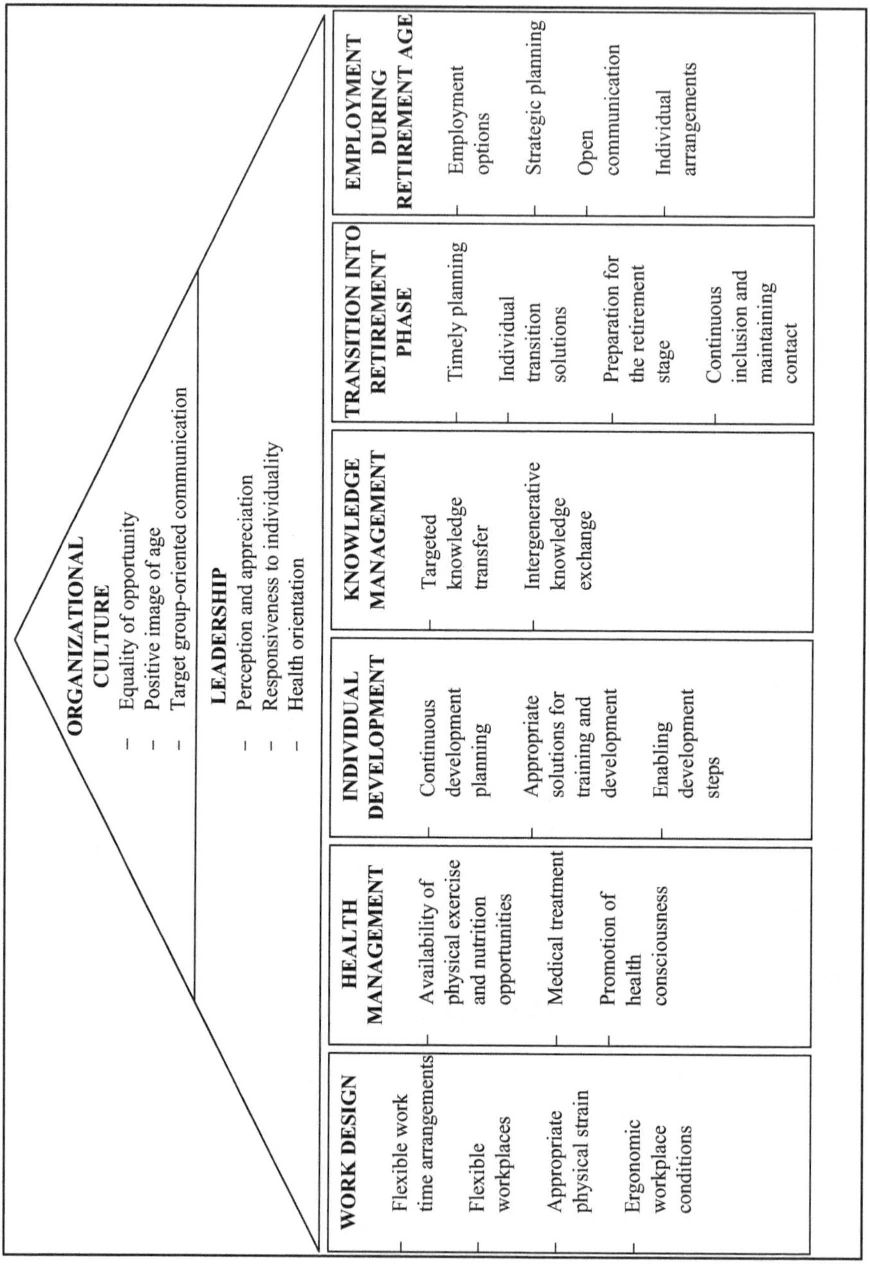

Fig. 5.3 House of SWI

decision criteria in the design of general work-related conditions should be the individual situation, competencies and experience. Special management of employees based on age is "a sign for a prevailing age discrimination as well as reservations and barriers" (from the interview with P14). Examples of equality of opportunity are recruitment from all age groups (P16) and age-independent employee management, which includes instruments such as performance appraisals and target agreements (P03). A sign of real equality of opportunity is also that in times of necessary downsizing, reduction is not solely conducted at the senior end (P19). The positive image of age indicator specifies that prevailing beliefs and attitudes regarding employees nearing retirement age or beyond are in fact shaped by a positive basic attitude.

A positive image of age is a "picture of an older person that does not place special emphasis on reduced performance but, on the contrary, the wish to profit from experience" (P03). Ageing should be understood as an individual process of change in competencies, motivation, values and behaviour. Specific situations and opportunities arising from this process should be recognized and implemented, for example, by providing jobs that consider the specific competencies and performance possibilities of the individual (P11). The target group-oriented communication indicator includes statements made by interviewees that point to good management of employees nearing retirement age or beyond as being recognizable through the external and internal representation of the organization. For example, explicitly addressing all age groups in job ads, in the employee magazine, or on the intranet should be positively valued (P07).

The leadership dimension includes the task performed by organizational executives of raising and utilizing the potential of employees nearing retirement age or beyond. The interviewed experts emphasized the aspect of appreciation and the necessity of considering the individuality of every employee. Indicators were perception and appreciation, responsiveness to individuality, and health orientation. The perception and appreciation indicator states that managers of an organization should have a basic attitude of appreciation that is manifested in respect and cordiality towards employees of all age groups. Managers should listen to employees and appreciate their experience and what they have achieved for the organization (P23), for example, by granting scope for action in the fulfilment of tasks (P20). The responsiveness to individuality indicator suggests that managers of an organization should be sensitized regarding the characteristic features of different life stages. They should also take into account each individual's personality, needs and performance capability. The interviewees mostly disavowed any special leadership methodology to be used with employees nearing retirement age or beyond. P27 stated that "leadership aligned to demography is nothing but 'individual consideration.' Where does he stand? What does he need? Where does he have development potential? Where does he have barriers to development?" The task of managers is therefore to recognize, raise, and utilize individual potential regardless of age and to create performance-enhancing general conditions in the workplace. This includes dealing constructively with a possibly changing physical capacity or performance constraints that can occur as one ages (P18). Also, the consideration of individual life circumstances, such as the need for family care (P10) or the necessity of arranging individual support and advisory services (P02) falls under this indicator. The aspect of health promotion is perceived as an essential leadership task by experts as well (P15). The health orientation indicator therefore includes the question of whether managers of an organization fulfil their responsibilities of providing a health-promoting work design and serving as a role model in terms of healthy lifestyle choices. For example, if they themselves take part in health-promoting offers, this would be evaluated positively (P09).

The work design dimension encompasses the adaptation of work to the individual needs and performance abilities of employees so as to positively affect job satisfaction and performance as well as relieve strain. Respective indicators are maintaining flexibility with regard to schedules as well as work locations, striving for improved work-family balance, having knowledge of appropriate levels of physical strain and implementing ergonomic workplaces. The flexible working time arrangements indicator summarizes expert statements about offering working schedules that enable employees to influence the duration, location, and distribution of work time according to their individual needs. "Flexible working time arrangements are important. However, I believe that older employees might need a different flexibility than younger employees" (P07). Different organizational options were mentioned, such as part-time

(P15), flex-time (P13) and job-sharing (P25) scheduling. The flexible workplaces indicator includes offers that can sometimes enable employees to determine the worksite location themselves according to their individual needs. This arrangement could be realized through the facilitation and technical support of home-office solutions (P04) as well as the installation of silent workplaces (P12). The appropriate physical strain indicator states that employees should have an adequate workplace that corresponds to their individual physical performance capability and resilience (P11). According to P04, this objective could be realized through a permanent move to another workplace that demanded less physical strain. Also, swapping workplaces in terms of job rotation (P07, P08), as well as carefully reconsidering and adapting workflow, should be taken into consideration (P25). The ergonomic workplace conditions indicator comprises interviewee statements regarding workplace design in terms of ergonomics and at the same time the consideration of individual circumstances, for example, occupational safety measures (P03) or providing supporting work equipment and tools (P08, P11).

The health management dimension contains all of the organizational activities aimed at the maintenance and promotion of employee health and work capability. Indicators are availability of physical exercise and nutrition opportunities, existing arrangements for medical treatment, and information geared to the promotion of health consciousness and nutrition, all of which work to strengthen employees' overall health-related behaviours as well as to ensure that they seek medical care when necessary. The interviewees emphasized that health management should be characterized by a holistic approach that addresses not only specific measures but also health-promoting work design and leadership (P18). The availability of physical exercise and nutrition opportunities indicator poses the question of whether any measures to strengthen employee health and work capability are being offered. According to the experts' statements, these are specifically opportunities for exercise, such as company sports activities (P23), active breaks (P21) and healthy nutrition options (P20). The medical treatment indicator outlines strategies for helping employees to prevent avoidable medical conditions and in general promotes recovery of employee health. Examples of this indicator might be having doctors on staff (P15), medical check-ups provided by the company (P03), and physiotherapeutic opportunities offered on site (P11). The promotion of health consciousness indicator states that measures should be taken to convey health-relevant knowledge and to increase employees' decision-making ability and responsibility regarding health questions. This goal could be realized by sensitization of employees through information on healthy living (P06) or in the context of "action days" (P20).

The individual development dimension includes supporting employees in the area of professional and personal development during their entire working lives. Development is "not an age specific phenomenon but something that has to be offered and has to take place until old age, until the end actually" (P03). Further, it would be a fatal sign if, "[as of] a certain age, participation in further training and occupational qualification ends, or that in potential analyses and the question on how someone is going to develop in a company, they say: 'We are not dealing with older employees anymore'" (P14). A special emphasis was placed on the idea of lifelong learning through continuous further education and training as well as through the demonstration and realization of further developmental steps. Determined indicators are continuous development planning, appropriate solutions for training and development, and enabling of developmental steps. The continuous development planning indicator involves activities that target developmental and qualification needs of employees of all age groups and their qualification needs in the upcoming years. This could be accomplished using development interviews (P01) or workshops on professional assessment that would reflect an employee's own abilities, competencies, and goals (P12). The appropriate solutions for training and development indicator comprises statements made by interviewees on the necessity of conducting further training and education aligned with individual life, professional and learning experience, and organizational goals. For that reason, subject matter and methods should be target group specific (P22). A reasonable composition of learning groups should be considered (P04). The enabling of developmental steps indicator states that modifications of the current position or job parameters should be made possible in order to reflect the specific competencies and interests of the employees. For example, this could be done by broadening of responsibilities, inclusion in projects (P18), or horizontal or vertical change of position (P24).

The knowledge management dimension includes the targeted design of work conditions and procedures for the transfer and exchange, and thus the conservation, of the knowledge and expertise of different employee generations. The determined indicators are targeted knowledge transfer and intergenerative knowledge exchange. The targeted knowledge transfer indicator reflects the necessity for measures to exist in organizations that enable a targeted transfer of knowledge from experienced employees to successors. In this context, the interviewees mentioned mentoring and "buddy" models (P16), as well as a systematic knowledge transfer process that would occur before employees of the company retire (P10). The intergenerative knowledge exchange indicator, on the other hand, is defined by a mutual transfer of knowledge and experience between generations, such as generational tandems (P15) or age-mixed teams (P19).

The transition into retirement phase dimension outlines modes of individual planning for and design of entry into the retirement phase. This encompasses the indicators timely planning, individual transition solutions, preparation for retirement stage, continuous inclusion and maintaining contact. The timely planning indicator includes the idea that organizations should talk with older employees about their individual planning for entry into retirement and also have active discussions of potential transition scenarios, for example, in the context of annual employee interviews (P11, P15). The individual transition solutions indicator reflects expert statements about the adaptation of generic transition scenarios tailored to employees' individual needs. A special emphasis was put on enabling "sliding" into retirement through a reduction of work hours (P03). The preparation for retirement stage indicator indicates how organizations should support their employees in the preparation for retirement. Proposals that can sensitize employees to the need for an actively designed retirement stage should be made. Furthermore, organizations should help individuals prepare for this stage, for example, by utilizing a structured approach that would reflect individual expectations and plans (P13). The continuous inclusion and maintaining contact indicator suggests introducing tools to maintain contact with employees even after their retirement. Examples of an active management of relationships are alumni networks (P16) and invitations to organizational events (P14).

The employment during retirement age dimension offers these employees various employment options as well as the possibility of designing specific work conditions themselves. Indicators are employment options, strategic planning, open communication and individual arrangements.

The employment options indicator reflects that providing employment opportunities for individuals at retirement age is to be evaluated positively, especially since this practice is not very common yet in Germany (P09). Options are continued employment or re-employment of current or former employees, as well as initial employment of individuals of retirement age who are new to the organization (P24). The strategic planning indicator reflects that an organization should strategically align policies regarding the employment of individuals of retirement age and systematically frame employment options (P09), for example, through a definition of areas and activities suited for employment for those of retirement age (P11), or through surveying the interest of employees in work after retirement age (P12). The open communication indicator reflects a transparent and open communication of employment options in retirement age. This could be effected through informing employees about options for employment in retirement age (P03). The individual arrangements indicator summarizes experts' statements about the necessity of individual solutions regarding content of tasks as well as working hours. Possible arrangements are consulting or mentoring activities (P18), work on specific projects (P19), holiday replacement possibilities (P01) and the taking over of voluntary tasks (P04). The majority of the experts held the opinion that the arrangements should be temporary and should involve fewer hours than a full-time position (P22).

5.5 Discussion and Implications

The SWI is intended to be an instrument that will make the management of and the employment framework conditions for employees nearing retirement age or beyond measurable and comparable. The aim of the current study therefore is to determine suitable dimensions and corresponding indicators of the index that will allow for such measurement and comparison.

The current study shows that good organizational management of employees nearing retirement age or beyond is a complex subject. It goes beyond historically accepted courses of action relating to age management and, more specifically, explores the topics of transition into retirement and employment in retirement age. Our results indicate that aspects of good organizational management of employees of this particular age group are often beneficial for employees of all age groups. However, several indicators specific to age were identified. This is reflected in the dimensions transition into retirement phase and employment during retirement age.

Further, the analysis of the data revealed that the aspects of "individuality" and "appreciation" were emphasized by the interviewees as being the core of good organizational management of employees nearing retirement age or beyond. According to the experts, flexibility in designing work conditions and the willingness to consider individual needs are the most important levers for retaining employees in the organization who are nearing retirement age or older (P09, P10). The interviewees' responses argue for the necessity of individual solutions, especially in light of the target group's heterogeneity (P22). In line with some gerontological findings, and against the backdrop of individual ageing processes and differing professional and life biographies, these results indicate that the older employee is essentially non-existent (Lehr and Kruse 2006), which is also the context for research findings that flexibility and individually tailored working conditions are the most important instruments of employee retention (Deller and Maxin 2009; Wang et al. 2013). In addition, the experts' statements in this study show that chronological age should definitely not be a criterion for differentiation. Good organizational management should be guided by competence, personality, and the living situation of the individual (P26), and "a positive aspect would be not to talk about age at all any more" (P14).

Appreciation and recognition are central to leadership culture and organizational culture. Both should be conveyed through different facets of organizational management. For example, providing the possibilities of reducing work hours (P05), or changing the focus of the work content, or working on other projects (12), expresses management's appreciation of the employee. And recognizing experience is central to being open to sharing knowledge (P23). Moreover, if the transition to retirement is arranged in a manner that communicates appreciation, that may be an impetus for positively motivating an employee until actual retirement (P11). This is consistent with the results of several studies (Armstrong-Stassen 2008; Armstrong-Stassen and Schlosser 2011; Hennekam and Herrbach 2013; Cheung and Wu 2013).

5.5.1 Implications

The Silver Work Index will complement the AAI by focusing on both organizational and individual levels. In a way, the concept of SWI delineates those organizational policies and processes that lay the basis for and support the development of higher AAI scores. Therefore, AAI and SWI interface. Both aim at tapping the potential of older people. While AAI focuses on the international, national and subnational levels, SWI will be applied within organizations by defining appropriate work conditions. The interface can be exemplified using the SWI dimension of employment during retirement age (cf. Fig. 5.3). Based on an organization's strategic planning, good practices include the identification of employment options and an open communication with employees. These aspects, together with individual arrangements between employer and employee, support a higher workforce participation rate of older people within the organization. Such good organizational practices in the employment during retirement age dimension are ultimately reflected in the employment rate indicators of the AAI domain employment.

Further, the SWI will be useful in supporting research on successful personnel practices regarding older employees and employees at retirement age. It may also modify the status quo in existing organizations regarding the management of employees nearing retirement age or beyond as well as contribute to the public debate regarding an improvement of the employment situation of individuals nearing retirement age or beyond in organizations. It may raise the awareness that employees, organizations, and society may benefit from the use of the available potential of this group.

The responsible parties in an organization will be able to use the SWI as a tool to determine how well that organization handles ageing issues.

It may help to reflect the attitudes of management towards the organization's current employees who are nearing retirement age or beyond in that organization. The courses of action determined in this study can already provide indications as to the identification of strengths and the development potential of the organization's personnel policies. It can also give first hints for implementation possibilities. Additionally, the results show that it may not be necessary to develop entirely new instruments in order to attain better management of employees nearing retirement age or beyond. Instead, existing instruments can be adapted to this target group. An appreciative attitude and a positive image of age are important prerequisites. In the overall course of the project, we plan to develop comprehensive recommendations for action regarding each dimension. These could support organizations in developing and implementing targeted measures for an improvement of the situation of employees nearing retirement age or beyond.

5.5.2 Limitations and Future Research

The current study completes the definition of relevant dimensions and indicators of the SWI. Through the selection of a heterogeneous group of interviewees, a variety of viewpoints regarding the subject matter was achieved. Data analysis was done systematically. Dimensions and indicators of the SWI were determined in a data-oriented manner and were therefore generated empirically. The experts' statements on the field of employment for older individuals resulted in many and diverse clues as to the important aspects of good organizational management. They confirm the aspects identified as important in a preliminary literature search in preparation for these interviews. In this respect, it can be assumed that essential dimensions and sub-dimensions of the construct were identified and are content-valid. However, future studies need to test whether results are country-specific, as rules and laws may differ across countries. As the final revision of the category system resulted in changes of the dimensions and indicators, it should be subjected to another reliability testing in order to scientifically validate the categories. With such testing, an improvement of the coefficients is expected. While following the steps of the development of the index, suitable items and parameters for the different dimensions should be developed as well as a calculation scheme. To give an example for the work ahead, several items will be used for the measurement of the dimension "individual development." Examples are "To what extent is continuous development planning realised in the organization," "To what extent are appropriate solutions for training and development applied in the organization," and "To what extent are development opportunities enabled in the organization." Answers can be given on multiple-point Likert scales, including the ratings "to a great extent," "somewhat," "very little" and "not at all." The index development will be completed by the examination of the scientific validity in pilot projects.

References

Armstrong-Stassen, M. (2008). Organisational practices and the post-retirement employment experience of older workers. *Human Resource Management Journal, 18*(1), 36–53.

Armstrong-Stassen, M., & Schlosser, F. (2011). Perceived organizational membership and the retention of older workers. *Journal of Organizational Behavior, 32*(2), 319–344.

Büsch, V., Dorbritz, J., Heien, T., & Micheel, F. (2010). Weiterbeschäftigung im Rentenalter, Wünsche—Bedingungen—Möglichkeiten [Continued employment in retirement age, wishes—conditions—options]. *Materialien zur Bevölkerungswissenschaft 129*. Wiesbaden: Bundesinstitut für Bevölkerungsforschung. Viewed January 17, 2016. http://www.ssoar.info/ssoar/bitstream/handle/document/33082/ssoar-2010-busch_et_al-Weiterbeschaftigung_im_Rentenalter__Wunsche.pdf?sequence=1

Cheung, F., & Wu, A. M. (2013). 'Older workers' successful aging and intention to stay. *Journal of Managerial Psychology, 28*(6), 645–660.

Deller, J., Kern, S., Hausmann, E., & Diederichs, Y. (2008). *Personalmanagement im demografischen Wandel: ein Handbuch für den Veränderungsprozess* [Personnel management in times of demografic change: A handbook for the change process]. Heidelberg: Springer.

Deller, J., Liedtke, P. M., & Maxin, L. M. (2009). Old-age security and Silver Workers: An empirical investigation identifies challenges for companies, insurers, and society. *Geneva Papers on Risk and Insurance, 34*(1), 137–157.

Deller, J., & Maxin, L. (2009). Berufliche Aktivität von Ruheständlern [Professional activity of retirees]. *Zeitschrift für Gerontologie und Geriatrie, 42*(4), 305–310.

Deller, J., & Pundt, L. (2014). Flexible transitions from work to retirement in Germany. In C. M. Alcover, G. Topa, M. Depolo, & F. Fraccaroli (Eds.), *Bridge employment: A research handbook* (pp. 167–192). London: Routledge.

Hennekam, S., & Herrbach, O. (2013). HRM practices and low occupational status older workers. *Employee Relations, 35*(3), 339–355.

Klaffke, M. (Ed.) (2014). *Generationen-Management: Konzepte, Instrumente, Good-Practice-Ansätze* [Generations management: Concepts, tools, good practice approaches]. Springer: Wiesbaden.

Krippendorff, K. (2013). *Content analysis: An introduction to its methodology* (3rd ed.). Thousand Oaks: Sage.

Lehr, U., & Kruse, A. (2006). Verlängerung der Lebensarbeitszeit—eine realistische Perspektive [Extension of working life—A realistic perspective]? *Zeitschrift für Arbeits- und Organisationspsychologie, 50*(4), 240–247.

Mayring, P. (2010). *Qualitative Inhaltsanalyse: Grundlagen und Techniken* [Qualitative content analysis: Fundamentals and techniques] (11th ed.). Weinheim: Beltz.

Naegele, G., & Walker, A. (2006). *A guide to good practice in age management*. Dublin: European Foundation for the Improvement of Living and Working Conditions.

Schuett, S. (2014). *Demografie-management in der Praxis: Mit der Psychologie des Alterns wettbewerbsfähig bleiben* [Demography management in practice: Remain competitive with the psychology of aging]. Heidelberg: Springer.

Wang, M., Olson, D. A., & Schultz, K. S. (2013). *Mid and late career issues: An integrative perspective*. New York: Routledge.

Zaidi, A., Gasior, K., Hofmarcher, M. M., Lelkes, O., Marin, B., Rodrigues, R., Schmidt, A., Vanhuysse, P., & Zolyomi, E. (2013, March). *Active Ageing Index 2012. Concept, methodology, and final results*. Research Memorandum/ Methodology Report. European Centre Vienna. Retrieved from www.euro.centre.org/data/aai/1253897823_70974.pdf

Anne Marit Wöhrmann is a researcher at the Federal Institute for Occupational Safety and Health (BAuA) in Dortmund, Germany. Her research focuses on late careers and retirement as well as work-life balance and working time.

Jürgen Deller is a university professor at the Institute of Management & Organization, Leuphana University Lüneburg, Germany, and the research director of the Silver Workers Research Institute, Berlin, Germany. Previously, he worked in senior HR functions of the Daimler Group and at Commerzbank. His research focuses on individual and organizational perspectives of work, ageing and retirement. He is also involved in research projects in the fields of international human resource management and labour market integration of refugees.

Leena Pundt is a professor of human resources management at the SiB School of International Business, City University of Applied Sciences, Bremen, Germany. Previously, she worked as diversity manager for Otto Group, Hamburg, Germany. Her research interests include the fields of ageing, labour market integration of refugees and international human resource management.

Part II

Subnational Adaptations of the AAI

6

The Active Ageing Index in a Southern European Region (Biscay): Main Results and Potentials for Policymaking

Amaia Bacigalupe, Yolanda González-Rábago, Unai Martín, Sergio Murillo, and Alfonso Unceta

6.1 Introduction

Biscay is a one of the three provinces of the Basque Country in Northern Spain. In 2014, it had nearly 1.2 million inhabitants, 20.7% of whom were above 65 years old (23.5% of women and 17.7% of men). The dependency rate is 0.30, which means that for every 100 people aged 16–64, there are 30 people over 65. Life expectancy is high in Biscay, reaching 85.1 and 78.2 years in women and men, respectively. The foreign population represents 8% of the total population.

Socio-economically, Biscay's gross domestic product (GDP) per capita reached 29.249 € in 2013, slightly lower than the average for the Basque Country but higher than in the Spanish data. In 2014, the unemployment rate was 16.1% for men and 15.8% for women. Important differences exist in the activity rate, since it reached 63.1% in men but 49.5% in women.

A. Bacigalupe • Y. González-Rábago (✉) • U. Martín • A. Unceta
University of the Basque Country, Biscay, Spain

S. Murillo
Government of Biscay, Biscay, Spain

A. Zaidi (eds.), *Building Evidence for Active Ageing Policies*,
https://doi.org/10.1007/978-981-10-6017-5_6

As part of the activities developed by the Observatory for the Elderly of Biscay—a participatory organization that brings together social, economic and political agents related to the elderly in Biscay—the Provincial Council of Biscay proposed and promoted in 2013 the calculation of the Active Ageing Index (AAI) as a tool for the diagnosis and monitoring of active ageing among the old population of Biscay. In collaboration with the Department of Sociology of the University of the Basque Country, the index has been calculated to support policymaking and evidence-based decision-making in this area, and help develop adequate policies to improve the quality of life of the population. Ageing is an extraordinary social and political challenge in Europe, which has to be faced in each region not only according to their demographic and social situation but also taking into account the opportunities for an active ageing of each context (González-Rábago et al. 2015).

The objective of this chapter is to describe the methodology used for the calculation of the AAI for Biscay, as an example of how to do it in the subnational level using two different kinds of sources—primary and secondary data. The results provide useful information about the situation of the elderly in a small region and highlight the potentials of the AAI for policymaking to improve active ageing.

6.2 Methodology

In order to calculate the Active Ageing Index for the region of Biscay, we followed the AAI methodology regarding indicators and weighted values for the calculation of the overall value and for each of its domains (Zaidi et al. 2013). Two strategies have been developed to obtain the scores of the indicators as well as the final result of the AAI for Biscay, combining two data sources: on the one hand, we produced primary data through a survey directed to the population of 55 years and older living in Biscay. The survey was carried out exclusively for the calculation of the index due to the absence of suitable secondary data. On the other hand, we also used data from secondary sources to gather information for those indicators related to other groups of population or whose calculation needed additional information, or when we needed a higher reliability than we had using the primary data source.

6.2.1 Calculation of the AAI Through a Combined Methodology: Primary and Secondary Data

The main source of information for the calculation of the AAI was the survey, whose design was based on the original questions of the AAI methodology. The sample was selected from a file of phone numbers of 15,000 households in Biscay. The selection was made by sex and age quotas taking into account the distribution of the population aged 55 and older. A total of 1362 telephone interviews were conducted using the Computer Assisted Telephone Interview (CATI) system. The survey was conducted between November 6 and 18, 2014. The fieldwork for the survey was coordinated by the Gaia-Research and Consulting Company.

As a complementary strategy, we collected secondary data from differences sources, most of them supplied by the Eustat-Basque Institute of Statistics. All its statistical operations are directed to the population of the Basque Country, so data for the population of Biscay had to be selected. We analysed the comparability of the indicators supplied by secondary sources of information and concluded that most of them were directly or indirectly comparable with the original indicators. Indeed, comparability reached 86%. We calculated all the values for the available indicators through the secondary data sources in order to compare and validate the primary data obtained through the specific survey designed for this study.

As a result of the combined methodology, the final calculation of the Active Ageing Index has been carried out using primary data for 12 out of 22 indicators, and secondary data for the rest. The selection of the values that we finally employed was based on comparability and reliability criteria. Thus, we used secondary data in those cases where primary data were not possible to calculate, or additional information was needed ('remaining life expectancy', 'share of healthy life expectancy'), or information about other age groups was needed to calculate the indicator ('relative median income' and 'no poverty risk'). Secondary sources were also employed when quality of data was better not only for comparability but also for reliability of the source ('employment rates').

The five secondary data sources used were the following:

1. Survey on the Population in relation to activity (Labour Force Survey) of the Basque Country 2012–2013: we collected the values related to

occupation rate and educational level (indicators 1.1–1.4 and 4.6, see Table 6.1 and Annex).

2. Demographic Survey of the Basque Country 2011: we collected the value related to household type (indicator 3.3, see Table 6.1 and Annex).
3. Survey of Poverty and Social Inequalities of the Basque Country 2012: we used data to calculate the indicators 'relative median income' and 'no poverty risk' (indicators 3.4 and 3.5, see Table 6.1 and Annex).
4. Mortality Statistics of the Basque Country: we used data to calculate the remaining life expectancy at age 55 (indicator 4.1, see Table 6.1 and Annex).
5. Health Survey of the Basque Country 2013: we collect data to calculate the share of healthy life expectancy at age 55 (indicators 4.2, see Table 6.1 and Annex).

In Appendix 1, all the information regarding the data used, years and sample sizes for the population of Biscay is summarized for each of the 22 indicators.

Although the primary data was collected in 2014, most of the available secondary data referred to the years 2012–2013, so the comparison of Biscay with the European countries was done with reference to AAI 2012.

6.3 Results

Biscay's overall score in the Active Ageing Index in 2012–2014 is 35.9 points. This score places the region of Biscay in an advantageous position in comparison with the 28 countries of the European Union (EU). The score of Biscay is 1.9 points higher than the EU28 average. The Active Ageing Index in Biscay shares the sixth position along with Germany and Luxembourg. However, it is also remarkable that there is a significant gap between the score of Biscay and the countries that are in a better situation (Fig. 6.1).

Analysing the four domains separately, it is observed that Biscay's position in each of them is different, highlighting the better rank in the second and fourth domains, that is, participation in society and capacity for active ageing. On the contrary, the situation of Biscay in relation to the

Table 6.1 Results for each indicator and dimension in Biscay, EU28 average and the country with the highest score (2012–2014)

	Biscay	EU28	Highest score
1.1. Employment rate for the age group 55–59	58.7	62.2	82.0 Sweden
1.2. Employment rate for the age group 60–64	30.4	31.6	64.2 Sweden
1.3. Employment rate for the age group 65–69	2.5	11.6	27.0 Estonia
1.4. Employment rate for the age group 70–74	1.2	6.1	20.4 Romania
Total employment	**23.2**	**27.9**	**43.4 Sweden**
2.1. % of older population aged 55+ providing unpaid voluntary work through the organizations	20.0	14.8	32.7 Austria
2.2. % of older population aged 55+ providing care to their children, grandchildren (at least once a week)	37.4	32.5	53.7 Italy
2.3. % of older population aged 55+ providing care to elderly or disabled relatives (at least once a week)	21.5	12.9	17.1 Finland
2.4. % of older population aged 55+ taking part in the activities of meeting of a trade union, a political party or political action group	14.7	12.0	26.5 Sweden
Total participation in society	**23.7**	**18.1**	**25.2 Ireland**
3.1. % of people aged 55 years and older undertaking physical exercise or sport at least 5 times a week	54.3	15.6	48.9 Finland
3.2. % of people aged 55 years and older who report no unmet need for medical and dental examination or treatment during the last 12 months	76.7	88.2	97.7 Netherland
3.3. % of people aged 75 years and older who live in a single household alone or in a couple household	54.2	84.2	99.5 Denmark
3.4. The relative median income ratio is defined as the ratio of the median equivalized disposable income of people aged above 65 to the median equivalized disposable income of those aged below 65	76.8	86.7	110.2 Luxemburg
3.5. % of people aged ≥65 who are not at risk of poverty	90.8	93.0	98.6 Czech Rep

(continued)

Table 6.1 (continued)

3.6. % of people aged ≥65 who are not severely materially deprived	94.1	90.0	100.0 Luxemburg
3.7. % of people aged ≥55 who are not worried about becoming a victim of violent crime	83.7	69.3	94.6 Slovenia
3.8. % of people aged 55 to 74 who stated that they received education or training in the four weeks preceding the survey	16.9	4.5	22.9 Denmark
Total independent and secure living	**67.8**	**70.6**	**79.0 Denmark**
4.1. Remaining life expectancy achievement of 50 years at age 55	60.2	53.8	59.4 France
4.2. % of healthy life years in the remaining life expectancy at age 55	61.9	53.2	76.1 Sweden
4.3. Mental well-being in people ≥55	87.9	64.6	87.3 Denmark
4.4. % of people aged 55–74 using the internet at least once a week	46.4	40.8	78.0 Sweden
4.5. % of people aged 55 or more that meet socially with friends, relatives or colleagues several times a week or every day	67.7	49.0	73.6 Portugal
4.6. % of people aged 55–74 with upper secondary or tertiary educational attainment	33.1	59.7	84.3 Czech Rep
Total capacity for active ageing	**63.5**	**54.4**	**69.2 Sweden**
Active Ageing Index	**35.9**	**34.0**	**44.8 Sweden**

other two domains—employment and independent and secure living—is relatively worse than in other countries. This is observed in both men and women, although with some differences that are described below. In Appendix 2, a figure for each domain is shown comparing all the countries of the European Union.

Thus, as it is shown in Fig. 6.2, in the first domain—employment—Biscay has a lower score than the EU28 average, both among men and women. Moreover, there is a remarkable inequality between the occupation rate in men and women in Biscay, being 10 points lower among women (19.2%). Among men, the activity rates in the 55–59 and 60–64 age groups are similar and slightly higher than the EU mean, but for those aged 65 and older, Biscay's occupation rates fall to very low levels. This explains the worse score in the domain, standing over 3 points away from the EU28 and very far from Sweden, the leading country in men's

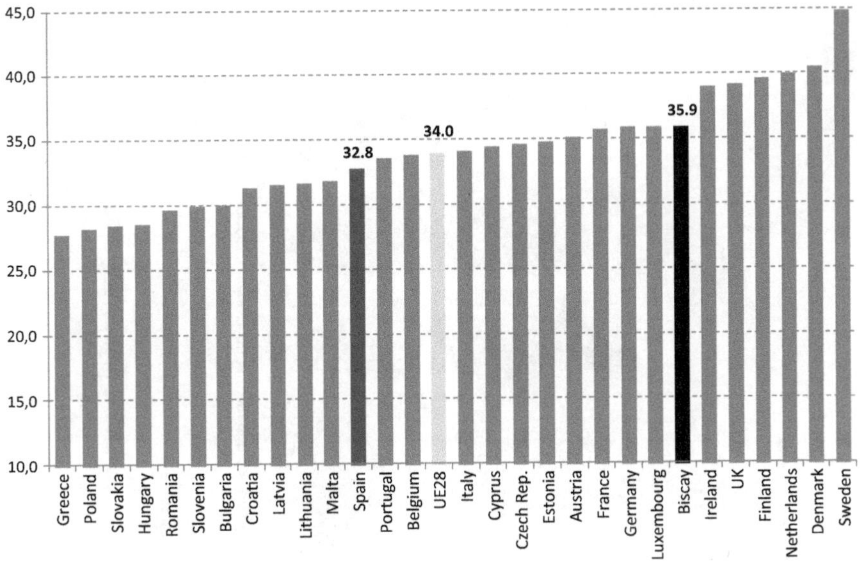

Fig. 6.1 AAI results for the European countries, the EU28 average and Biscay: 2012–2014

employment. Among women, the occupation rate in the 55–59 age group is lower than the EU average, as well as among those aged 65 or older, while it is slightly higher in the 60–64 age group. Thus, occupation in women over 54 in Biscay is 4 points below the EU28 mean and far away from Estonia, the country with a higher female employment rate.

Secondly, in relation to participation in society of people above 54, Biscay reaches a high score in every indicator. Indeed, social participation in Biscay is significantly higher than the average of the European Union, both among men and women. The score in social participation is higher among women (24.1%) than among men (23.2%). Elderly people in Biscay provide voluntary work through organizations more frequently than in other countries of the EU (19% in men and 20.8% in women). Furthermore, population in Biscay stands out compared to the rest of the countries in the indicators of provision of care (2.2 and 2.3), especially in that provided to older adults or to disabled relatives, in which Biscay's score, both among men and women, is the highest of the EU28 (18.2% and 24%, respectively). However, these data highlight the existence of an

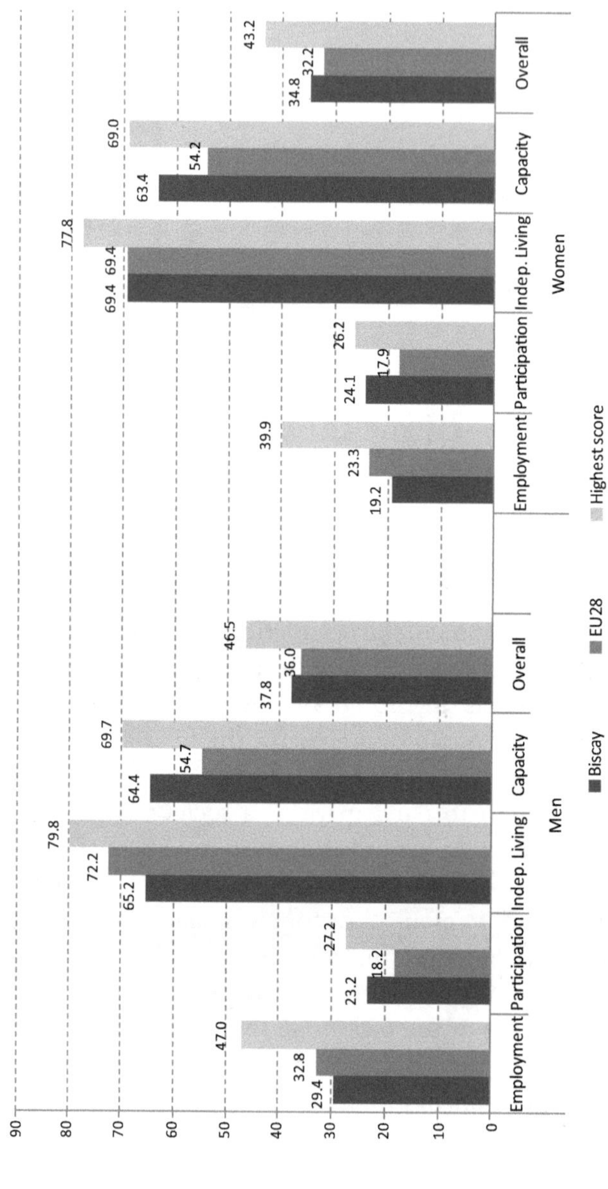

Fig. 6.2 Results by domain and sex in Biscay, EU28 average and the country with the highest score: 2012–2014

important gap between sexes. In addition, the political and trade union participation is remarkable, being also higher than the EU28 average, and with a special interest in women as leaders (11%) in the EU.

Thirdly, the independent and secure living domain shows values slightly lower than the EU28 average among men (7 points less), but not among women, whose score coincides with the EU28 mean. In men, the general worse position of Biscay in this domain is mainly due to the low proportion of men that live independently, that is, alone or in a couple (39.2%) compared to the EU28 average (84.2%). In women, this indicator is also lower than the EU28 mean, but the proportion is more similar to that of the EU28 (62.7% in Biscay vs. 84.2%). Moreover, the percentage of people who report no unmet need for medical and dental examination or treatment is lower in Biscay (77.1% in men and 76.5% in women) than in the EU28 (88.6% and 87.8%, respectively). In the rest of indicators on this domain (related to physical exercise, income, security and training), Biscay obtains similar scores or even higher than the EU28 average. However, the smaller weight given to these indicators compared to the two in which Biscay has a lower result (unmet need for medical and dental treatment, and independent living) produces a decrease in the global score in this domain.

Finally, in relation to the fourth domain, capacity for active ageing, Biscay is on a very good position compared to the rest of the countries. The higher life expectancy—Biscay has the highest result of all countries both among men and women—and the share of healthy life years, the two indicators with the largest weight in this domain, provide a very positive idea of the capacity for active ageing of elderly people in Biscay. Especially relevant is the distance in the score of the share of healthy life years in men between Biscay and the EU28 average, nearly 12 points higher. On the other hand, the distance in the share of healthy life years in women is also relevant, although slightly lower than that in men (7.4 points). It is also remarkable that there is better mental well-being of elderly people in our region, being higher than the EU28 average in nearly 25 points among women and 21.4 among men. Besides, other indicators such as the use of the internet or social connectedness show a better score for the elderly in Biscay.

In Table 6.1, the values of each indicator are presented as well as the score for each dimension and the overall AAI. The results for Biscay are compared with the EU28 average and the highest value reached by any country.

6.4 Conclusions

6.4.1 What Is the Position of Biscay in Relation to Other Countries in the European Union?

The AAI for the province of Biscay in 2012–2014 was 39.5, which means 1.9 points above the average of the EU28. Biscay is placed in a high position within the 28 countries of the EU. After a first group of countries with a very high score in the overall index, with Sweden leading, Biscay is positioned within the second group of countries with a high score, at the level of France, Germany or Luxembourg.

The relatively better position of Biscay compared to the rest of the countries was evident in the Participation domain. Indeed, the limited available data comparing European regions confirm the existence of differentials in involvement in political protest activities and the positioning of the Basque Country among the regions with highest values (Valencia et al. 2010). Regarding the data of clearly fewer people aged 75 and over who live in a single household alone or in a couple, the explanation could be a stronger familistic ideology and a family-centred welfare system in Southern European countries (Minas et al. 2014; Moreno 2014), where much of the care (to older people, children and also adult offspring) is provided within the family sharing the same household (Bianchi-Pernasilici and González-Rábago 2016). Another indicator where Biscay's result is lower than the EU28 score is related to the relative median income in people above 65 compared to those below that age. In Biscay, retirement pensions are fixed by the Spanish Government, which means that similar income level is observed among the retired. However, the relatively higher income of the rest of the population in Biscay compared to the Spanish median could imply that, finally, the ratio is lower in Biscay than in Spain, and also than in other countries of the EU28.

As regards the employment domain, the relatively lower score of Biscay in the employment rates of those above 65 could be due to the specific characteristics of the productive structure of this region. The higher relative weight of manual workers employed in the industrial sector in the Basque Country compared to the EU28 could explain that the preretirement policies applied in the 1990s during the industrial restructuring in Biscay affected more workers in this province, compared to the whole EU28 (Eustat 2014). Finally, the characteristics of the history of the educational structure of Biscay (Delgado Criado 1994) can also explain partly the lower score in the indicator 4.6, especially compared to the EU28. Later universalization of education in Biscay entails that the proportion of older population with secondary or tertiary education was clearly lower.

6.4.2 What Are the Methodological Challenges and Limitations for the Calculation of AAI at Subregional Level?

The calculation of the AAI at the subregional level is of highest interest for better policymaking on ageing. Most of the competencies regarding ageing (including active ageing) are located at the province level in the Basque Country, which at present generates an important mismatch between the lack of reliable and high-quality data and the area where most of these policies are implemented. It is, therefore, necessary to build advocacy for a progressively higher availability of data in those geographical levels where policies take place. That is the only way to advance in a more rigorous assessment of policymaking and better accountability to citizens.

The calculation of the AAI for the province of Biscay is, therefore, of great relevance since it allows testing the possibilities of calculation in the subregional level. However, this calculation is limited by some methodological challenges, affecting not only Biscay, but also any other region at the same level. Most of the challenges were related to the fact that no statistical data is available for such a local area [a province (NUTS 3) of a region (NUTS 2) inside a country (NUTS 1)], with sufficient sample size and level of comparability with the rest of the countries. Even though a regional Institute of Statistics supplies data for the Basque Country

(EUSTAT), most of the indicators for Biscay showed some degree of difficulty for their full comparability with the rest of the countries.

Besides, in relation with the results shown in this chapter, it is necessary to mention a limitation about the period of calculation. Despite the primary data that all referred to 2014, most of the available secondary data were gathered for the years 2012 and 2013, depending on the data source. This is the reason why we compared the results for Biscay with those for EU28 in the AAI round 2012.

6.4.3 Current Limitations and Future Proposals for Advancing with the AAI

Taking into account the main limitations of the calculation of the AAI at the subregional level as well as the relevance of continuing to do so, we propose some ways to advance a better calculation of the AAI in Biscay and in any other subregion.

On the one hand, it would be highly advisable that, at least in highly decentralized countries such as Spain, the European Commission-Eurostat would encourage existing regional institutes of statistics to offer comparable data, as it does with national institutes, where European-level statistics are directly applied to national-level samples. These data should also be representative for smaller areas within the regions such as provinces, with sufficient sample size to be able to carry out analyses for different population subgroups defined by gender, socioeconomic position or residence in smaller areas within the province. However, when the previous strategy is not possible, it would be worthwhile to promote a specific survey that can offer the necessary data to calculate the AAI and to explore the reasons explaining (some of) their scores.

On the other hand, the Provincial Council of Biscay has also an interest in building a more locally based AAI, which considers citizens' opinions about what active ageing should be. This will require exploring which are the dimensions and specific aspects that the index should consider according to the elderly in Biscay and the weight each of them should have. In its current version, 70% of the weights assigned to the AAI constituent indicators correspond to paid and unpaid work, while the weight of each of the other dimensions hardly reaches 20%. Weights could be differently distributed according to local needs.

Appendix 1. Data Sources, Years and Sample Sizes for Each Indicator of the AAI (2012–2014)

Domain	Indicator	Data source	Years	Sample size
1. Employment	1.1. Employment rate for the age group 55–59	Survey on the population in relation to activity (Labour Force Survey) of the Basque Country	2012–2013	*n* = 531
	1.2. Employment rate for the age group 60–64	Survey on the population in relation to activity (Labour Force Survey) of the Basque Country	2012–2013	*n* = 531
	1.3. Employment rate for the age group 65–69	Survey on the population in relation to activity (Labour Force Survey) of the Basque Country	2012–2013	*n* = 531
	1.4. Employment rate for the age group 70–74	Survey on the population in relation to activity (Labour Force Survey) of the Basque Country	2012–2013	*n* = 531
2. Participation in society	2.1. % of older population aged 55+ providing unpaid voluntary work through the organizations	Survey on Active Ageing in Biscay	2014	*n* = 1362
	2.2. % of older population aged 55+ providing care to their children, grandchildren (at least once a week)	Survey on Active Ageing in Biscay	2014	*n* = 1362
	2.3. % of older population aged 55+ providing care to elderly or disabled relatives (at least once a week)	Survey on Active Ageing in Biscay	2014	*n* = 1362
	2.4. % of older population aged 55+ taking part in the activities of meeting of a trade union, a political party or political action group	Survey on Active Ageing in Biscay	2014	*n* = 1362

(*continued*)

Appendix 1 (continued)

Domain	Indicator	Data source	Years	Sample size
3. Independent and secure living	3.1. % of people aged 55 years and older undertaking physical exercise or sport at least five times a week	Survey on Active Ageing in Biscay	2014	*n* = 1362
	3.2. % of people aged 55 years and older who report no unmet need for medical and dental examination or treatment during the last 12 months	Survey on Active Ageing in Biscay	2014	*n* = 1362
	3.3. % of people aged 75 years and older who live in a single household alone or in a couple household	Demographic Survey of the Basque Country	2011	
	3.4. The relative median income ratio is defined as the ratio of the median equivalized disposable income of people aged above 65 to the median equivalized disposable income of those aged below 65	Survey of Poverty and Social Inequalities of the Basque Country	2012	*n* = 616
	3.5. % of people aged ≥65 who are not at risk of poverty	Survey of Poverty and Social Inequalities of the Basque Country	2012	*n* = 616
	3.6. % of people aged ≥65 who are not severely materially deprived	Survey on Active Ageing in Biscay	2014	*n* = 1362
	3.7. % of people aged ≥55 who are not worried about becoming a victim of violent crime	Survey on Active Ageing in Biscay	2014	*n* = 1362
	3.8. % of people aged 55 to 74 who stated that they received education or training in the four weeks preceding the survey	Survey on Active Ageing in Biscay	2014	*n* = 1362

(*continued*)

Appendix 1 (continued)

Domain	Indicator	Data source	Years	Sample size
4. Capacity for active ageing	4.1. Remaining life expectancy achievement of 50 years at age 55	Mortality Statistic of the Basque Country	2012	Population register
	4.2. % of healthy life years in the remaining life expectancy at age 55	Basque Health Survey	2013	$n = 1801$
	4.3. Mental well-being in people ≥55	Survey on Active Ageing in Biscay	2014	$n = 1362$
	4.4. % of people aged 55–74 using the internet at least once a week	Survey on Active Ageing in Biscay	2014	$n = 1362$
	4.5. % of people aged 55 or more that meet socially with friends, relatives or colleagues several times a week or every day	Survey on Active Ageing in Biscay	2014	$n = 1362$
	4.6. % of people aged 55–74 with upper secondary or tertiary educational attainment	Survey on the population in relation to activity (Labour Force Survey) of the Basque Country	2012–2013	$n = 1.345$

Source: Appendix in Zaidi et al. (2013)

Appendix 2. Results in Biscay, EU28 Average and Each EU Country by Domain

Employment

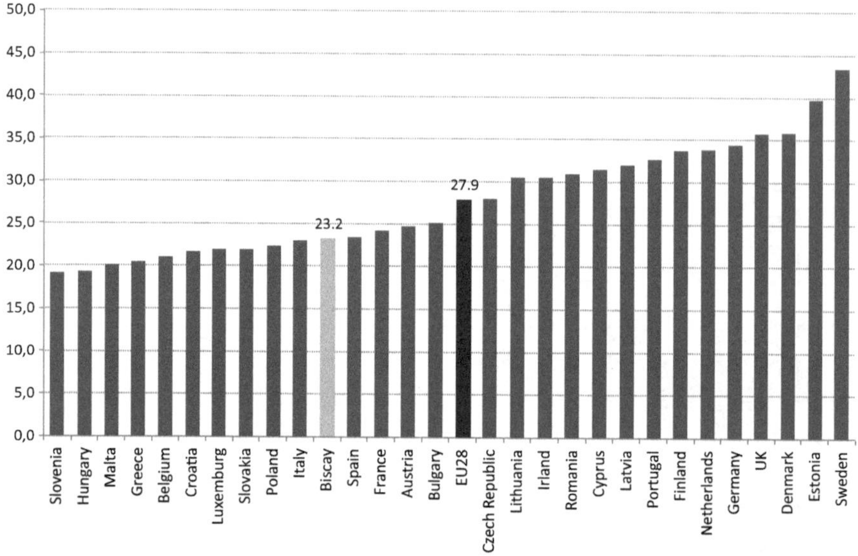

Participation in Society

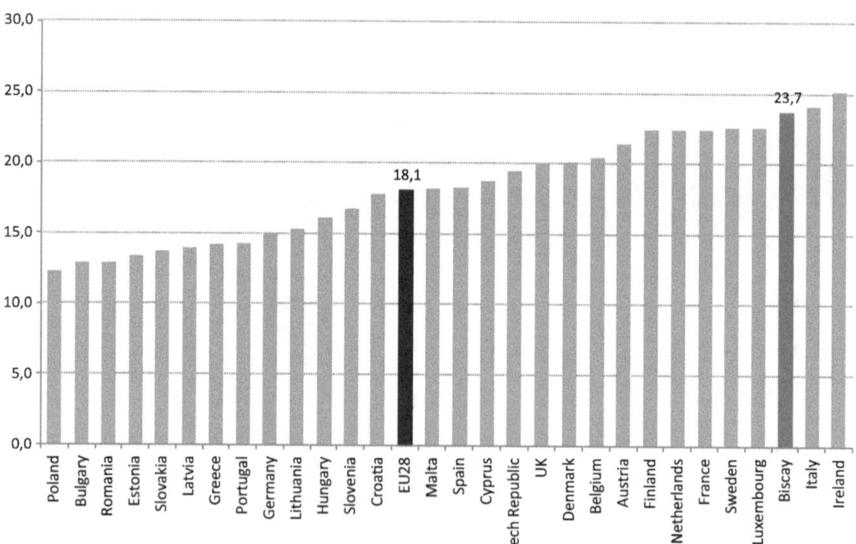

Independent and Secure Living

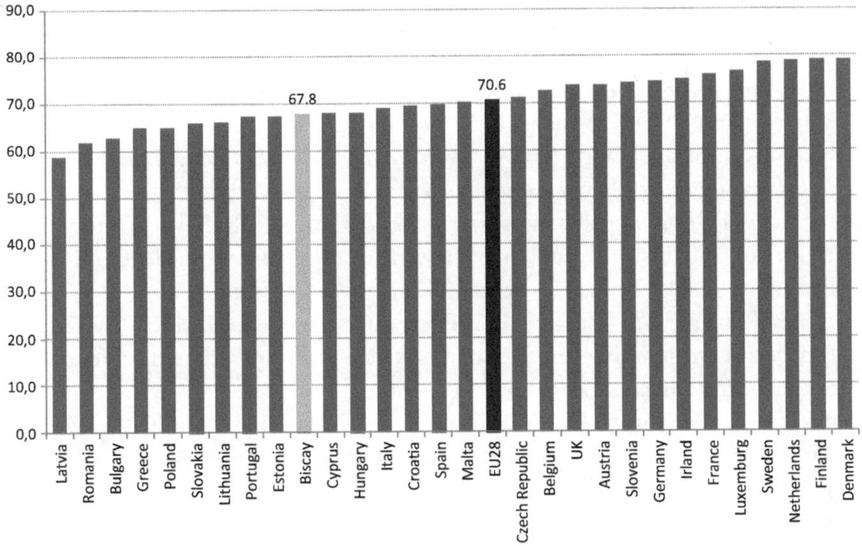

Capacity for Active Ageing

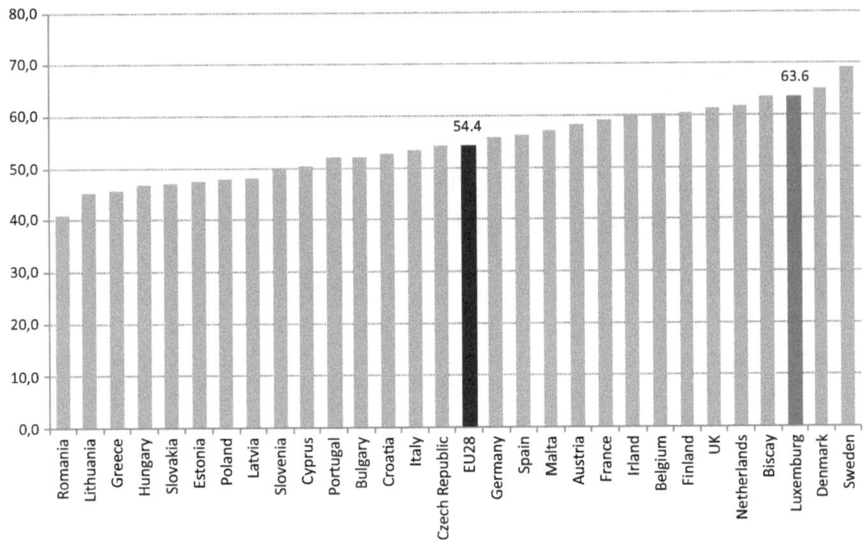

References

Bianchi-Pernasilici, G., & González-Rábago, Y. (2016). El cuidado de personas en la familia. In M. Legarreta (Ed.), *Dos décadas de cambio social en la CAE a través del uso del tiempo: Encuesta de Presupuesto de Tiempo 1993–2013*. Vitoria-Gasteiz: Eustat-Instituto Vasco de Estadística.

Delgado Criado, B. (1994). *Historia de la Educación en España y América. Vol. 3: La educación en la España Contemporánea*. Madrid: Morata.

Eustat. (2014). *Basque industry overview 2014* [Online]. Retrieved October 19, 2015 in Spanish from http://www.eustat.es/elementos/ele0011300/ti_Panorama_de_la_Industria_Vasca_2014/inf0011343_c.pdf

González-Rábago, Y., Martin, U., Bacigalupe, A., & Murillo, S. (2015). Envejecimiento activo en Bizkaia: situación comparada en el contexto europeo. *Revista Zerbitzuan, 59*, 145–159.

Minas, C. H., Jacobson, D., Antoniou, E., & Mcmullan, C. (2014). Welfare regime, welfare pillar and Southern Europe. *Journal of European Social Policy, 24*(2), 135–149.

Moreno, L. (2014). Spain's transition to new welfare: A farewell to superwomen. In P. Taylor-Gooby (Ed.), *New risks, new welfare: The transformation of the European welfare*. New York: Oxford University Press.

Valencia, J. F., Cohen, E. H., & Hermosilla, D. (2010). Social trust and political protest. The mediating role of the value of Power Distance. *Psicologia Política, 40*, 61–80.

Zaidi, A., Gasior, K., Hofmarcher, M. M., Lelkes, O., Marin, B., Rodrigues, R., Schmidt, A., Vanhuysse, P., & Zolyomi, E. (2013, March). *Active Ageing Index 2012. Concept, methodology, and final results*. Research Memorandum/Methodology Report. European Centre Vienna. Retrieved from www.euro.centre.org/data/aai/1253897823_70974.pdf

Amaia Bacigalupe has a position as a lecturer at the University of the Basque Country. She holds a PhD in Public Health and a Master's in Public and Social Policies. For years, she has combined university teaching with research work in the Research Unit of the Department of Health of the Basque Government, developing projects about the social factors that influence population health, with a special focus on social inequalities in health and health care, Health Impact Assessment and Health in All Policies. She is member of the Research Group on Social Determinants of Health and Demographic Change-OPIK.

Yolanda González-Rábago has a position as a lecturer at the University of the Basque Country. She holds a PhD in Sociology and a Diploma in Public Health. Currently, her main research lines include sociology of migration, sociology of health, with special focus on social inequalities in health, and demographic change, especially the analysis of population ageing. She is a member of the Research Group on Social Determinants of Health and Demographic Change-OPIK.

Unai Martín has a position as a lecturer at the University of the Basque Country teaching Demography and Statistics. He holds a PhD in Sociology and Advances Studies Diploma in Public Health. Before his position at the university, he worked as a researcher in the Basque Foundation for Health Innovation and Research. Currently, his main research lines are the sociology of health, especially the analysis of social inequalities in health, and demographics, especially the analysis of health and mortality, and population ageing. He is Director of the Research Group on Social Determinants of Health and Demographic Change-OPIK.

Sergio Murillo is Director General for the Promotion of Personal Autonomy, in the Government of Biscay. He is a member of the Expert group on the Active Ageing Index, Treasurer and member of the Board of Directors of the European Covenant on Demographic Change. He holds a Bachelor's degree in Economics and Business Administration from the University of Deusto (Bilbao) and a Master's in Management Development. His professional background is related to the leadership and management of ageing-related social services.

Alfonso Unceta is a Professor of Sociology at the University of the Basque Country. Researcher in numerous R&D competitive projects and main R&D contracts; he has published more than fifty different papers and books on academic change in education, social responsibility, participation and social innovation. He is also the Director of SINNERGIAK, Social Innovation Centre at Euskampus—Campus of International Excellence—in the Basque Country.

7

Evaluating Socio-economic Impact of Age-Friendly Environments

Mireia Ferri Sanz, Jorge Garcés Ferrer, Willeke van Staalduinen, Rodd Bond, and Menno Hinkema

7.1 Introduction

According to the World Health Organization (WHO) (2007), physical and social environments are key determinants of whether people can remain healthy, independent and autonomous. Health and well-being are not only determined by our personal characteristics but also by the environment where we were born and where we live throughout our life (WHO 2015). In fact, promoting age-friendly environments (AFE) is the most effective approach for responding to demographic change

M. Ferri Sanz (✉) • J. Garcés Ferrer
University of Valencia, València, Spain

W. van Staalduinen
Age-friendly Nederland, Gouda, Netherlands

R. Bond
Netwell Centre at the Dundalk Institute of Technology, Dundalk, Ireland

M. Hinkema
TNO (Netherlands), The Hague, Netherlands

A. Zaidi (eds.), *Building Evidence for Active Ageing Policies*,
https://doi.org/10.1007/978-981-10-6017-5_7

because they empower elderly people to age in better health, they promote their social inclusion and active participation, and they help them to live autonomously into old age (Parent 2012). AFE foster health, well-being and the participation of people as they age; AFE are accessible, equitable, inclusive, safe, secure and supportive; and they provide services centred on people and support citizens to enable recovery or to compensate the loss of function so that people can continue to do things that are important to them (WHO 2015).

Moreover, environment initiatives that promote ageing in place have the potential to lead to positive outcomes for the entire population (Metlife Mature Market Institute and Stanford Center on Longevity 2013). Ageing in place is also believed to be less expensive than institutional long-term care for elderly people, their families and governments (Sixsmith and Sixsmith 2008). This idea is directly linked with the Social Sustainability Theory promoted by one of the authors of this chapter that consists in a joint reorganisation of the health and social care systems to provide an answer to the necessities of people requiring long-term care (Garcés and Ródenas 2012; Garcés et al. 2011).

Many cities and communities are already implementing initiatives to become more age-friendly. For them, the WHO has launched a dedicated website,[1] Age-friendly World, that supports them providing guides and tools, practices and information about existing initiatives. In parallel, several WHO initiatives including the "Healthy Cities movement"[2] and the "WHO Strategy and Action Plan for Healthy Ageing in Europe" (WHO 2012) are framing and reinforcing efforts to make Europe more age-friendly. At the European level, the European Commission launched an Action Group on AFE inside the European Innovation Partnership on Active and Healthy Ageing (EIP AHA), the D4 Action Group on Age-Friendly Buildings, Cities and Environments. This group works on: setting up the mechanisms to engage older people and ensure their participation, adapting and developing principles on AFE for the EU context, understanding the role that Information and Communication Tecnologies (ICTs) can play in this field, exploring new ways to promote active and healthy ageing with AFE, and running pilots in this line (EIP AHA 2012).

On the occasion of the EU Summit on Active and Healthy Ageing in 2013, mayors and senior political representatives of European cities, municipalities, communities and regions signed the Dublin Declaration on Age-Friendly Cities and Communities in Europe 2013, highlighting the importance of promoting AFE. Signatories committed themselves to ensure that the views and opinions of older people are heard and valued; to adopt measures to develop urban and public spaces that are inclusive, can be shared and desirable for all; and to promote the development of neighbourhoods and communities for all ages, among other actions. Moreover, in December 2015, the Covenant on Demographic Change was launched aiming to gather all local, regional and national authorities, and other stakeholders, that commit to cooperate and implement (evidence-based) solutions to support active and healthy ageing as a comprehensive answer to Europe's demographic challenge.[3] Practical, pragmatic, based on user experience, and context-sensitive initiatives are welcomed inside the Covenant.

Thus, many solutions in this area have been implemented, but they remain isolated and unknown. Moreover, there is a need to raise the awareness of local and regional authorities of the importance of self-assessing the initiatives implemented so they can evaluate their efficacy at different levels (social, economic and environmental) and make them transferable to other regions or municipalities. In this context, the authors have proposed the Socio-Economic and Environmental Impact Tool (SEE-IT) to help local and regional authorities assess the socio-economic and environmental impact of their interventions in this field and adopt evidence-based policies. This tool has been developed within the work-plan of the European project "Thematic Network Innovation for age-friendly environments in the European Union—AFE-Innovnet" funded by the CIP programme of the European Commission (ref.: 620978),[4] following a participatory methodology involving municipalities and regional authorities that provided their feedback in an active way.

The SEE-IT is presented in this chapter as a framework for local and regional authorities to evaluate their age-friendly initiatives and, consequently, to better evaluate the costs and benefits of their initiatives and to take informed decisions on AFE. In this sense, the creation of a common framework to evaluate the AFE initiatives will support these authorities to develop new initiatives with a stronger evidence base, it will support the scaling replication-transfer and adaptation of innovations across

Europe, it will facilitate the creation of new business opportunities, and it will provide greater insight into return on investment in AFE innovations in terms of economic, social and environmental aspects.

The AAI arose as a tool for policymakers to enable them to design informed strategies for ageing and to measure their impacts on the society (Zaidi et al. 2013). Moreover, it is referred as a tool to support the implementation of social investment by the European Commission in its Social Investment Package (European Commission 2014a). The SEE-IT uses some indicators of the AAI and makes the tool comprehensive and usable at the local and regional levels where the data required for the AAI is not always available. In this sense, the SEE-IT is designed as a friendly tool to local and regional authorities that can be developed for collecting own data at the local and regional levels and linking with national and European levels. Consequently, it provides an added value to local and regional authorities who can use the tool to evaluate their initiatives with their own resources. Moreover, this tool is recommended to be complemented using the guidelines for co-production developed in the framework of the AFE-Innovnet project in order to involve older citizens and relevant stakeholders in all the AFE initiative processes (Ferri et al. 2015).

7.2 Methodology

The SEE-IT to evaluate local and regional initiatives in the field of AFE is based on the application of the AAI model to higher resolution spatial scales at the regional and local levels (Zaidi et al. 2013), the WHO-Europe revised guidelines on AFE (WHO 2007), the Canada Mortgage Housing Corporation model (MacLeod et al. 2005), the Global AgeWatch Index (Gorman and Zaidi 2013; Know-Vydmanov et al. 2014), the liveable communities (Kihi et al. 2005) and the Research and Evaluation Framework (Manchester) (Handler 2014). Moreover, an extensive bibliography review was carried out to complete the models and to offer a complete overview of the dimensions and their corresponding indicators that should be considered and assessed to achieve a comprehensive evaluation of the socio-economic and environmental impact of an AFE initiative. Among others, the SEE-IT's design and development has been influenced by: the UN Madrid International Plan of Action on Ageing (2002), the URBACT programme promoting sustainable development (European Commission 2014b), the European Commission (2009) Impact assessment guidelines and the Environmental Protection Agency (2002) social, cultural, economic impact assessment literature review.

The SEE-IT was developed following a participatory methodology involving municipalities and regional authorities that provided their feedback in an active way. Concretely, the draft of the protocol with the selected dimensions was shared with the 29 partners of the AFE-Innovnet project in a project meeting. The 29 partners[5] participating in the meeting were from 16 countries: 13 cities, 6 regions, 5 large EU networks active in the field of ageing or representing large numbers of local authorities or seniors' organisations, 4 research centres specialised in ICT and ageing policies, and a communication agency with vast EU experience. Most of the consortium partners are actively involved in the EIP AHA D4 action group on AFE.

Then the draft of the completed protocol (dimensions, indicators and proceeding) was shared with the members of the AFE-Innovnet thematic network in an online webinar, held on 15 January 2015, to discuss their impressions and with the objective to make it practical and useful for them. Furthermore, an AAI expert was invited to the webinar to offer his point of view and to provide scientific comments to the draft protocol; 27 members of the thematic network attended the webinar and gave their feedback to build the final version of the SEE-IT.

7.3 Results: The SEE-IT

As mentioned, the tool was designed on the basis of the AAI, the WHO guidelines on AFE, the CMHC Logic Model, the Global AgeWatch Index, the liveable communities, and the Research and Evaluation Framework (Manchester). As a result, the SEE-IT uses several indicators from the AAI and it is extended with other variables related to economy and environment in order to make it usable at regional and local levels. After the review, the tool is composed of the following three high-level domains: social, economic and environmental (see Fig. 7.1).

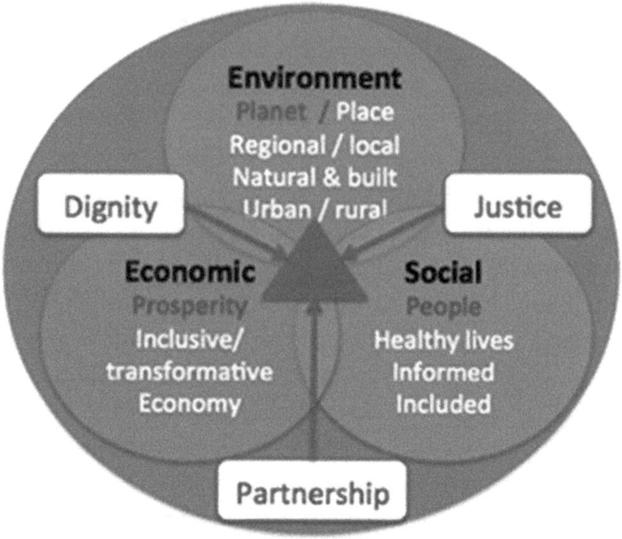

Fig. 7.1 High-level domains to the European AFE initiative. Derived from United Nations (2014)

These three high domains are related to the six essential elements for delivering the Sustainable Development Goals established by the United Nations (2014): Dignity, People, Planet, Prosperity, Justice and Partnership. Furthermore, the core of the SEE-IT is characterised by the following features:

- A holistic framework with the three high-level domains mentioned (social, economic and environmental) that considers older people in a society for all and in different timelines
- An iterative process with multiple entry points depending on the use-context and status of the initiative that covers from overview to detailed plans
- A co-design partnership with multiple stakeholders and a collaborative process of decision-making and assessment
- A complete set of performance criteria: flexibility to adapt to broad range of use-case scenarios, time as an important dimension, applicability in specific and general AFE policies with different approaches, responsible and reinforcing to involving WHO-AFE guidelines, simply for use across different scales and stakeholders

The tool defines a generic impact assessment process with the steps to be followed by the local and regional authorities in order to evaluate their AFE initiatives from the three high-level domains perspective. The relationships between the five stages are illustrated in the Fig. 7.2, and the main aspects of the SEE-IT process are described below. Depending upon the development and robustness of the team involved and the AFE initiative, strategy or policy, development teams may join the assessment process at stages 2 or 3.

7.3.1 Step 1: Aim of the SEE-IT

The first stage of the process is the most critical as it sets the overall direction and extent of the evaluation exercise. Before starting, the initiative and the *core problem or challenge* addressed should be identified. The initial identification of problems can come from many sources, including national, regional and local strategies and their reviews, consultations with experts, citizens' councils and representative NGO organisations,

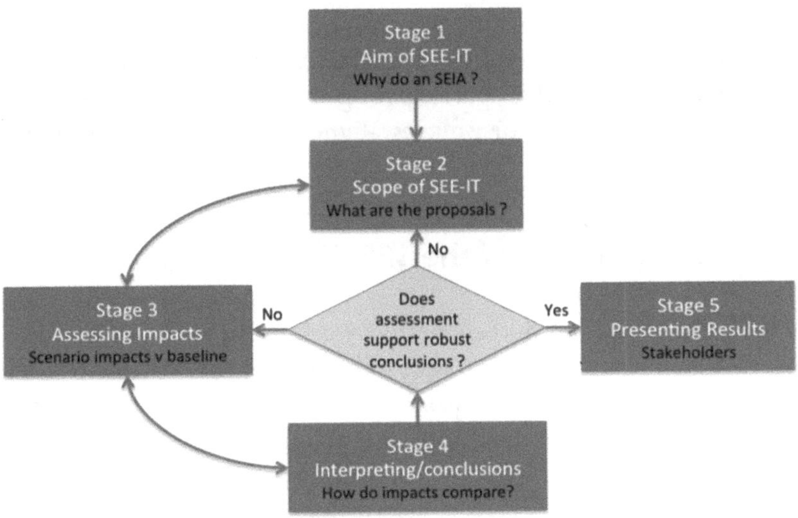

Fig. 7.2 The overall SEE-IT process (Bond et al. 2015)

and mainstream and specialised media reporting. In relation to AFE, they can come from sources such as:

- A review or audit of community experiences and considerations for improvements across the eight domains of the WHO age-friendly cities guide
- Consideration of cross-cutting themes in the AFEE guide in areas such as physical activity, falls and factures, diet and nutrition, social connectedness, loneliness and isolation, and universal access and inclusive design
- Higher-level systemic problems such as market failures (inequities/access to care and assistive technologies), regulatory failures (right, responsibilities, benefits and entitlements), and public awareness and attitudes (education, better and different choices, personal planning and greater tolerance, etc.)

When identifying the core problems to be addressed, it is also important to consider who the *targeted or affected populations* are. Also the *objectives* should be established ranging from general to specific and operational. Local and regional authorities should ensure that the detailed objectives are coherent with other local/regional/national strategies, or should consider if these strategies can influence the specific objectives. Moreover, the government representatives participating in the SEE-IT process should identify the programmes or policies aligned with the AFE initiative to support and reinforce it.

7.3.2 Step 2: Scope of the SEE-IT

The bulk of the creative and generative work of the evaluation process is under this step where the work-plan is developed, the baseline scenario is set out, the boundaries of the work are defined, and several future state scenarios and actions are developed and described:

> Much of the success of the process is based on the *team*. The key is to balance a mix of ownership, passion and commitment to improvement with objectivity, impartiality and tolerance. It is recommended that the team in

charge of the process is composed by a multi-stakeholder committee and a leader capable of directing and managing a group formed by people from different sectors and disciplines. It is important to involve an adequate representation of older persons in the team who can help further to engage elderly people when exploring options and co-producing with them (Ferri et al. 2015).

At this stage, the *baseline* should be defined, that is, the current situation where no explicit intervention is introduced and that represents the cost of doing nothing. For that, different techniques can be developed: community audits, existing surveys, new data collection through questionnaires, and so on. Additionally, and although AFE initiatives are often connected to other policy domains, themes and actions, it is important that the *boundary* conditions on the problem are as clear as possible in terms of direct and indirect outcomes:

- Vertical scale: ranging across community, neighbourhood, district, city, county and/or region
- Horizontal thematic areas: social, economic and/or environmental focus
- Horizontal organisational: local authority, health service, industry, commercial and retail, education and learning, rural development, crime prevention and/or community groups
- Domain specific: housing, transport, education, employment, and so on
- Cross-cutting health and well-being: physical activity, diet and nutrition, stress, chronic conditions, social isolation, dementia, and so on
- Timelines for action: short, medium or long term

With all previous information, the team should define the potential *scenarios* that can be achieved by the AFE initiative which is being assessed and identify cross-links and establish a short list of valid/desirable options. In this process, *risks* should be predicted and their related solutions agreed. Some of the potentials risks during the SEE-IT process can be:

- Lack of involvement of representatives or target groups
- High regional bureaucracy
- Lack of communication with local authorities who finally implement the initiative

- High geographic coverage
- Reluctant attitude of target groups
- Legislative barriers at different levels
- Budget constraints

7.3.3 Step 3: Assessing Impacts

In this step, the team analyses the AFE initiative from the three high-level domains: social, economic and environmental. First, the *direct and indirect impacts* should be identified; for that, the experience and knowledge of the stakeholders involved in the process is crucial. Below, some examples of AFE initiatives impacts are listed:

- Social impacts: health and longevity, safety, lifelong learning, quality and social integration, and so on
- Economic impacts: investment flows, public budgets, market mechanisms, innovation, property rights, and so on
- Environmental impacts: natural environment, culture, housing, sustainable transport, and so on

Moreover, the team should consider to gather additional data from the previous stages, from the AFE impact framework where the initiative, strategy or policy is implemented, data from checklists and data from models/simulations (where available). Then, the AFE impacts should be assessed using *indicators* that respond to the following characteristics: to represent relative importance, to provide an appropriate level of detail, to respond to the timeline agreed at the beginning of the process and to be reliable.

Some of the proposed indicators to measure the above high-level domains are taken from the AAI (Table 7.1). However, the data required in the AAI is not always available at regional/local level, so local and regional authorities should collect their own data following the comprehensive process detailed in this chapter. Moreover, not all the impacts can be measured in monetary terms, for those impacts (such as impact on professionals' skills or technological benefits), the SEE-IT team should express each impact in the more suitable unit of measure (Monacciani et al. 2012).

Table 7.1 Impacts for consideration in the SEE-IT process (Bond et al. 2015)

	Domains of impact	Possible impacts for consideration/indicators
Economic impacts	Economic prosperity	Regional Gross Domestic Product (GDP)/capita Employment rate for age group* Disparities by sub-areas Household savings—older people
	Investment flows	Cross-border flows—import/export/access to age-related products/services Cross-border jobs/mobility
	Public budgets/ services	Public agency-specific budgets (hospital/ community services/older people services/ housing/transport/welfare benefits, etc.) Service quality/person/citizen-centeredness Cross-agency budget readjustments/sharing/ pooling
	Market mechanisms	Effects on private sector business opportunities/small and medium enterprises (SMEs) Effects on private social enterprise opportunities & structures Balance/transactions across sectors Private Public Partnerships (PPP)
	Innovation, Research & Development (R&D)	Investment in R&D related to active and healthy ageing Pre-commercialisation/intellectual property Accelerated time to market
	Sustainable consumption & production	Consumption: Household expenditure—Structure of older households—energy use, car ownership Production: Persons at work by sectors—industry/manufacturing, professional/services, agriculture/farming Food production/supply: organic farming/ intensity/urban farming
	Property rights	Home ownership/private renting/social housing/residential/nursing homes Legacy/transfers/asset release/leans

(*continued*)

Table 7.1 (continued)

	Domains of impact	Possible impacts for consideration/indicators
Social impacts	Health and longevity	Remaining life expectancy achievement* Healthy life years in the remaining life expectancy* Physical activity*, nutrition, mental well-being* Causes of death/chronic diseases, places of death, Accidental injuries/falls Independent living*
	Safety	Criminality, safety of social environment Protection of older people against abuse Protection against risks/in emergencies Feelings of safety Physical safety*
	Productive and valued activities	Employment opportunities (also, e.g. market-volunteering) Absence of discrimination Working conditions/quality of work environments Amount of leisure Care for others (informal care)*
	Standard of living of older people	Principal status Private consumption, median income* Poverty and deprivation* Mobility (also for disabled/impaired) Choice/control over where/how one lives—tenure status
	Education/ lifelong learning	Literacy Use of ICT* Opportunity to go to higher education Educational attainment* Lifelong learning*
	Quality of social interaction	Social connectedness* Social participation—volunteering* Migration, ethnicity, languages Religious participation Political participation*
	Private and family life	Families by family cycle Marital status—widow/widower One person households
	Personal data	Access to information—availability/restrictions Identity/identifiers Protection of data/sharing information
	Basic rights and responsibilities	European Charter on Fundamental Rights: Human dignity, Equality, Freedoms, Justice, Solidarity, Citizens' rights Moral outlook/responsibilities

(*continued*)

Table 7.1 (continued)

	Domains of impact	Possible impacts for consideration/indicators
Environmental impacts	Natural environment	Quality of local environment—air, water, etc. (toxins) Nature of environment—urban/suburban/rural Landscape: Typological features—hills/mountains Waterways/lakes/coast Biodiversity—common/rare flora/fauna
	Culture, heritage and leisure	Availability and quality of cultural assets (music/arts/drama/literature) Heritage sites Events/festivals Opportunities to participate: artists/performance
	Land use	Geographic context Zoning—agricultural, forestry, marine, industrial, retail, residential, recreational, educational, health service, mixed use Conservation—natural, archaeological, architectural
	Climate & energy	Energy conservation Alternative energy sources Environmental/energy control Seasonal variation/weather/drainage Climate change hazards—sea levels/heat/cold/floods
	Renewable resources/waste	Re-use Reduce Recycling
	Settlement—urban/rural	Spatial hierarchy—connection and access Zoning/mixed use Density/proximity to amenities/retail/services Public spaces: streets/squares/parks Furnishings: lights/seats
	Housing	Households by type of accommodation Quality/age of housing/parking/garages Accessibility Gardens/pets Heating/Water/Drainage/Electricity/waste Broadband/Internet
	Sustainable transport	Means/modes of travel—availability/frequency Time leaving/returning to work/amenities, etc. Journey times Safety/accidents

Indicators marked with * are part of the AAI

7.3.4 Step 4: Interpreting/Conclusions

In this stage, the team should compare the impacts achieved in the three different domains (social, economic and environmental) between the situation before and after the AFE initiative/policy/strategy. Also the distribution of affects is relevant (e.g. "Which target is the most affected?" "How is the impact on different geographies?") and the strength of conclusions (e.g. "Is there a need of more data?" "Do we need greater detail?" "Should we refine scenarios?"). The team can use Table 7.2 to interpret the results obtained considering the relevance of each domain for the framework in which the AFE initiative/policy is implemented:

7.3.5 Step 5: Presenting Results

The final step consists of presenting the report of the SEE-IT process that should consider all the aspects that have been detailed along the different steps: aims and scope, team, impacts, comparison of results, conclusions and appendices, if needed. Moreover, the team should internally analyse the effectiveness and the quality of the work done and propose improvements if needed.

7.4 Discussion and Conclusions

As defined throughout the chapter, promoting AFE is justified as an essential and innovative strategy to face the demographic change as they empower elderly people to enjoy better health, they promote their social inclusion and active participation, and they help them to live autonomously into old age (Parent 2012). In this framework, the need for a tool that supports regional and local authorities in Europe to evaluate the impact of their AFE initiatives has been highlighted from different organisms as the EU2020 vision, the EIP AHA, and the WHO. Thus, the creation of a common framework to evaluate the AFE initiatives entails different benefits: supporting local and regional authorities to develop new initiatives with a stronger evidence base, supporting the scaling replication-transfer and adaptation of innovations across Europe, facilitating the

Table 7.2 Table for comparing scenarios (Bond et al. 2015)

	Policy/Proposal/Scenario A		Policy/Proposal/Scenario B	
Economic domains of impact	**Nature of impacts on resources/ assets**	**Impact assessment**	**Nature of impacts on resources/assets**	**Impact assessment**
Economic property	Advantages/disadvantages/ improvements	+	Advantages/disadvantages/ improvements	+
Investment flows	Advantages/disadvantages/ improvements	+	Advantages/disadvantages/ improvements	+
Public budgets/services	Advantages/disadvantages/ improvements	++	Advantages/disadvantages/ improvements	++
Market mechanisms	Advantages/disadvantages/ improvements	++	Advantages/disadvantages/ improvements	0
Innovation, R&D	Advantages/disadvantages/ improvements	0	Advantages/disadvantages/ improvements	0
Sustainable consumption & production	Advantages/disadvantages/ improvements	+	Advantages/disadvantages/ improvements	+
Property rights	Advantages/disadvantages/ improvements	0	Advantages/disadvantages/ improvements	0
Social domains of impact	**Nature of impacts on capabilities**		**Nature of Impacts on capabilities**	
Health and Longevity	Advantages/disadvantages/ improvements	+	Advantages/disadvantages/ improvements	+
Safety	Advantages/disadvantages/ improvements	+++	Advantages/disadvantages/ improvements	0
Productive and valued activities	Advantages/disadvantages/ improvements	+++	Advantages/disadvantages/ improvements	+
Standard of living of older people	Advantages/disadvantages/ improvements	++	Advantages/disadvantages/ improvements	0

(continued)

Table 7.2 (continued)

	Policy/Proposal/Scenario A		Policy/Proposal/Scenario B	
Education/life-long learning	Advantages/disadvantages/ improvements	+++	Advantages/disadvantages/ improvements	++
Quality of social innovation	Advantages/disadvantages/ improvements	++	Advantages/disadvantages/ improvements	++
Private and family life	Advantages/disadvantages/ improvements	+	Advantages/disadvantages/ improvements	+
Personal data	Advantages/disadvantages/ improvements	–	Advantages/disadvantages/ improvements	–
Basic rights and responsibilities	Advantages/disadvantages/ improvements	0	Advantages/disadvantages/ improvements	0
Environment domains of impact	**Nature of impacts on environment**		**Nature of impacts on environment**	
Natural environment	Advantages/disadvantages/ improvements	+	Advantages/disadvantages/ improvements	+
Culture, heritage and leisure	Advantages/disadvantages/ improvements	+	Advantages/disadvantages/ improvements	+
Land use	Advantages/disadvantages/ improvements	0	Advantages/disadvantages/ improvements	0
Climate & energy	Advantages/disadvantages/ improvements	0	Advantages/disadvantages/ improvements	0
Renewable resources/ Waste	Advantages/disadvantages/ improvements	–	Advantages/disadvantages/ improvements	–

(*continued*)

Table 7.2 (continued)

	Policy/Proposal/Scenario A		Policy/Proposal/Scenario B	
Settlement—urban/ rural	Advantages/disadvantages/ improvements	+	Advantages/disadvantages/ improvements	–
Housing	Advantages/disadvantages/ improvements	++	Advantages/disadvantages/ improvements	++
Sustainable transport	Advantages/disadvantages/ improvements	+	Advantages/disadvantages/ improvements	+
Overall evaluation		++ to +++[a]		+ to ++[a]

[a]The overall evaluation requires consideration—not simple aggregation as impacts may need to be weighed

creation of new business opportunities and providing greater insight into return on investment. Moreover, initiatives that promote ageing in place have the potential to lead to positive outcomes for the entire population (Metlife Mature Market Institute and Stanford Center on Longevity 2013) and they are also believed to be less expensive than institutional long-term care for elderly people, their families and governments (Sixsmith and Sixsmith 2008), as defended by the Social Sustainability Theory promoted by Garcés and Ródenas (2012) and Garcés et al. (2011).

In this context, this chapter presents the SEE-IT developed within the work-plan of the project "Thematic Network Innovation for age-friendly environments in the European Union—AFE-Innovnet" funded by the CIP programme of the European Commission (ref.: 620978). It offers the framework to be followed by local and regional authorities in order to evaluate their AFE initiatives, policies and strategies considering three high dimensions: social, economic and environmental. These three high domains are related to the six essential elements for delivering the Sustainable Development Goals established by the United Nations (2014): Dignity, People, Planet, Prosperity, Justice and Partnership.

With this tool, authors pretend to offer a comprehensible framework for local and regional authorities in order to encourage them to self-assess their AFE initiatives and raise awareness among them about their related social, economic and environmental benefits. For that, the design of the SEE-IT was developed following a participatory methodology that involved local and regional authorities along Europe. As a result, the SEE-IT proposes a simple process that enables authorities to evaluate their initiatives even when external data is not available, making usable the AAI at low levels and replicable at European level. Indeed, the AAI offers an analytical framework for further extensions, for example, at the regional and local analysis (Zaidi 2014) as made with the SEE-IT and presented in this chapter.

Nevertheless, additional research is required to validate the SEE-IT in different fields and approaches with the aim of evaluating the tool and its effectiveness for local and regional authorities. A common validated tool as the SEE-IT will support local and regional authorities to develop new initiatives with a stronger evidence base that will increase innovations across Europe in terms of economic, social and environmental gains. This will encourage governments to build AFE and the quality of life and well-being of older people will increase, as health and well-being are not only determined by personal characteristics but also the environment where we were born and where we live throughout our life (WHO 2015).

Acknowledgements The SEE-IT is developed under the Project "Thematic Network Innovation for age-friendly environments in the European Union—AFE-Innovnet" funded by the CIP programme of the European Commission (ref.: 620978).

Notes

1. http://www.agefriendlyworld.org/
2. http://www.euro.who.int/en/health-topics/environment-and-health/urban-health/activities/healthy-cities
3. http://afeinnovnet.eu/content/towards-covenant
4. http://afeinnovnet.eu/
5. Age Platform Europe AISBL, Conseil Des Communes et Regions D'Europe-Association, Polibienestar Research Institute, Nederlandse Organisatie Voor Toegepast Natuurwetenschappelijk Onderzoek—TNO, PAU Education SL, Louth County Council, Azienda per I Servici Sanitari n 5 Bassa Friulana, Office of Commissioner for Older People in Wales, Manchester City Council, Academisch Ziekenhuis Groningen, Tamperen Kaupunki, Stockholms Stad, Ville de Bruxeless-Service Seniors, Miasto Stoleczne Warszawa, Municipality of Ljubljana, Eurohealthnet ASBL, Alzheimer Europe, Dundalk Institute of Technology, Conseil regional de Franche-Comté, European Social Network, INOVAMAIS-Servicos de Consultadoria em Inovacao Tecnologica SA, Urzad Miasta Krakowa, Frederica Kommune, Vereniging Voor Vlaamse Steden en Gemeente, Kuldiga District Municipal Agency "The Social Service", Fundaçao para o Desenvolvimento Social do Porto, Consiglio Nazionale Delle Ricerche, JZ SOCIO, Javni Zavod za Sozialnovardstvene Dejavnosti Celje and Tallinna Linn

References

Bond, R., Ferri, M., van Staalduinen, W., Garcés, J., & Hinkema, M. (2015). A protocol for European regions, local authorities, and communities. Social, Economic and Environment Impact Tool (SEE-IT). D.4.2-Methodology and indicators for LRAs to assess socio-eco impact of investing in AFE developed in coordination with WHO Europe technical work. AFE-Innovnet project funded by the European Commission within the ICT policy Support Programme.

EIP AHA. (2012). Action plan on innovation for age-friendly buildings, cities & environments. Conference of partners, Brussels.

Environmental Protection Agency. (2002). *Social, cultural, economic impact assessments: A literature review*. Albuquerque: Galisteo Consulting Group.

European Commission. (2009). *Impact assessment guidelines*. European Commission.

European Commission. (2014a). Policy roadmap for the 2014 implementation of the social investment package. *Employment, social affairs and inclusion*. http://ec.europa.eu/social/main.jsp?catId=1044&langId=es&newsId=1807&moreDocuments=yes&tableName=news

European Commission. (2014b). *URBACT programme 2014–2020* [Online]. Retrieved February 15, from http://urbact.eu/en/about-urbact/urbact-2014-2020/#overview

Ferri, M., et al. (2015). Participatory method to involve end-users (older people): In co-production of AFE solutions by LRAs and older people, to be used for future Covenant. Guidelines for co-producing Age Friendly Environments with older people. AFE-INNOVNET project co-funded by the European Commission within the ICT policy Support Programme.

Garcés, J., Carretero, S., & Ródenas, F. (2011). *Reading of the social sustainability theory: Applications to long-term care field*. Valencia: Tirant Lo Blanch.

Garcés, J., & Ródenas, F. (2012). La teoría de la Sostenibilidad Social: aplicación en el ámbito de cuidados de larga duración. *Revista Internacional de Trabajo Social y Bienestar, 1*, 49–60.

Gorman, M., & Zaidi, A. (2013). *Global AgeWatch Index 2013: Insight report*. London: HelpAge International.

Handler, S. (2014). *A research and evaluation framework for age-friendly cities*. UK Urban Ageing Consortium.

Kihi, M., Brennan, D., Gabhawala, N., List, J., & Mittal, P. (2005). *Livable communities: An evaluation guide*. Washington: AARP Public Policy Institute.

Know-Vydmanov, C., Mihnovits, A., & Zaidi, A. (2014). *Global AgeWatch Index 2014. Methodology update*. London: HelpAge International.

MacLeod S., Davidson B., Ricketts, B., & Wilson C. (2005). *A framework for cost-benefit analysis of HASI and PRAP-D. Research highlight* (Socio-economic Series 05-038).

Metlife Mature Market Institute and Stanford Center on Longevity. (2013). *Livable community indicators for sustainable aging in place*. Study, New York.

Monacciani F., Passani, A., Bellini, F., & Debicki, M. (2012). Deliverable D.3.3b—SEQUOIA self-assessment how-to guide. WP3: Socio-Economic Impact Assessment of Research Projects. SEQUOIA project "Socio-Economic Impact Assessment for Research Projects" project funded by the 7°FP.

Parent, A. S. (2012, October 29–30). *Age friendly goods and services—An opportunity for social and economic development. Toward an age-friendly EU by 2020.* Stakeholder Paper AGE, Poland.

Sixsmith, A., & Sixsmith, J. (2008). Ageing in place in the United Kingdom. *Ageing International, 32*, 219–235.

United Nations. (2002). *Political declaration and Madrid international plan for action on ageing.* New York: United Nations.

United Nations. (2014). *The road to dignity by 2030: Ending poverty, transforming all lives and protecting the planet.* Synthesis Report of the Secretary-General. On the post-2015 agenda, New York.

WHO. (2007). *Global age-friendly cities: A guide.* France: WHO.

WHO. (2012). *Strategy and action plan for healthy ageing in Europe 2012–2020.* Malta: Regional office for Europe.

WHO. (2015). *Age-friendly environments* [Online]. Retrieved January 12, 2015, from http://www.who.int/ageing/projects/age-friendly-environments/en/

Zaidi, A. (2014). AAI 2014: Active Ageing Index for 28 EU countries. Policy Brief presented at the UNECE/European Commission stakeholders event "The Active Ageing Index—Potential for Evidence-Based Policy Development: A Dialogue with Policymakers" 14 November, Brussels.

Zaidi, A., K. Gasior, M. M. Hofmarcher, O. Lelkes, B. Marin, R. Rodrigues, A. Schmidt, P. Vanhuysse and E. Zolyomi (2013). Active Ageing Index 2012: Concept, Methodology and Final Results. EC/UNECE, Active Ageing Index Project, UNECE Grant ECE/GC/2012/003, European Centre for Social Welfare Policy and Research, Vienna. http://www.euro.centre.org/data/aai/1253897823_70974.pdf

Mireia Ferri Sanz is a PhD in Social Sciences, Labour and Human Resources. She is an external expert for the Spanish Service for the Internalisation of the Education for the evaluation of funded projects under the ERASMUS+ programme. In 2012–2015, she co-coordinated the groups related with accessible spaces and ICT in the Action Group D4 of the International Partnership on Active and Healthy Ageing and now coordinates the group on age-friendly tourism. She has recently been nominated member of the Board of Directors of the Covenant on Demographic Change.

Jorge Garcés Ferrer is a Full Professor at Universitat de València, Príncipe de Asturias Professor at Georgetown University, Washington DC, and the Director of Polibienestar Research Institute (Universitat de València). His research has focused on comparative European social policy, especially on ageing and social innovation and on the efficiency of long-term care policies in Europe. He has participated in the Committee on Employment and Social Affairs at the Spanish Parliament as an expert and also in different committees at the European Commission.

Willeke van Staalduinen is a political scientist and was a nurse by profession. Currently, she owns the private company Age-friendly Nederland. She is assigned by the European Commission as promoter of Action Group D4 age-friendly environments of EIP on AHA. She co-founded the AFEdemy: Academy for capacity building on the implementation of age-friendly environments. Her former professional assignments were as researcher and coordinator at TNO and Netherlands Institute of Healthcare Institutions and as health policy assistant of the social liberal party in Dutch parliament D66.

Rodd Bond is the Director of the Netwell Centre at the Dundalk Institute of Technology (Ireland) and an architect by his background. Rodd has a wide range of interests including creating and sustaining healthy communities and homes; the interrelationship between the environment and peoples' health and well-being; organic design methods; and community decision-making frameworks. Rodd is a Partner in the EIP AHA's D4 Specific Action Group on Age Friendly Environments and has acted as a rapporteur for the D4 group.

Menno Hinkema is a senior research scientist at TNO (Netherlands). Menno has over 15 years' senior-level experience in health systems and health property planning. His activities in ageing research have focused on the role of the physical environment in supporting independent living and empowerment of older individuals. His current focus is on ageing-resilient housing solutions and map-based instruments for policymaking support.

8

The Active Ageing Index: A Tool to Develop the Strategy of Active Ageing in the Basque Country (Spain)

Elena del Barrio, Angel Rodriguez-Laso, Olga Mayoral, Mayte Sancho, and Lide Amilibia

8.1 Introduction

Demographic changes experienced in recent years and projections of the future population make ageing of the population of the Basque Country a pressing issue, both for society as a whole and for its governance.

Currently, the Basque Country population is slightly above 2 million (Table 8.1, referring to year 2014): out of them, 732,611 are people of age 55+, constituting more than one-third of the total population (33.7% of the men population, 36.4% of the women population). For comparison, 19.4% are young people (16–34 years old). In Spain and the rest of the 28 developed countries of the European Union (EU-28), the percentage of the population over 55 years is 29.6% (INE 2014a, b, c) and 35.0% (Eustat 2014a, b), respectively. The Basque Country is the fourth most aged region of Spain.

E. del Barrio (✉) • A. Rodriguez-Laso • O. Mayoral • M. Sancho
Matia Institute of Gerontology, San Sebastián/Donosti, Basque Country, Spain

L. Amilibia
Basque Government, Basque Country, Spain

A. Zaidi (eds.), *Building Evidence for Active Ageing Policies*,
https://doi.org/10.1007/978-981-10-6017-5_8

Table 8.1 Population by age and sex, Basque Country, 2014

	Total		Men		Women	
2014	Total	%	Total	%	Total	%
Total	2,172,877	100.0	1,057,455	100.0	1,115.422	100.0
16–34 years	420,494	19.4	213,981	20.2	206,513	18.5
35–54 years	699,365	32.2	352,210	33.3	347,155	31.1
55 and over	732,611	33.7	326,780	30.9	405,831	36.4

Source: Compiled from data of Eurostat. Estadística Municipal de Habitantes a 1 de enero 2014

Table 8.2 Projections of population by age and sex, Basque Country, 2029

	Total		Men		Women	
2029	Total	%	Total	%	Total	%
Total	2,056,975	100.0	989,187	100.0	1,067,788	100.0
16–34 years	390,284	19.0	198,486	20.1	191,796	18.0
35–54 years	508,312	24.7	251,587	25.4	256,723	24.0
55 and over	923,010	44.9	417,588	42.2	505,416	47.3

Source: Compiled from data of INE: Proyecciones demográficas en base 2014 a 1 de enero

Population projections estimate that by 2029 people of age 55+ will be 44.9% of the total population (923,010; Table 8.2). This number will almost double the adult population (35–54 years).

Population pyramids (Fig. 8.1) summarize the demographic past and future of the Basque Country. From 1996 to 2029, we can observe the ageing of the more populated cohort. In 1996, that cohort was 20–40 years old. In 2014, it was 35–55 years old. The largest cohort in 2029 will be aged 50+ years. On the contrary, in 2029, the population between 25 and 45 years old will decrease dramatically. Another important age group that will increase its numbers is that of the centenaries, especially women. All these changes are stressful and will further pressurize the sustainability of public pensions and social and healthcare services. This situation can be only overcome through an active contribution of older people themselves and by fostering lifestyles throughout the life course that will allow a healthy and fulfilling old age (Zaidi and Stanton 2015).

In this context, the Department of Employment and Social Policy of the Basque Country in collaboration with Matia Institute of Gerontology, decided to design the 'Strategy of Active Ageing of the Basque Country

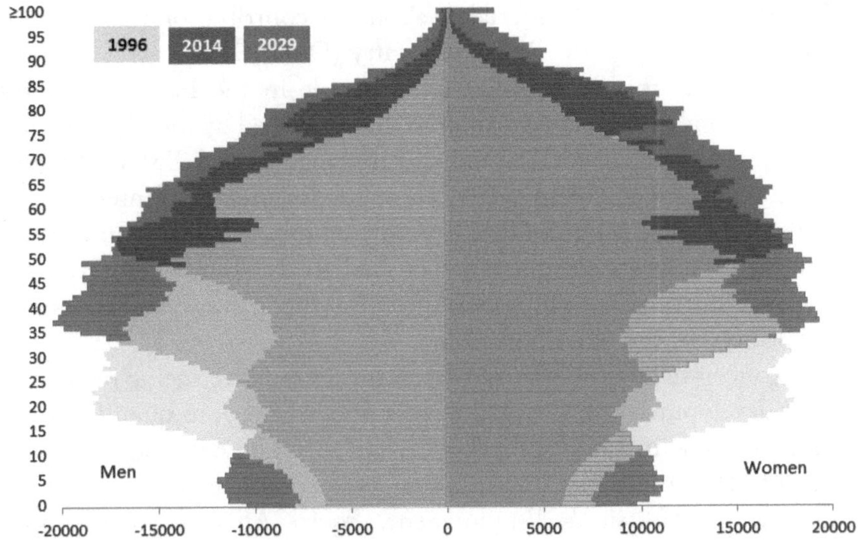

Fig. 8.1 Population pyramids of the Basque Country. 1996, 2014, 2029. Source: Compiled from data of Eustat. Estadística Municipal de Habitantes a 1 de enero 2014. Proyecciones demográficas en base 2014 a 1 de enero. INE: INEBASE: Proyecciones de población 2009

2015–2020' (SAABC), with the aim to contribute to the identification of a model of governance that gives voice to ageing people and helps building a society for all ages.

8.2 Research Context

In recent years, Basque policymakers have adopted an approach to developing plans to meet the needs of the population based on the information provided by several quantitative and qualitative studies. This follows the motto: 'knowing for intervening'. The first survey with this objective was carried out in 1993 (Survey of the Elderly, CAPV 1993) with the aim to launch a gerontological plan in 1994. In 2010, the first 'Living Conditions Survey of people of 60+ in the Basque Country' was implemented and used as the basis for the development of the document: '100 Proposals for advancing wellbeing and good treatment for people growing old. Basis for an Action Plan' (Sancho 2010).

In 2012, research was undertaken about the contribution to older people to the economy of the Basque Country (Durán 2014), with the purpose to analyze the situation of older people in the Basque Country economy, from a perspective similar to that proposed by the Commission on the Measurement of Economic Performance and Social Progress (Stiglitz et al. 2009) and advocated in many documents of international organizations, especially the Platform for Action of the UN Conference (Beijing 1995). It was the first effort of the Basque Country Government to measure the quality of life, social progress, inequalities, and economic contributions of the older population.

The conceptual basis of the SAABC lays on the active ageing paradigm. Recent developments in the field have produced a tool to quantify it following the guidelines of health, activity, and autonomy: the Active Ageing Index (AAI) is the result of a research project managed jointly by the European Commission's Employment, Social Affairs and Inclusion Directorate General and the United Nations Economic Commission for Europe (UNECE).

The AAI is a toolkit comprising 22 individual indicators grouped in four domains: employment, social participation, independent living, and capacity for active ageing. The first three domains measure achievements, while the fourth is a measure of the starting conditions for achieving positive active ageing outcomes (Zaidi and Stanton 2015). In April 2013, the European Commission issued this system of ageing indicators to monitor and compare countries and regions around the world and to promote the active role and the autonomy of older people in ageing societies. According to Echeverria (2014), the main objective was to contribute with empirical data to the development of public policies in Europe that were not oriented only to the welfare of older persons but also to their possible contributions to the economy and society. The overall goal of the AAI project is to identify areas in which different policies and programs can promote the contribution and potential of older people (Zaidi and Stanton 2015).

The development of the AAI tool shows a different way to treat active ageing. The concept of active ageing starts from the definition of the WHO, which is based on the well-known three pillars: health, security, and participation. The AAI incorporates an economic dimension that involves the insertion of older persons in the labor market and other productive activities. That transcends the individual perspective to a societal one. Active ageing is 'the situation where people are able to live healthy, independent and secure lives as they age and thus continue to participate in the formal labour market as well as engage in other unpaid productive activities (such as volunteering and care provision to family members)' (Zaidi et al. 2013, p. 6).

This change in the perspective of social studies and public policy on ageing based on the AAI should provide relevant social innovations, due to the use of a measuring system that attaches considerable weight to socio-economic activities, understood as capabilities and ways of participation in society. It is expected that the ageing policies arising from the AAI will be very different to the traditional policies, based on the WHO paradigm of 2002 (Echeverría 2014). The added value of the AAI is that it encourages policymakers to look at active ageing in a comprehensive way (Zaidi et al. 2013, p. 5).

These innovative aspects have led the leading institutions of the SAABC to adopt this tool, broadly used at the national level and over the European Union, as the basis for it. There was a need for obtaining regional scores to inform a plan that has a regional scope. The AAI project allows policymakers to base their interventions on the comparative and substantive evidence of active ageing indicators and composite indices. It aims to help in identifying priority areas of policy development in the near future (Zaidi and Stanton 2015).The wealth of quantitative information obtained in recent years for the development of the successive gerontological plans allows the calculation of the AAI, although an ad hoc survey was needed to measure some of its items.

8.3 Objectives

The calculation of the AAI in the Basque Country has been carried out with two goals: the first and most important is to detect those areas in which it is necessary to influence public policies in order to further promote the quality of life of the older citizens; the second is to compare the situation of the Basque Country with Europe and Spain in relation to active ageing.

The working hypothesis is that there are differences in the results of AAI between Spain and the Basque Country.

8.4 Methodology

The calculation of the AAI in the Basque Country was performed using different statistical sources: the National Institute of Statistics of the Basque Country (Eustat), the Department of Statistics of the Department of Employment and Social Policy (DESP) of the Basque Country, and the Living Conditions Survey (LCS+55) +55 of the DESP (Barrio et al. 2015).

Eustat supplied data and indicators to calculate different domains of the AAI:

Domain 1: Employment: Population in Relation to Activity survey 2013 (PRA 2013) (Eustat 2017)

Domain 4: Capacity and enabling environment for active ageing: Data from the Mortality register of the Department of Health of the Basque Country, corresponding to the year 2013, Health Survey of the Basque Country 2013 (HSBC 2013) (Osakidetza 2017), and the statistical operation 'Survey on the Information Society-ESI-Families' (SISF 2014) (Eustat 2017)

The Department of Statistics provided data on:

Domain 3: Independent, healthy, and secure living: Survey on Poverty and Social Inequalities (SPSI 2014) (Departamento de Empleo y Políticas Sociales 2017)

In addition, a cross-sectional study of the Basque Country community-dwelling population aged 55 years or more was performed ad hoc in 2014 (The LCS+55). It included indispensable indicators needed to build the AAI of the Basque Country and its provinces.

The sample design of this survey was multi-staged. Clusters (census tracts) were randomly selected proportional to its population over 55 years in strata defined by provinces and size of habitat (capitals, >50,000, 20,001–50,000, 5001–20,000 and <5000 inhabitants).

Households and one individual per household were randomly selected. The final sample included 2469 subjects. The survey was structured on nine thematic topics: sociodemographic data; household composition; family networks and exchange of care; housing, environment, and friendliness; activities and active ageing, health, and dependency; retirement; well-being and social resources.

This survey was developed as a continuation of the Survey of the Elderly 1993 (CAPV 1993) and the Living Conditions Survey 60+ (Barrio et al. 2011). In the LCS +55 a topic about active ageing was included to address the SAABC, and the following questions directly taken from the AAI were added:

Domain 2: Participation in society

Voluntary activities
Care of children, grandchildren
Care of older adults
Political participation

Domain 3: Independent, healthy, and secure living

Physical exercise
Access to health and dental care
Independent living arrangements
No severe material deprivation
Physical safety
Lifelong learning

Domain 4: Capacity and enabling environment for active ageing

Mental well-being
Social connectedness
Educational attainment of older persons (Table 8.3)

From the total 22 items included in the AAI, nine (40.9%) came from existing sources and 13 from the ad hoc survey (LCS+55; 59.1%). Taking into account the specific weight of each domain, the data used to calculate the AAI in the Basque Country came 47.5% from secondary sources and 52.5% from the ad hoc survey (Fig. 8.2).

Table 8.3 Domains, indicators, source, and year of the Active Ageing Index of the Basque Country

Domains and indicators	Source	Year
1. Employment		
1.1. Employment rate in the population 55–59 years old	PRA	2013
1.2. Employment rate 60–64	PRA	2013
1.3. Employment rate 65–69	PRA	2013
1.4. Employment rate 70–74	PRA	2013
2. Participation in society		
2.1. Voluntary activities	LCS+55	2014
2.2. Care of children, grandchildren	LCS+55	2014
2.3. Care of older adults	LCS+55	2014
2.4. Political participation	LCS+55	2014
3. Independent, healthy, and secure living		
3.1. Physical exercise	LCS+55	2014
3.2. Access to health and dental care	LCS+55	2014
3.3. Independent living arrangements	LCS+55	2014
3.4. Relative median income	SPSI	2014
3.5. No poverty risk	SPSI	2014
3.6. No severe material deprivation	LCS+55	2014
3.7. Physical safety	LCS+55	2014
3.8. Lifelong learning	LCS+55	2014
4. Capacity and enabling environment for active ageing		
4.1. Remaining life expectancy achievement of 50 years at age 55	Mortality register, HSBC	2013
4.2. Share of healthy life years in the remaining life expectancy at age 55	Mortality register, HSBC	2013
4.3. Mental well-being	LCS+55	2014
4.4. Share of people aged 55–74 using the Internet at least once a week	SISF	2014
4.5. Social connectedness	LCS+55	2014
4.6. Educational attainment of older persons	LCS+55	2014

Note: PRA Population in Relation to Activity survey, *LCS* Living Conditions Survey +55, *SPSI* Independent, healthy, and secure living: Survey on Poverty and Social Inequalities, *HSBC* Health Survey of the Basque Country, *SISF* Survey on the Information Society-ESI-Families

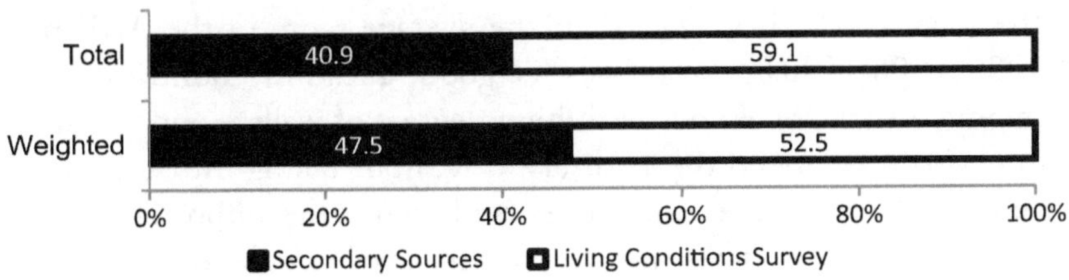

Fig. 8.2 Items included in the Basque Country Active Ageing Index by source

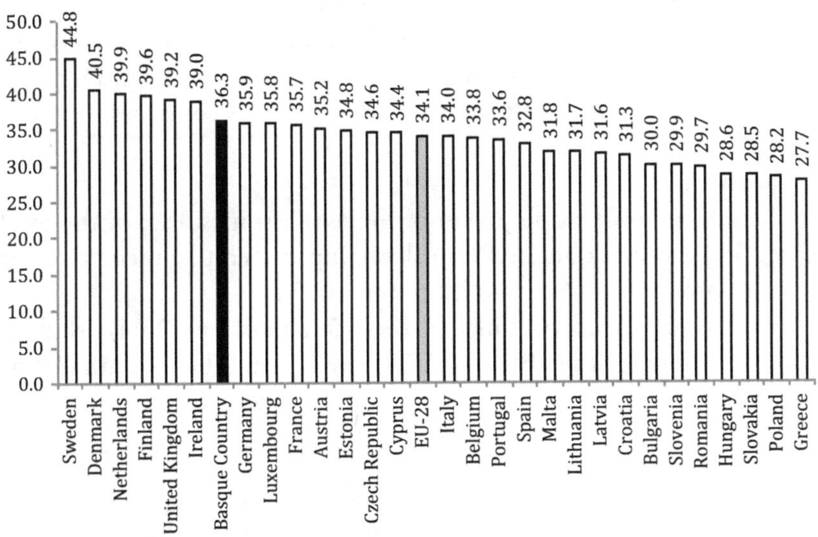

Fig. 8.3 Ranking in the Active Ageing Index of 28 countries of the European Union and the Basque Country. Source: Own elaboration based on: Population in Relation to Activity survey (2013), Living Conditions Survey +55 (2014), Independent, healthy, and secure living: Survey on Poverty and Social Inequalities (2014), Health Survey of the Basque Country (2013), and Survey on the Information Society-ESI-Families (2014)

8.5 Results

1. Overall Index

According to the methodology proposed in the AAI project, the calculation of the global index places the Basque Country in the seventh position of the ranking of the EU-28 (Fig. 8.3).

The score (36.6) is greater than the average score in the AAI in the EU-28 (34.05). This position is very good, ahead of countries such as Germany, Austria, or France with high degrees of welfare and social protection. Spain is ranked the 18th (17th without taking into account the Basque Country). These differences underscore the importance of the territorial level analysis beyond the country-level calculation.

The same comparison by domain of the AAI is presented in Fig. 8.4. The Basque Country fares slightly better than Spain as a whole and the EU-28 countries in the Independent living and Participation in society domains, and clearly better in the Capacity for active ageing domain. In the Employment domain the Basque Country obtains slightly better results than Spain, but slightly worse than the EU-28 countries.

2. Employment

   The sub-index of employment of the AAI assumes 35% of the total score. The score is obtained by calculating the employment rate for each age group (55–59; 60–64; 65–69; 70–74).

   The Basque Country ranks in the 15th position (25.4), slightly below the average of the EU-28 (27.9) and slightly above Spain (19th) (Fig. 8.5).

Fig. 8.4 Comparison of the results of the Active Ageing Index domains between the Basque Country, Spain, and 28 countries of the European Union (EU-28). Source: Own elaboration based on: Population in Relation to Activity survey (2013), Living Conditions Survey +55 (2014), Independent, healthy, and secure living: Survey on Poverty and Social Inequalities (2014), Health Survey of the Basque Country (2013), and Survey on the Information Society-ESI-Families (2014)

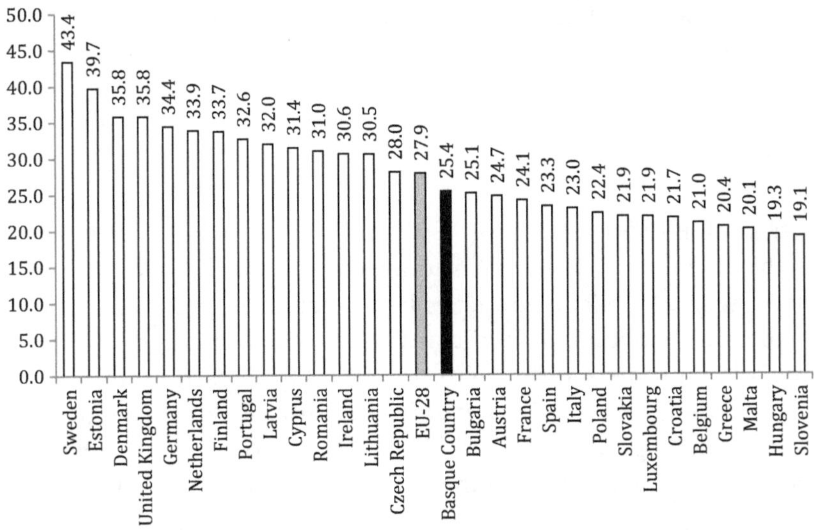

Fig. 8.5 Ranking in the employment domain of the Active Ageing Index of 28 countries of the European Union and the Basque Country. Source: Own elaboration based on: Eustat: Population in Relation to Activity survey (2013)

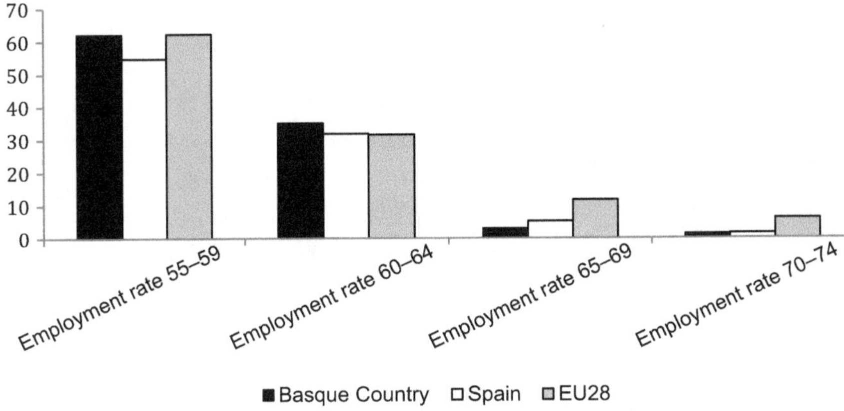

Fig. 8.6 Comparison of the results of the employment domain of the Active Ageing Index between the Basque Country, Spain, and 28 countries of the European Union (EU-28) by age groups. Source: Own elaboration based on: Eustat: Population in Relation to Activity survey (2013)

There are some differences between the employment rates of the Basque Country with regard to the Spanish rates. The percentage of people who work between the ages of 55 and 64 is greater in the Basque Country, while the rates of the group 65 and older are slightly lower (Fig. 8.6).

3. Participation in Society

   The domain of social participation accounts for another 35% of the overall score. Its calculation takes into account unpaid activities such as voluntary activities; those that involve caring of children, grandchildren, older people and the disabled; and political participation.

   The Basque Country is located in the ninth place with a score (21) above the average of the EU-28 (18.1). Spain is positioned in the 15th place, with a similar score to the European average (18.3) (Fig. 8.7).

In comparison to Spain, the Basque Country achieves high scores in voluntary activities, political participation, and taking care of children and grandchildren; however, there is a lower percentage of the Basque population that at least once a week cares for old or disabled people (Fig. 8.8).

4. Independent, Healthy, and Secure Living

   The sub-index Independent, healthy, and secure living represents 10% of the total score. This is the component with less weight on the AAI. It includes information on physical exercise, health needs (medical and dental), type of accommodation, median income, poverty risk, material deprivation, physical security, and lifelong learning.

   The Basque Country is ranked 14th (72.3), standing slightly above the average score of the EU-28 (70.6) and Spain (69.8) (Fig. 8.9).

The much higher percentage of Basque people over 55 years who practice physical exercise more than five times per week (understanding this as practicing sports or any other physical activity such as biking, walking, gardening, etc.; 60.9%) is remarkable (Fig. 8.10). Access to health and dental care is slightly below the Spanish and the EU-28, respectively.

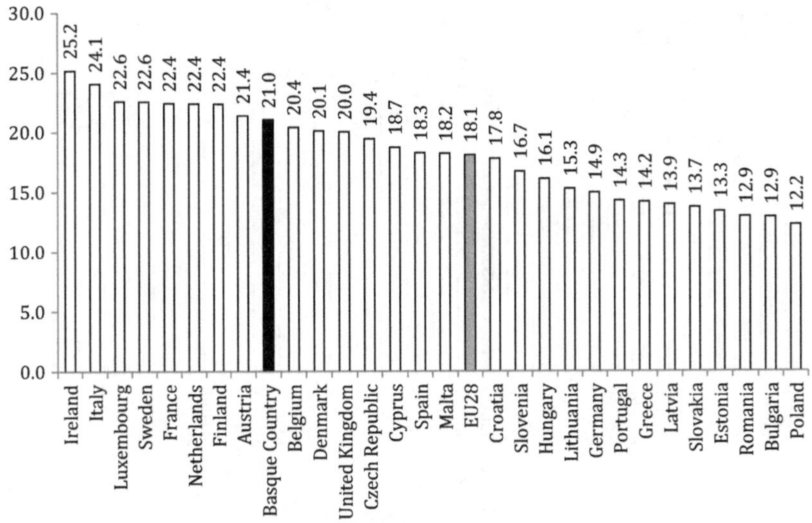

Fig. 8.7 Ranking in the participation in society domain of the Active Ageing Index of 28 countries of the European Union and the Basque Country. Source: Own elaboration based on: Living Conditions Survey +55 (2014)

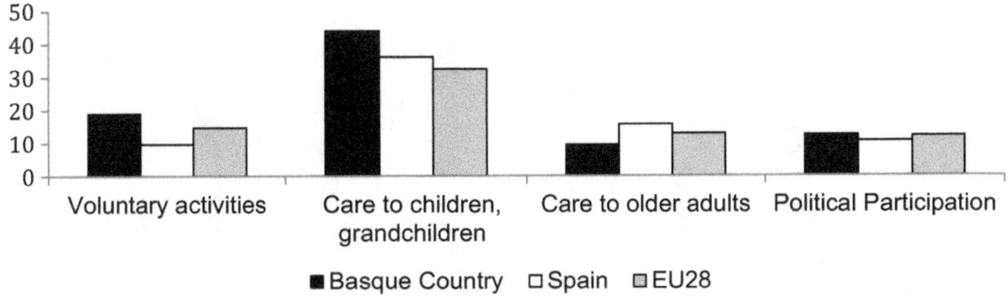

Fig. 8.8 Comparison of the different indicators of the participation in society domain of the Active Ageing Index between the Basque Country, Spain, and 28 countries of the European Union (EU-28). Source: Own elaboration based on: Living Conditions Survey +55 (2014)

There is a much lower prevalence of independent living. On the contrary, the indicators of median income, risk of poverty, and physical safety are better in the Basque Country.

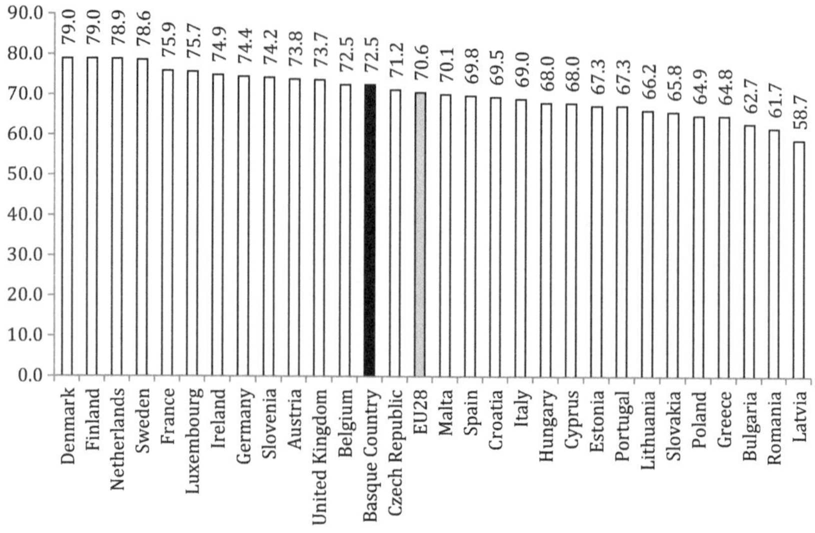

Fig. 8.9 Ranking in the independent, healthy, and secure living domain of the Active Ageing Index of 28 countries of the European Union and the Basque Country. Source: Own elaboration based on: Living Conditions Survey +55 (2014), Independent, healthy, and secure living: Survey on Poverty and Social Inequalities (2014) and Health Survey of the Basque Country (2013)

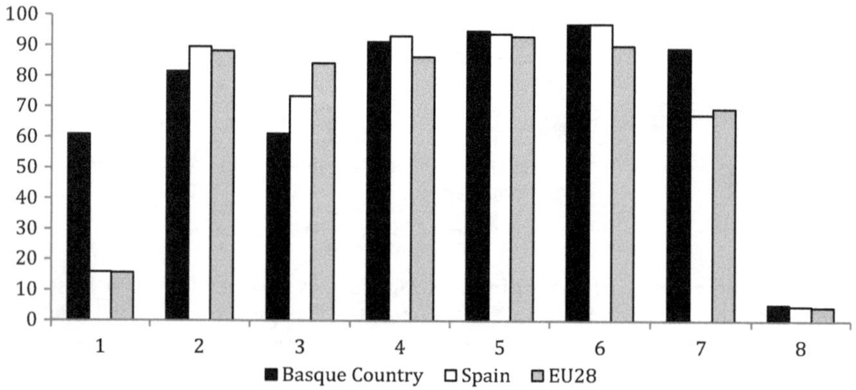

Fig. 8.10 Comparison of the different indicators of the independent, healthy, and secure living domain of the Active Ageing Index between the Basque Country, Spain, and 28 countries of the European Union (EU-28). Note: 1 = Physical exercise; 2 = No unmet needs of health and dental care; 3 = Independent living arrangements; 4 = Relative median income; 5 = No poverty risk; 6 = No material deprivation; 7 = Physical safety; 8 = Lifelong learning. Source: Own elaboration based on: Living Conditions Survey +55 (2014), Independent, healthy, and secure living: Survey on Poverty and Social Inequalities (2014) and Health Survey of the Basque Country (2013)

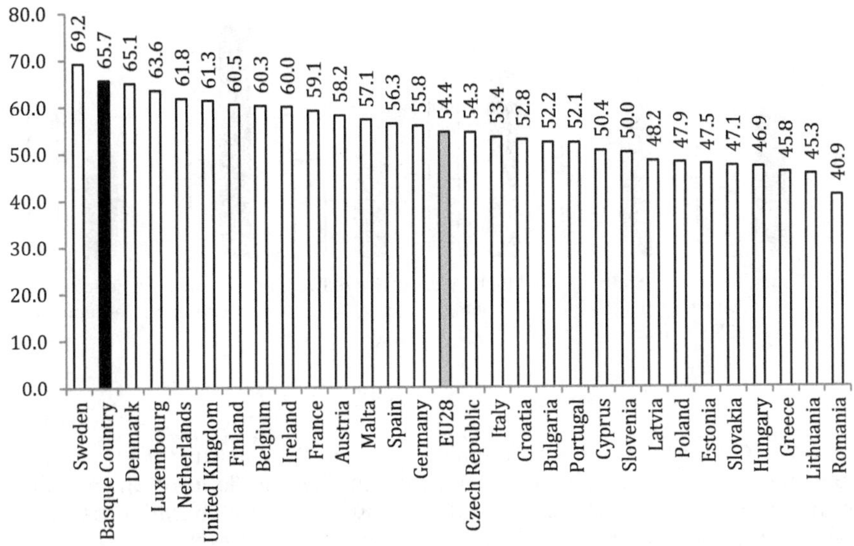

Fig. 8.11 Ranking in the active and healthy ageing domain of the Active Ageing Index of 28 countries of the European Union and the Basque Country. Source: Own elaboration based on: Living Conditions Survey +55 (2014), Health Survey of the Basque Country (2013), and Survey on the Information Society-ESI-Families (2014)

5. Capacity for Active and Healthy Ageing

   The sub-index of capacity for active ageing is the domain that measures if the environment facilitates active ageing. It accounts for 20% of the overall score and includes aspects such as life expectancy, life expectancy in health, mental health, the use of information and communication technologies, social connectivity, and level of education.

   In this sub-index, the Basque Country is ranked second compared to the EU-28, only after Sweden (Fig. 8.11).

The Basque Country gets better scores than Spain and the EU-28 in remaining life expectancy achievement of 50 years at age 55, share of healthy life expectancy at the same age (10 years above the figure of Spain and EU-28), mental well-being, and social connectedness. It is above the Spanish but below the European figures for use of information and communication technologies (ICTs) and educational attainment (Fig. 8.12).

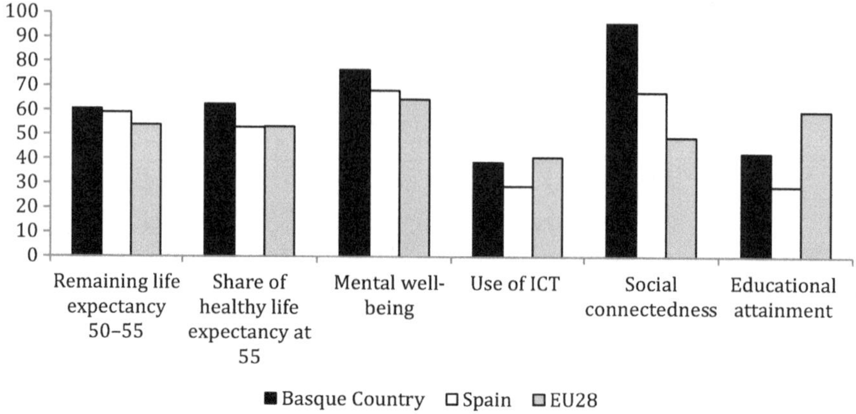

Fig. 8.12 Comparison of the different indicators of the capacity for active ageing domain of the Active Ageing Index between the Basque Country, Spain, and 28 countries of the European Union (EU-28). Source: Own elaboration based on: Living Conditions Survey +55 (2014), Health Survey of the Basque Country (2013), and Survey on the Information Society-ESI-Families (2014)

8.6 Discussion

In relation to the validity of the AAI in the Basque Country, data used from secondary sources (Eustat) were harmonized with the international corresponding surveys (Labor Force Survey -LFS-, Statistics on Income and Living Conditions -SILC-, European Quality of Life Survey -EQLS-, European Social Survey -ESS-.) and questions and answers used in LCS+55 were directly copied from the AAI indicators. Therefore, the results of the 22 items of the AAI in the Basque Country are directly comparable with other countries that use them.

The results of the Basque Country in the AAI are very good, standing out in almost all the domains, with higher scores than the average of the EU-28. They are related to its sociodemographic and economic structure and its culture but also to the actual context of an economic crisis. That is the main reason for the low results of the first domain (employment). In relation to the scores in the fourth domain, favorable environment for healthy ageing, it should be noted that the Basque Country is emerging as one of the best territories for active and healthy ageing.

All the information provided for the calculation of the AAI, obtained in 2014, has proven to be of great value to develop the proposals included in the SAABC, launched in September 2015.

The proposed measures are aimed at further improving the present conditions in the Basque Country in the field of ageing, making the Basque society increasingly adapted to an ageing population. The resulting strategy has three main areas: (1) 'Adaptation to an ageing society', (2) 'Anticipation and prevention for ageing better', and (3) 'Friendly environments and participation in the construction of a welfare society'. Each one has been structured into 12 operative objectives and actions.

A total of 100 measures are included in the SAABC with the aim to strengthen active ageing. A total of 62% of those measures are directly related to the AAI and their different domains (Table 8.4). This figure is not close to 100% because the AAI was not the only source of information to decide on the design of the SAABC: other elements, such as interviews with experts in the ageing field, were taken into account. Analyzing the number of measures related to the AAI included in the SAABC by domain, we can see that 40.3% of those measures are referred to the third domain, 'independent, healthy, and secure living'; 37.1% to the second domain, 'participation in society'; and 11.3% to the first and the fourth domains, respectively, 'employment' and 'capacity to active ageing' (Fig. 8.13). It is noticeable that the domain where the Basque Country fares worst, 'employment', is one of the domains that gets fewer measures. This is due to the abovementioned utilization of different sources of information for the design of the SAABC and to the actual economic crisis that makes difficult to increase the elderly employment when there are high rates of youth unemployment.

In the following paragraphs, we provide an explanation for the findings in the different domains of the AAI and mention the activities included in the SAABC to address them.

1. Employment

   In recent years, the employment rate of people aged 55 and over has increased in the Basque Country, from 10.5% in 2002 to 17.1% in 2014 (fourth quarter). This increase is in accordance with the goal of Active Ageing and the tendency reviewed by the European

Table 8.4 Strategy of Active Ageing of the Basque Country (SAABC) measures related to the domains of the Active Ageing Index (AAI)

Main areas of the SAABC	SAABC measures	Measures related to the domains of the AAI	AAI domain number addressed
1. Adaptation to an ageing society	**33**	**17**	
1.1. Improve Governance	6	4	2
1.2. Ensuring the rights of and combat discrimination against the elderly	7	4	3
1.3. Ensuring security and social inclusion of aged people	12	9	1, 2, 3
1.4. Promote economic development on ageing	8	0	
2. Anticipation and prevention for successful ageing	**32**	**22**	
2.1. Promote responsible decision-making and empowerment of older people	2	1	3
2.2. Promote a positive transition from work to retirement	6	3	1
2.3. Develop housing and lifetime homes	7	3	3
2.4. Promote healthy ageing	11	9	3
2.5. Promote lifelong learning	6	6	3, 4
3. Promote age-friendly environments and participation in the construction of a welfare society	**35**	**23**	
3.1. Promote volunteerism and community involvement	15	13	2, 4
3.2. Promote transfers of care and support provided by older people in the family	3	3	2
3.3. Promote age-friendly programs	17	7	2, 3, 4
SAABC 2015–2020	**100**	**62**	

Commission in 2009 (The 2009 Aging Report). Even so, data from the Basque Country are still low compared with the rest of EU-28.

The legal retirement age in Spain is 65 years. The average retirement age of people over 55 years in the Basque Country is 61 years (Barrio

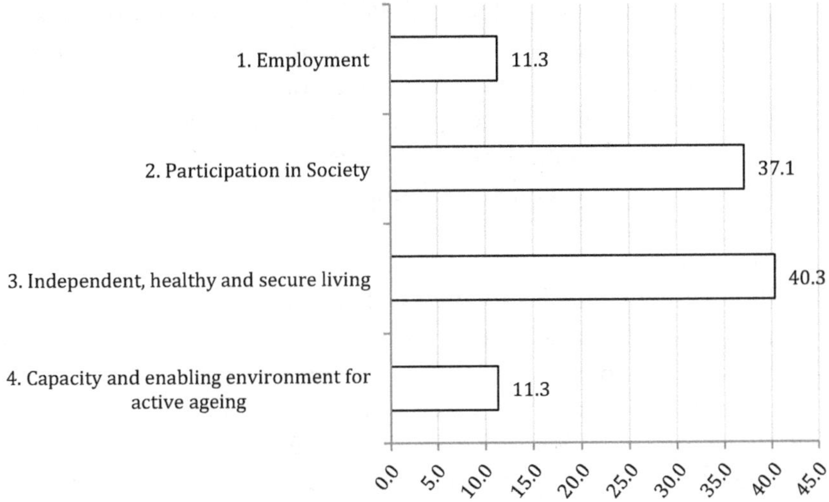

Fig. 8.13 % of Measures related to the AAI included in Strategy of Active Ageing of the Basque Country 2015–2020 by domain

et al. 2015). Retirement in Spain is regarded as a social achievement (Bazo 2001). Very few people agree with increasing the retirement age (8.0% of those aged 55+ in the Basque Country). A significant cultural change is required to accept employment as an achievement in older generations. This is going to prove difficult in the current times of crisis, when unemployment rates among the young are very high.

One of the possible leverages to be pulled is the adoption of flexible work times. In the report of the Economic and Social Center (CES 2011), it was noted that in the Basque Country, there were a few innovative practices to promote it, which would allow older people to remain in the labor market. Employment policies designed to increase the replacement of old workers by young individuals have promoted early retirement, however. Early retirement has taken over in the social and business culture, and expectations of early retirement have spread among people and companies.

The results in the employment domain of the AAI highlight the need for public policies that encourage the permanence in the labor market of older people. The SAABC includes some measures about employment mainly in its area 1 ('Adaptation to an ageing society'). These measures are related to encouraging the permanence in the labor market of older people through establishing flexible forms of retirement, adapting the jobs to the elderly, reducing the wage gap, ensuring continuous training, and committing companies to address this issue.

2. Participation in Society

   Providing informal support is characteristic of the Basque and Spanish society, as occurs in other Mediterranean countries. The family is an important complement of the Welfare State and, usually, it is the family who takes care of children, grandchildren, or older adults. A total of 32.9% of the population aged 55+ in Basque Country lives with their children at home. Children and parents live together for a long time. At the same time, 53.6% of the sons and daughters who do not share accommodation with their parents live within one kilometer and 67.4% have face-to-face contact with them every week. Family networking is very important to the daily life of the people in the Basque Country. Exchange of help inside the family is a consequence of that (Barrio et al. 2015).

   The most common voluntary activity in the Basque Country is participation in associations, for example, 6.9% of people 55+ participate in associations. Regarding political participation, we found that 8.8% of people have attended a protest or demonstration. This figure is probably related to the economic crisis, when the number of demonstrations and protests has increased.

   The scores in the participation in society domain ask to work harder to involve older people in society. Programs like Age-Friendly Cities and Communities of the WHO (OMS 2007) are working toward this aim. In the Basque Country, 16 cities, municipalities, and rural communities are involved in the WHO Global Network of Age-friendly Cities and Communities (Barrio et al. 2014).

   Measures of the SAABC about Participation in Society include improving the structures of participation of the elderly; promotion of voluntary and civil and social participation; transfers of care; and development of the age-friendly program in the Basque Country initiated in 2012.

3. Independent, Healthy, and Secure Living

   The percentage of Basque people over 55 years who practice physical exercise more than five times per week is much higher than the Spanish and European average. In the Basque Country, physical exercise is a cultural issue: taking walks or hiking are common activities in the population of all ages (Eustat 2015). Needs are not fully met in health and dental care in the Basque Country probably due to the main private provision of dental care. We do not have a satisfactory explanation for the finding of lower levels compared to Spain, where dental care is also mainly private. The Basque Country population is wealthier than the Spanish average, EUR 33,053 (household income/year) in the Basque Country and EUR 26,092 in Spain (INE 2015). Nevertheless, the question used to evaluate this item ('In the last 12 months, was there any occasion when you really needed to visit the dentist but you did not do it?') may measure other reasons for not going to the dentist, apart from economic circumstances. If, for example, the individual knows that he has to go to an annual review, but he does not go because of different non-economic reasons, such as lack of time, needs not satisfied in this item could be representing other reasons for lack of access to preventive care, apart from the financial barriers. The explanation of the differences with the EU-28 is even more difficult because, in addition, there are a variety of combinations of private and public provision of dental and health care in the different countries.

   In relation to the low prevalence of independent living, it has been mentioned above that there are a large number of children living with their parents at home (32.9%). Tradition, the economic crisis, and the unemployment rate as well as low salaries of the young could be contributing to this phenomenon. Almost 35% of people aged 55+ live in multigenerational households in the Basque Country (Barrio et al. 2015).

   On the other hand, the percentages of people aged 55 years and older who are not worried about becoming a victim of violent crime are higher in the Basque Country in comparison with Spain and EU-28. The Basque Country is one of the safest places in Europe, with an average crime rate of 42.23/1000 inhabitants in 2012 (Ertzaintza 2012), almost six points lower than the average for Spain (48 in 2012, Departamento de Seguridad Nacional 2017), and over 12 points lower than the EU-27 average, set at 54.72 (Gobierno Vasco 2017).

The measures referred to this domain in the SAABC include encouraging the autonomy and independency values, promoting accessible homes for the whole life, promoting making decisions on time, helping to manage loneliness, ensuring opportunities for personal development, developing prevention programs for making decisions on time, promoting physical exercise, and lifelong learning.

4. Capacity for Active and Healthy Ageing

The good results in Domain 4 are due mainly to Social Connectedness. As already mentioned, this area is very strong in the Basque Country: every week, 67.4% of the elderly see their children, 86.1% see their friends or neighbors, and 45.2% see other relatives. In the Mediterranean culture, it is very important to keep physical contact with others (Abellán et al. 2006).

Also noteworthy are the good results in life expectancy as a consequence of a good, free at the point of health system delivery, and the high levels in welfare of the region. The SAABC intervenes on this domain through the promotion of lifelong learning and social connectedness and the use of ICT, one of the items where the Basque Country fares worse than the EU-28. The measures about health are dealt with more extensively in the Plan for Health 2013–2020 (Departamento de Salud 2014).

8.7 Conclusions

The AAI proved to be a useful tool for policymaking. It helped to identify priority areas of policy development in the near future and will be used to evaluate the results of the implementation of the SAABC: the aim is to increase it by 5% (38.4 in 2020 compared to 36.6 in 2014). The AAI's analysis by sub-index shows both a challenging scenario for employment of older people and a very favorable environment for healthy ageing.

The results obtained for the Basque Country and its differences with the overall score for Spain show the need to use the AAI at a territorial level. Further research is needed.

Scores of the Active Ageing Index (AAI) and its domains in the Basque Country, Spain, and the 28 countries of the European Union (EU-28)

	Basque Country	Spain	EU-28
Overall index			
Employment	25.38	23.33	27.9
Participation in society	21.03	18.26	18.1
Independent living	72.3	69.8	70.6
Capacity for active ageing	65.7	56.28	54.5
AAI total score	36.3	32.8	34.1
Rank	7	18	
Domain 1: Employment			
Employment rate 55–59	62.1	54.8	62.2
Employment rate 60–64	35.1	31.8	31.6
Employment rate 65–69	3	5.2	11.6
Employment rate 70–74	1.3	1.5	6.1
Value	25.4	23.3	27.9
Rank	15	19	
Domain 2: Participation in society			
Voluntary activities	19	9.6	14.7
Care to children, grandchildren	44	36.1	32.5
Care to older adults	9.4	15.7	12.9
Political participation	12.3	10.6	12
Value	21	18.3	18.1
Rank	9	15	
Domain 3: Independent, healthy, and secure living			
Physical exercise	60.9	15.8	15.6
Access to health and dental care	81.5	89.6	88.2
Independent living	61	73.4	84.2
Relative median income	91.3	93.1	86.3
No poverty risk	94.8	93.8	93
No material deprivation	97	97.1	90

(*continued*)

(continued)

	Basque Country	Spain	EU-28
Physical safety	89.1	67.4	69.3
Lifelong Learning	5.3	4.8	4.5
Value	72.3	69.8	70.6
Rank	13	16	
Domain 4: Capacity and enabling environment for active ageing			
Remaining life expectancy 50–55	60.5	59.1	53.8
Share of healthy life expectancy at 55	62.5	52.9	53.2
Mental well-being	76.6	68.2	64.6
Use of ICT	38.8	29	40.8
Social connectedness	95.8	67.5	49.0
Educational attainment	42.5	28.9	59.7
Value	65.7	56.3	54.4
Rank	2	13	

References

Abellán, G. A., Puga, G. M., & Sancho, C. M. (2006). Mayores y familia en la sociedad actual. In Fundación Encuentro (Ed.), *Informe España 2006: Una interpretación de su realidad* (pp. 262–333). Madrid: Centro de Estudios del Cambio Social.

Barrio, E., Mayoral, O., & Sancho, M. (2015). *Estudio sobre las condiciones de vida de las personas de 55 y más años en Euskadi, 2014*. Vitoria: Gobierno Vasco.

Barrio, E., Sancho, M., & Yanguas, J. (2011). *La realidad sobre las personas de 60 y más años en Euskadi*. Gobierno Vasco: Departamento de Empleo y Asuntos Sociales.

Barrio, E., Tomasena, A., Indart, A., Elortza, G., & Sancho, M. (2014). *Euskadi amigable: Guía práctica para la implantación y uso en municipios*. Vitoria: Departamento de Empleo y Políticas Sociales, Gobierno Vasco.

Bazo, M. T. (2001). *La Institución social de la jubilación: de la sociedad industrial a la postmodernidad*. Valencia: Nau Libres.

CAPV. (1993). Encuesta Tercera Edad en el País Vasco, 1993.

CES. (2011). *El impacto económico y social en la CAPV de la evolución demográfica prevista* (Colección Estudios e Informes 8). Bilbao: Consejo Económico y Social Vasco.

Departamento de Empleo y Políticas Sociales. (2017). Informes de resultados de la Encuesta de Pobreza y Desigualdades Sociales. Retrieved September 6, 2017, from http://www.euskadi.eus/gobierno-vasco/-/informe-encuesta-pobreza-desigualdades-sociales/

Durán Heras, M. A. (2014). *La contribución de los mayores a la economía del País Vasco*. Vitoria: Gobierno Vasco.

Echeverría, J. (2014). Envejecimiento activo e innovación social. In M. Moteagudo, J. Cuenca, & R. San Salvador (Coords.), *Aportaciones al Envejecimiento satisfactorio* (Documentos de estudios de Ocio, 50). Bilbao: Universidad de Deusto.

Ertzaintza. (2012). *Estadísticas Delictivas*. Vitoria: Gobierno Vasco. http://www.ertzaintza.net/wps/portal/ertzaintza/!ut/p/c5/04_SB8K8xLLM9MSSzPy8xBz9CP0os3gXDydTo2AzN0tLE9dAdxNzDxdzAwjQDwfpMIv39_EMMw00dnR3NAl19vIJDHQyhKkAyxvgAI4G-n4e-bmp-gXZ2WmOjoqKAATK2bg!/dl3/d3/L2dJQSEvUUt3QS9ZQnZ3LzZfSUYzSFVWRjAwR1VVMDBJN0ozUjdQQU9NNjI!/.

European Commission. (2009). *The 2009 Aging Report. Economic and budgetary projections for the EU-27 Member States (2008-2060)*. Luxembourg.

European Commission. (2013). Active Aging Index. *Policy Brief*.

Eustat. (2014a). *Demographic indicators*. Vitoria: Instituto Vasco de Estadística.

Eustat. (2014b). *Estadística Municipal de Habitantes, 2014*. Vitoria: Instituto Vasco de Estadística.

Eustat. (2015). *Encuesta condiciones de vida 2014*. Vitoria: Análisis de resultados. Instituto Vasco de Estadística.

Eustat. (2017a). Encuesta sobre la sociedad de la información. Familias. Retrieved September 6, 2017, from http://www.eustat.eus/estadisticas/tema_15/opt_1/ti_Encuesta_sobre_la_sociedad_de_la_informacion_Familias/temas.html

Eustat. (2017b). Población en relación con la actividad (PRA). Retrieved September 6, 2017, from http://www.eustat.eus/banku/id_2307/indexLista.html

Gobierno Vasco. (2017). Tasa de Infracciones penales en la UE. Año 2012. Retrieved September 11, 2017, from http://opendata.euskadi.eus/w79-contgen/es/contenidos/estadistica/infrac_penales_ue_2012/es_def/adjuntos/doc_comparados.shtml

INE. (2014a). *Encuesta de Población Activa* (4° Trimestre). Madrid: INE.

INE. (2014b). *Padrón municipal de Habitantes*. Madrid: INE.

INE. (2014c). *Proyecciones demográficas en base 2014 a 1 de enero*. Madrid: INE.

INE. (2015). *Encuestas de Condiciones de vida. Renta por hogar por comunidades autónomas*. Madrid: INE.

OMS. (2007). *Ciudades amigables con las personas mayores*. Ginebra: OMS.

Osakidetza. (2017). Encuesta de salud 2013. Retrieved September 6, 2017, from https://www.osakidetza.euskadi.eus/informacion/encuesta-de-salud-2013/r85-ckosag01/es/

Sancho, M. (2010). *100 Propuestas para avanzar en el buen trato en las personas que envejecen. Bases para un plan de Acción*. Gobierno Vasco: Departamento de Empleo y Políticas Sociales.

Stiglitz, J., Sen, A., & Fitoussi, J.-P. (2009). Report by the Commission on the Measurement of Economic Perfomance and Social Progress. Paris: CMEPSP.

Zaidi, A., Gasior, K., Hofmarcher, M. M., Lelkes, O., Marin, B., Rodrigues, R., Schmidt, A., Vanhuysse, P. & Zolyomi, E. (2013). *Active Aging Index 2012: Concept, methodology and final results. EC/UNECE, Active Aging Index Project*(UNECE Grant ECE/GC/2012/003). Vienna: European Centre for Social Welfare Policy and Research. Retrieved from http://www.euro.centre.org/data/aai/1253897823_70974.pdf

Zaidi, A., & Stanton, D. (2015). *Active aging index 2014: Analytical report.* Report produced at the Centre for Research on aging, University of Southampton, under contract with UNECE (Geneva), co-funded by European Commission, Brussels. Retrieved from http://www.southampton.ac.uk/assets/sharepoint/groupsite/Administration/SitePublisher-document-store/Documents/aai_report.pdf

Elena del Barrio is Research Project Manager from Matia Institute of Gerontology. She has a BA in Sociology and PhD Candidate. She works in ageing projects relating to living conditions, time use research, active ageing, social policies, and Age-Friendly Cities and Communities. She is the coordinator of the Age-Friendly Basque Country Project from 2012. Previously, she was a member of the Social Area of the Spanish Society of Geriatrics and Gerontology.

Angel Rodriguez-Laso of Matia Institute of Gerontology is a Family Physician, Public Health Officer, and Researcher at the Biomedical Research Foundation of the Getafe University Hospital. At present, he works as technical advisor in the coordination team of the ADVANTAGE European Joint Action on frailty. Previously, he worked for Matia Institute of Gerontology, where he researched on inequalities in health and the operationalization of healthy ageing, and collaborated in the development of the Spanish Longitudinal Study on Aging (ELES).

Olga Mayoral of Matia Institute of Gerontology is a Sanitary Psychologist. She has a Master's in General Health Psychology and BA in Psychology. She was working in Matia Institute of Gerontology during 2013–2015 as researcher. She is a Social Entrepreneur—Co-Founder of Being Inclusive. Her work focuses on consultation with applied research.

Mayte Sancho is Scientific Director at Matia Institute of Gerontology. She has a long career including working at the IMSERSO Body of Social Security Management, and being the OECD Expert on Social Protection and Aging, member of the Technical Committee of the Spanish Society of Geriatrics and Gerontology, Director of the IMSERSO Observatory on Aging and Dependence, Co-Director of the Senior Portal, and Coordinator of the 'Mistreatment of Elderly' project at the World Health Organization.

Lide Amilibia is Deputy Minister of Social Policies at the Department of Employment and Social Policies of the Basque Government. Her priority objectives are to create and maintain a normative framework that guarantees the right of citizens to access basic and quality social services and the maintenance of a policy of minimum income and aid to families.

9

Greying Italy Across Time, Space, and Gender

Luciana Quattrociocchi, Daria Squillante, and Mauro Tibaldi

9.1 Introduction

Italy is one of the oldest countries in Europe, with a population of about 59 million in 2012, out of which 20.8% are aged 65 years and over (23.2% of women and 18.2% of men). Many demographic indicators confirm that the ageing process is taking place: during the years 2007 and 2012, the ageing index[1] passed from 142.3 to 148.6, and the demographic old age dependency ratio[2] from 30.5 to 32.0, healthy life expectancy at birth increased from 73 to 74.[3] Overall GDP per capita reached €26,760 in 2012, while the richest region, Bolzano, reached €40,043, and the poorest, Calabria, reached €16,628. In 2012, the overall employment rate of the population aged 15–64 was 56.6%, while the unemployment rate was 9.8% for men and 11.8% for women. In 2012, important differences existed in the employment rates between regions (71.8% in Bolzano and 41.3% in Calabria; 66.3% for men and 47.1% for women).

L. Quattrociocchi (✉) • D. Squillante • M. Tibaldi
Italian National Institute of Statistics, Rome, Italy

A. Zaidi (eds.), *Building Evidence for Active Ageing Policies*,
https://doi.org/10.1007/978-981-10-6017-5_9

In Italy, there is a strong recommendation for a shared commitment among different administrative levels in defining policies for all active ageing domains. In fact, the three policy levels (national, regional, local) have intertwined responsibilities, tasks, and activities to achieve the common goal of social well-being—for the elderly as well as for other subgroups—though each level has its particular roles and powers. In fact, the 2001 reform of the fifth title of the Italian Constitution (Parlamento Italiano, 2001) clearly bolstered the autonomy of policymaking at the regional level. The political power of these three levels differs depending on the specific domain: for example, the *employment* domain falls rather under the national level, whereas the *health* domain is governed more by local authorities.

The Active Ageing Index (AAI) in Italy is not yet used as a common and standardized indicator to evaluate the condition of the ageing population, neither at the national nor at the local level. In fact, Italy has not yet passed a national law dedicated categorically to active ageing, though there are policies promoting an ageing strategy, which focus on employment and care.

This chapter shows that the way of "becoming" and of "being" "older people" is strictly tied to temporal, territorial, and gender specificities of the country. Therefore, this chapter is aimed at calculating the AAI in 2007 and 2012, focusing on the following specific aspects:

1. Analysis of national results, to detect the impact of the crisis on the AAI over time
2. Analysis of regional differences, since Italy is still affected by pronounced social, economic, and cultural disparities by regions
3. Analysis of gender differences in regions and overall for the country to measure how females fall short in comparison to men in the active ageing experiences.

In order to calculate the Active Ageing Index for all regions of Italy, we followed the AAI methodology regarding indicators and weighted values for the calculation of the overall value and for each of its domains (Zaidi et al. 2013: 9–20). We used data from secondary sources, both surveys harmonized at European level (LFS and European Union Statistics on

Income and Living Conditions—EU-SILC) and surveys delivered by the Italian National Institute of Statistics. To ensure a higher reliability, we used national surveys with a sample size which allows to calculate indicators at a subnational level also by gender.

We have used data from the following secondary sources in the calculation of the AAI at subnational level in Italy:

1. *Labour Force Survey in 2007 and in 2012*
2. *Aspects of Daily Life in 2007, 2008, 2009 and in 2012*
3. *Family and Social Subjects in 2003 and in 2009*
4. *EU-SILC in 2007 and in 2012*

9.2 Results

9.2.1 Domain 1: Employment

In 2007, the index ranges from a minimum of 16.5 points in Puglia to a maximum of 24.9 points in Bolzano (Table 9.1). Although we are in the period of maximum expansion of the labor market in Italy (the total unemployment rate stands at 6.1%, the lowest figure since 2004, and the employ-

Table 9.1 Employment rates 55–74 years old in 2007, reporting top and bottom three ranked regions only

Employment (Total)	1.1 Employment rate 55–59	1.2 Employment rate 60–64	1.3 Employment rate 65–69	1.4 Employment rate 70–74	Index	
Region	LFS 2007	LFS 2007	LFS 2007	LFS 2007	Value	Rank
Bolzano	54.6	24.1	13.7	7.0	24.9	1
Emilia-Romagna	53.1	21.2	10.1	4.7	22.3	2
Lazio	50.6	23.0	8.1	3.1	21.2	3
Italy	**46.0**	**19.4**	**7.3**	**3.1**	**19.0**	**12**
Friuli-Venezia Giulia	42.3	16.5	6.8	3.5	17.3	20
Piemonte	43.1	13.6	7.0	3.5	16.8	21
Puglia	41.8	18.2	4.2	1.8	16.5	22

Source: Labour Force Survey

ment rate at the highest level with 58.6%), by scrolling the ranking we can see that some regions are still far away from achieving the policymaking objectives set at EU level (Lisbon 2000) to improve the employability of people and the inclusion in the labor market as major social inclusion elements.

At the bottom of the ranking are not only the regions of southern Italy, which are traditionally disadvantaged, but also some northern regions like Piemonte and Friuli-Venezia Giulia with an economy of a developed industrial structure. Indeed, in these northern regions, the low employment rate of older workers is due to a plurality of reasons:

- A developed economy that has allowed working life to start as early as between 15 and 18 years
- A smooth/seamless working career
- Social regulations which permitted leaving work before the age of 60 years, without any penalties, after having contributed for 35 years
- Incentives from employers who, in case of excess staff, have encouraged ageing workers to an early withdrawal rather than laying off younger ones
- Low effectiveness of the pension system's first measures to limit early retirement

In 2012, all employment indicators show significant improvements, leading to an increase in the value of the domain from 19.0 points to 23.0 points at Italy level (Table 9.2). Between 2007 and 2012, it changes

Table 9.2 Employment rates 55–74 years old in 2012, reporting top and bottom three ranked regions only

Employment (Total)	1.1 Employment rate 55–59	1.2 Employment rate 60–64	1.3 Employment rate 65–69	1.4 Employment rate 70–74	Index	
Region	LFS 2012	LFS 2012	LFS 2012	LFS 2012	Value	Rank
Bolzano	68.3	29.8	15.0	6.9	30.0	1
Marche	64.4	23.6	14.6	4.7	26.8	2
Liguria	64.5	24.2	11.4	5.8	26.5	3
Italy	**57.7**	**22.7**	**7.9**	**3.5**	**23.0**	**13**
Campania	48.0	24.1	6.0	1.4	19.9	20
Sicilia	47.6	22.4	5.9	2.5	19.6	21
Puglia	45.7	19.7	6.6	2.4	18.6	22

Source: Labour Force Survey

the weight that covers the diverse components (young, adult, mature) of the total workforce. While among young people (15–34) and adults (35–54) the employment rate decreased, respectively, −7.7 p.p.[4] and −2.0 p.p., on the contrary, it increased by 6.6 p.p. for people aged 55 and older (Table 9.3).

This increase can be explained by the entrance of the "baby boomers" in mature age (about 4,214,000 persons aged 50–54 years in 2007) which replace a lower number of labor force cohorts leaving the labor market force 3,284,000 persons aged 60–64 years in 2007).

In addition, in 2012, although modest, there are the first signs of the impact of Minister Fornero's pension reform. The reform led to higher retirement ages, the end of early retirement, and the immediate and widespread implementation of a contributory system of calculating the pension, the "indexed" retirement age to life expectancy.

Regarding the subnational level, Trento and Veneto have retrieved many positions.

Over the years, the gender gap in employment activity reflects the women's historically scant participation in the labor market, particularly pronounced in the southern regions of the country. From 2007 to 2012, the gender gap decreased from 14.2 p.p. to 13.1 p.p. (Table 9.4).

Table 9.3 Overall employment in thousands and in percentage—2007 and 2012 for Italy

	Employers (15 years old and over)		Variation 2012–2007		Employment rate (15–64 years)		Absolute variation employment rate (%)
	2007	2012	Absolute	%	2007	2012	2012/2007
Gender							
Men	13,812	13,194	−618	−4.5	70.6	66.3	−4.3
Women	9,083	9,372	289	3.2	46.6	47.1	0.5
Age							
15–34	7,082	5,638	−1,444	−20.4	50.8	43.1	−7.7
35–54	13,050	13,483	433	3.3	74.9	72.9	−2.0
55 and over	2,762	3,445	683	24.7	33.7	40.3	6.6
Italy	**22,894**	**22,566**	**−328**	**−1.4**	**58.6**	**56.6**	**−2.0**

Source: Labour Force Survey

Table 9.4 Employment rates 55–74 years old, ranking by gender in 2007 and in 2012, reporting top and bottom three ranked regions only

2007						2012					
Men			Women			Men			Women		
Region	Value	Rank	Region	Value	Rank	Region	Value	Rank	Region	Value	Rank
Bolzano	32.5	1	Bolzano	17.7	1	Bolzano	36.3	1	Bolzano	24.2	1
Emilia-Romagna	29.8	2	Umbria	16.3	2	Marche	32.9	2	Liguria	22.7	2
Lazio	29.2	3	Valle D'Aosta	15.6	3	Toscana	32.7	3	Marche	21.2	3
Italy	**26.3**	**9**	**Italy**	**12.1**	**13**	**Italy**	**29.8**	**12**	**Italy**	**16.7**	**13**
Valle D'Aosta	23.4	20	Sicilia	9.6	20	Sicilia	27.4	20	Sicilia	12.6	20
Friuli-Venezia Giulia	23.0	21	Sardegna	9.6	21	Piemonte	27.2	21	Campania	12.4	21
Piemonte	22.3	22	Puglia	8.7	22	Puglia	27.2	22	Puglia	10.7	22

Source: Labour Force Survey

Bolzano was at the first place in 2007 as in 2012, but here, for women, there is the highest increase in employment rates compared to men (+6.5 p.p.). In 2012, all the southern regions, except for Sardinia, have values higher than the national average of this indicator. One crucial explanation of the decrease in the gender gap consists of the differences in the impact of the crisis on the economic sectors: the economic crisis has hit more traditionally "male" economic sectors (construction/manufacturing) than "female" ones (health, social activities, care).

9.2.2 Domain 2: Participation in Society

In 2007, the active ageing indicators of political participation (overall value for Italy: 25.5%), and voluntary activities (overall value for Italy: 10.2%) are widespread activities among the older Italian population. Older people in Italy also play an important part in caring for children (9.4%) and adults (4.4%) although possibly not as high as could be expected (Table 9.5).

Also in this domain, the country is sharply divided into three areas: a northern area with high levels of participation, a central area with middle values, and a southern area which is bringing up the rear. The strength of the "informal care network" in different geographical areas is strictly tied

Table 9.5 Social participation indicators in 2007, reporting top and bottom three ranked regions only

Participation in society (Total)	2.1 Voluntary activities	2.2 Care to children, grandchildren	2.3 Care to older adults	2.4 Political participation	Index	
Region	AVQ2007	FSS2003	FSS2003	AVQ2008	Value	Rank
Trento	20.5	14.0	5.8	33.6	17.1	1
Bolzano	23.7	13.0	2.0	28.3	15.4	2
Emilia-Romagna	13.3	10.6	5.3	33.3	14.2	3
Italy	**10.2**	**9.4**	**4.4**	**25.5**	**11.3**	**11**
Puglia	5.5	6.1	2.6	23.2	8.3	20
Sicilia	5.3	6.6	3.6	19.2	7.9	21
Calabria	4.1	4.2	2.5	25.1	7.8	22

Source: Aspects of Daily Life 2007, 2008; Family and Social Subjects 2003

both to different labor market involvements and to the efficiency of local welfare systems.

Since 2011, government grants to families with young children begin to decline, and the number of children attending public kindergartens reduced. Informal support networks guaranteed by grandfathers/grandmothers become decidedly in central/northern regions, where working parents cannot rely on public childcare services. In the southern area, on the contrary, where female unemployment/inactivity is still high, care for children/adults can often be guaranteed by the mothers.

From 2007 to 2012, the overall domain's indicator rises from 11.3 points to 12.2 points. The improvement may be ascribed specially to "care for children" (+2.9 p.p.). The regions showing the most marked improvement are Lazio (+2.8 p.p.) and Trento (+2.4 p.p.). The first two places in the rankings are held by Trento (19.5 points) and Bolzano (16.9 points), Molise (7.8 points), and Sicily (8.2 points) hold the last two positions (Table 9.6).

Central and northern regions show greater increases than the average as far as "care activities" are concerned, while values above average are registered only in a scant number of southern regions. In 2012, one can notice a greater participation of older men in care activities, and that's why the gender gap for these indicators has become smaller (Table 9.7). Nevertheless, informal care still remains the only sphere women prevail in.

Table 9.6 Social participation indicators in 2012, reporting top and bottom three ranked regions only

Participation in society (Total)	2.1 Voluntary activities	2.2 Care to children, grandchildren	2.3 Care to older adults	2.4 Political participation	Index	
Region	AVQ 2012	FSS 2009	FSS 2009	AVQ 2012	Value	Rank
Trento	21.0	22.1	9.0	30.0	19.5	1
Bolzano	25.3	13.8	2.9	31.5	16.9	2
Veneto	14.4	15.5	8.2	26.8	15.3	3
Italy	**10.8**	**12.3**	**5.2**	**24.1**	**12.2**	**11**
Calabria	5.9	5.5	4.3	21.7	8.5	20
Sicilia	5.9	6.9	3.3	20.1	8.2	21
Molise	5.6	5.6	3.8	19.1	7.8	22

Source: Aspects of Daily Life 2012; Family and Social Subjects 2009

Table 9.7 Social participation overall value in top three and bottom three ranked regions by gender in 2007 and in 2012

2007						2012					
Men			Women			Men			Women		
Region	Value	Rank	Region	Value	Rank	Region	Value	Rank	Region	Value	Rank
Trento	18.6	1	Trento	15.6	1	Trento	20.0	1	Trento	19.1	1
Bolzano	17.0	2	Bolzano	14.3	2	Bolzano	17.9	2	Bolzano	16.0	2
Veneto	15.7	3	Emilia-Romagna	13.3	3	Veneto	16.9	3	Veneto	14.0	3
Italy	**13.2**	**10**	**Italy**	**9.8**	**10**	**Italy**	**14.0**	**9**	**Italy**	**9.4**	**14**
Basilicata	10.7	20	Sicilia	6.3	20	Basilicata	10.1	20	Basilicata	6.7	20
Calabria	10.0	21	Calabria	6.1	21	Calabria	10.1	21	Abruzzo	6.4	21
Sicilia	9.8	22	Puglia	6.0	22	Sicilia	9.6	22	Molise	6.2	22

Source: Aspects of Daily Life 2007, 2008; 2012 Family and Social Subjects 2003, 2009

9.2.3 Domain 3: Independent, Healthy, and Secure Living

In 2007, for the overall value of this domain, the central and northern regions are ranked higher than the southern regions, with Valle D'Aosta in the first place (82.2 points) and Sicily in the last (62.7 points). Sardinia is an exception with a value of 73.7 points (the third place in the ranking) (Table 9.8).

The indicators related to the economic conditions (like material deprivation) confirm the greater disadvantage of the southern regions in 2007.

After five years (2012/2007), the overall value of the sub-indicator of this domain does not show a meaningful progress (Table 9.9). Nevertheless, in spite of the economic crisis, the two indicators referring to economic conditions and poverty risk improved considerably, and the share of people without access problems to meet the needs of health and dental care has fallen. However, at regional level, the material deprivation indicator has lower values in 2012, which means that the situation between both periods became worse. Valle D'Aosta confirms its leading position followed by Bolzano, whereas the worsened situation of Friuli-Venezia Giulia is important.

The nationwide gender gap shows women lagging by 5.6 p.p., ranging from −7.9 p.p. in Marche to −0.5 p.p. in Valle D'Aosta, with a slight decline in 2012 compared to 2007 (−0.3 p.p.) (Table 9.10).

9.2.4 Domain 4: Capacity for Active Ageing

In 2007, the fourth domain's indicators trace a ranking on the map that starts from the north, passes through central Italy, and heads southward. The leading positions are all concentrated in the northern regions, and the last ones are all crowded in the southern regions (Table 9.11).

"The attainment of 105 years" (56.5%) combined with the "number of healthy life years we may expect to live after 55 years of age" (31.9 years) describes a positive landscape of quantity and quality of "elderly life". However, the depth of regional gaps persists: Marche promises the most

Table 9.8 Independent, healthy, and secure living in 2007, reporting top and bottom three ranked regions only

Independent, healthy, and secure living (Total)	3.1 Physical exercise	3.2 No unmet needs of health and dental care	3.3 Independent living arrangements	3.4 Relative median income	3.5 No poverty risk	3.6 No material deprivation	3.7 Physical safety	3.8 Lifelong learning	Index	
Region	AVQ 2007	SILC 2007	SILC 2007	SILC 2007	SILC 2007	SILC 2007	AVQ 2009	LFS 2007	Value	Rank
Valle D'Aosta	46.0	95.5	82.8	77.0	91.8	99.2	69.3	0.5	82.2	1
Trento	63.1	92.8	81.8	81.3	89.5	100.0	68.1	3.2	75.4	2
Sardegna	34.7	81.6	56.8	109.5	91.3	91.6	59.2	1.8	73.7	3
Italy	**35.0**	**86.2**	**73.5**	**85.8**	**87.9**	**93.7**	**46.0**	**1.5**	**66.9**	**17**
Basilicata	22.9	74.7	76.0	81.5	81.8	93.3	66.7	1.1	64.9	20
Puglia	18.4	77.0	75.3	100.1	85.4	86.9	46.2	0.9	64.3	21
Sicilia	17.0	74.2	78.5	100.9	75.7	84.3	43.2	0.5	62.7	22

Source: Aspects of Daily Life 2007, 2009; EU-SILC 2007; Labour Force Survey 2007

Table 9.9 Independent, healthy, and secure living in 2012, reporting top and bottom three ranked regions only

Independent, healthy, and secure living (Total)	3.1 Physical exercise	3.2 No unmet needs of health and dental care	3.3 Independent living arrangements	3.4 Relative median income	3.5 No poverty risk	3.6 No material deprivation	3.7 Physical safety	3.8 Lifelong learning	Index	
Region	AVQ 2012	SILC 2012	SILC 2012	SILC 2012	SILC 2012	SILC 2012	AVQ 2012	LFS 2012	Value	Rank
Valle D'Aosta	41.1	96.6	88.1	84.2	97.8	96.9	73.4	2.6	84.8	1
Bolzano	65.8	91.1	74.4	91.8	92.2	95.4	69.3	3.0	74.8	2
Sardegna	32.6	75.6	72.2	99.3	94.2	94.1	55.7	2.5	74.6	3
Italy	**36.1**	**82.6**	**74.5**	**94.8**	**93.0**	**87.0**	**47.7**	**2.3**	**67.5**	**16**
Campania	16.7	80.3	67.6	111.5	88.0	76.9	37.7	1.1	65.6	20
Sicilia	16.3	81.4	75.8	108.9	79.5	64.1	43.7	1.0	62.8	21
Puglia	20.2	63.5	71.7	107.5	91.9	69.8	44.9	1.2	60.6	22

Source: Aspects of Daily Life 2012; EU-SILC 2012; Labour Force Survey 2012

Table 9.10 Independent, healthy, and secure living overall value by gender in 2007 and in 2012, reporting top and bottom three ranked regions only

2007						2012					
Men			Women			Men			Women		
Region	Value	Rank	Region	Value	Rank	Region	Value	Rank	Region	Value	Rank
Valle D'Aosta	82.7	1	Valle D'Aosta	82.2	1	Valle D'Aosta	86.1	1	Valle D'Aosta	84.1	1
Trento	76.6	2	Trento	74.8	2	Sardegna	78.2	2	Trento	72.7	2
Bolzano	76.4	3	Sardegna	72.2	3	Bolzano	77.0	3	Bolzano	71.9	3
Italy	**70.1**	**15**	**Italy**	**64.5**	**17**	**Italy**	**70.5**	**16**	**Italy**	**65.2**	**16**
Calabria	66.6	20	Campania	63.8	20	Abruzzo	66.8	20	Campania	63.1	20
Sicilia	66.3	21	Puglia	62.1	21	Sicilia	64.2	21	Sicilia	61.9	21
Basilicata	65.7	22	Sicilia	60.1	22	Puglia	64.0	22	Puglia	57.9	22

Source: Aspects of Daily Life 2007, 2009, 2012; EU-SILC 2007, 2012; Labour Force Survey 2007, 2012

Table 9.11 Capacity for active ageing in 2007, reporting top and bottom three ranked regions only

Capacity and enabling environment for active ageing (Total)	4.1 RLE achievement of 50 years at age 55	4.2 Share of healthy life years in the RLE at age 55	4.3 Mental well-being	4.4 Use of ICT	4.5 Social connectedness	4.6 Educational attainment	Index	
Region	AVQ 2007	AVQ 2007	AVQ 2010	AVQ 2008	AVQ 2007	LFS 2007	Value	Rank
Trento	57.8	54.2	62.2	8.4	65.6	29.1	53.5	1
Bolzano	58.1	49.9	62.4	12.9	62.4	21.6	52.0	2
Veneto	57.3	34.7	46.8	9.8	61.8	22.7	47.0	3
Italy	**56.5**	**31.9**	**39.9**	**9.9**	**59.5**	**26.3**	**43.3**	**13**
Calabria	56.3	13.9	34.6	6.2	61.4	24.7	38.0	20
Puglia	56.4	19.3	30.1	6.3	57.2	21.2	37.8	21
Campania	53.6	26.1	26.3	6.3	58.7	23.6	35.0	22

Source: Aspects of Daily Life 2007, 2010; Labour Force Survey 2007

likely attainment of 105 years (58.5%), while Campania (53.6%) holds the last position. For the second indicator (4.2), 40.3 p.p. separate Trento (54.2%) and Calabria (13.9%).

The possibility to access a "digital citizenship" is now increasingly indispensable for social inclusion. Nevertheless, in Italy, digital citizenship is still not common for many people between 55 and 74 years of age: only 9.9% use the Internet at least once a week; Lazio leads the ranking (14.4%), whereas Molise covers the last place (5.5%).

In 2012, the fourth domain (44.6 points) shows an increase by 1.3 p.p. from 2007, thanks to the rise of three indicators: "share of healthy life years in the remaining life expectancy at age 55", "use of ICT", and "educational attainment" (Table 9.12). An increased percentage of people who expect a healthy old age is also the result of a greater awareness/prevention of healthy attitudes. In 2012, Trento and Bolzano lead the ranking again, whereas in the last positions, once again, we find Campania and Calabria.

This domain ranking of regions by gender shows the split between north and south (Table 9.13). A nationwide gap of 2.7 p.p. lies between the positive value difference of 2.8 p.p. in Bolzano and the negative gap (−4.4 p.p.) in Sicily. Overall, in 2012, for Italian women, this domain indicator equals to 43.3 points, while for men it is 46.3 points. In 13 regions out of the 22, the gender gap has increased in this domain. In 2012, the two regions with the worst gender imbalance are Abruzzo (−5.6 p.p.) and Puglia (−4.8 p.p.).

9.3 Overall Value of AAI

In five years, the overall index has grown from 25.9 points to 28.0 points (Table 9.14), thanks to the increased values of indicators of all domains, especially the domain of "employment". Generally, the northern regions, the most developed area of the country, have achieved the best records, while the lowest values are logged in the southern area. The highest value is observed for Bolzano (34.5 points), while the worse in Puglia (with the index value of 23.6 points).

Table 9.12 Indicators of the fourth domain of capacity for active ageing in 2012, reporting top and bottom three ranked regions only

Capacity and enabling environment for active ageing (TOTAL)	4.1 RLE achievement of 50 years at age 55	4.2 Share of healthy life years in the RLE at age 55	4.3 Mental well-being	4.4 Use of ICT	4.5 Social connectedness	4.6 Educational attainment	Index	
Region	AVQ 2012	AVQ 2012	AVQ 2012	AVQ 2012	AVQ 2012	LFS 2012	Value	Rank
Bolzano	58.9	47.8	62.0	22.0	62.3	32.1	53.0	1
Trento	60.0	52.9	43.5	22.0	59.2	37.3	51.5	2
Veneto	58.4	45.2	39.9	17.8	61.0	31.9	49.3	3
Italy	**57.5**	**38.7**	**32.5**	**17.8**	**56.4**	**34.2**	**44.6**	**15**
Puglia	58.0	25.0	26.1	11.1	54.9	26.6	39.4	20
Calabria	57.2	15.6	28.3	11.8	61.8	32.3	38.6	21
Campania	54.9	35.5	18.9	12.4	55.8	31.3	37.7	22

Source: Aspects of Daily Life 2012; Labour Force Survey 2012

Table 9.13 Capacity for active ageing, regions' ranking by gender in 2007 and in 2012, reporting top and bottom three ranked regions only

2007						2012					
Men			Women			Men			Women		
Region	Value	Rank	Region	Value	Rank	Region	Value	Rank	Region	Value	Rank
Trento	53.5	1	Bolzano	53.5	1	Bolzano	55.1	1	Bolzano	51.1	1
Bolzano	52.0	2	Trento	53.4	2	Trento	52.7	2	Trento	50.5	2
Friuli-Venezia Giulia	47.0	3	Veneto	46.8	3	Veneto	50.3	3	Veneto	48.5	3
Italy	**44.8**	**14**	**Italy**	**42.1**	**13**	**Italy**	**46.3**	**15**	**Italy**	**43.3**	**14**
Calabria	40.3	20	Puglia	36.5	20	Puglia	42.0	20	Calabria	37.3	20
Puglia	39.5	21	Calabria	36.2	21	Calabria	40.2	21	Campania	37.3	21
Campania	35.5	22	Campania	34.6	22	Campania	38.3	22	Puglia	37.2	22

Source: Aspects of Daily Life 2007, 2010, 2012; Labour Force Survey 2007, 2012

Table 9.14 Overall values of AAI in 2007 and in 2012 at subnational level of Italy

2007							2012						
Total	Indices						Total	Indices					
Region	Emp	Soc	Liv	Cap	Value	Rank	Region	Emp	Soc	Liv	Cap	Value	Rank
Piemonte	16.8	11.5	69.2	44.6	25.7	14	Piemonte	22.3	13.8	69.7	44.9	28.6	12
Valle D'Aosta	19.5	11.0	82.2	45.4	28.0	4	Valle D'Aosta	24.1	11.8	84.8	47.6	30.5	5
Lombardia	18.3	13.3	68.8	46.4	27.2	7	Lombardia	22.9	14.8	69.4	46.8	29.5	7
Bolzano	24.9	15.4	72.6	52.0	31.8	1	Bolzano	30.0	16.9	74.8	53.0	34.5	1
Trento	18.2	17.1	75.4	53.5	30.6	2	Trento	24.8	19.5	73.3	51.5	33.1	2
Veneto	18.1	14.1	67.9	47.0	27.5	6	Veneto	24.5	15.3	70.6	49.3	30.8	3
Friuli-Venezia Giulia	17.3	13.0	71.3	46.2	27.0	9	Friuli-Venezia Giulia	23.2	13.2	67.3	45.9	28.6	11
Liguria	20.1	10.2	69.9	44.7	26.6	10	Liguria	26.5	11.4	71.6	46.3	29.7	6
Emilia-Romagna	22.3	14.2	67.9	45.6	28.7	3	Emilia-Romagna	26.3	13.5	70.7	47.8	30.6	4
Toscana	20.8	12.9	71.0	45.4	27.9	5	Toscana	26.3	12.4	69.4	44.8	29.4	8
Umbria	20.1	11.4	68.7	45.8	27.0	8	Umbria	24.4	9.7	69.3	45.0	27.9	15
Marche	21.0	9.8	68.5	43.4	26.3	12	Marche	26.8	11.9	67.6	44.4	29.2	9
Lazio	21.2	9.4	68.9	40.6	25.7	15	Lazio	24.3	12.2	69.9	43.7	28.5	13
Abruzzo	19.5	10.4	67.3	42.1	25.6	16	Abruzzo	23.0	8.8	65.7	45.6	26.8	16
Molise	18.0	9.2	67.0	43.1	24.8	18	Molise	22.0	7.8	69.9	45.3	26.5	18
Campania	18.6	8.5	65.2	35.0	23.0	20	Campania	19.9	8.6	65.6	37.7	24.0	21
Puglia	16.5	8.3	64.3	37.8	22.7	22	Puglia	18.6	9.2	60.6	39.4	23.6	22
Basilicata	20.5	9.1	64.9	41.0	25.1	17	Basilicata	22.7	8.7	67.7	43.9	26.5	17
Calabria	19.1	7.8	65.4	38.0	23.6	19	Calabria	20.9	8.5	66.7	38.6	24.7	19
Sicilia	17.4	7.9	62.7	38.7	22.9	21	Sicilia	19.6	8.2	62.8	40.4	24.1	20
Sardegna	17.3	13.0	73.7	42.8	26.5	11	Sardegna	22.0	13.9	74.6	45.2	29.1	10
Italy	**19.0**	**11.3**	**66.9**	**43.3**	**25.9**	**13**	**Italy**	**23.0**	**12.2**	**67.5**	**44.6**	**28.0**	**14**

The regions that are proceeding at the fastest rate are Veneto (+3.3 p.p.) and Liguria (+3.1 p.p.). The regions that are still behind are Umbria and Puglia (+0.9 p.p.) and Campania (+1.0 p.p.).

At national level, the gender gap shows a more elevated female disadvantage (−7.3 p.p.) in the southern regions, furthermore without any signs of improvement (Table 9.15). Particularly critical is the gender gap in Puglia (9.1 p.p.) and Sardegna (8.8 p.p.), while there are less differences by gender in Piemonte (3.9 p.p.) and Valle D'Aosta (5.1 p.p.).

The distance between the first (Bolzano) and the last region (Puglia) evenly increased (from 9.1 p.p. in 2007 to 10.9 p.p. in 2012). It is interesting to highlight how, even though the overall ranking has not changed, all the Italian regions in 2012 obtained higher values of the AAI. This general result is not enough and Italian policy is called to increase the effort for enacting programs that promote active ageing with a renewed and shared awareness between the political responsibility, personal responsibility, age-friendly environments, and intergenerational solidarity.

9.4 Conclusions

The AAI adopts a holistic perspective and, for the first time, it operationalised the multidimensional concept of active ageing, which had been hitherto considered only conceptually (ActivAge Consortium 2008; Walker and Maltby 2012; WHO 2002).

Until now, the lack of a multidimensional approach has hindered to adopt incisive policymaking in Italy, in particular at the regional and the municipal governments' level.

This research shows that in order to understand what happens on a national level and to promote active ageing, it is necessary to know how welfare is being measured on a local level.

The observed variations in the AAI on the regional level show the specific social, economic, political, cultural, and demographic characteristics of every region. Furthermore, the AAI allows us to identify the areas where regions should intervene with appropriate measures to improve the regional context.

Table 9.15 Overall AAI by gender in 2007 and in 2012, reporting top and bottom three ranked regions only

2007						2012					
Men			Women			Men			Women		
Region	Value	Rank	Region	Value	Rank	Region	Value	Rank	Region	Value	Rank
Bolzano	35.1	1	Bolzano	28.9	1	Bolzano	37.7	1	Bolzano	31.5	1
Trento	33.4	2	Trento	27.9	2	Trento	35.8	2	Trento	30.6	2
Emilia-Romagna	32.5	3	Valle D'Aosta	25.9	3	Veneto	34.5	3	Valle D'Aosta	28.7	3
Italy	**29.8**	**13**	**Italy**	**22.5**	**14**	**Italy**	**31.6**	**12**	**Italy**	**24.3**	**15**
Calabria	27.3	20	Campania	19.2	20	Calabria	28.2	20	Sicilia	20.7	20
Puglia	27.3	21	Sicilia	18.9	21	Sicilia	28.0	21	Campania	20.6	21
Campania	27.3	22	Puglia	18.6	22	Campania	27.9	22	Puglia	19.4	22

Overall, the AAI reveals a substantial rigidity over the two years despite the economic crisis. Although some regions improve their position and climb the ranking, other regions get lower scores and lose the position occupied in 2007.

The gender differences reveal small improvements, such as older women's ability to have and maintain employment largely due to male-dominated economic sectors being penalized by the crisis. Also the role of elders in care work remains high because of the combined effect of the gradual deterioration of social protection systems and mechanisms of family assistance (informal assistance networks). The gender gap in the southern regions remains higher than in the northern regions.

Active ageing policy appears to be particularly difficult to implement. The main blocking reasons for not having reached better results in our country are:

- The excessively complex, fragmentary, and interdependent responsibilities of the policy authorities on central and local levels
- Political convenience, which has encouraged policymakers to implement active ageing policies with limited results but with high visibility
- Company strategy decisions focused on cost containment and production increment (early retirement schemes)
- The incapability, mainly in the southern regions, to profitably employ European funds to promote active ageing

Surely, the AAI will enable to plan, monitor, and analyze the results obtained from policies at national and subnational levels in a better way. For this reason, ISTAT has released the first information web system on the elderly (http://www.istat.it/it/anziani), which in the future should also include a section dedicated to the AAI.

Notes

1. The ageing index is calculated as the number of persons who are 60 years old or over per 100 persons under 15 years of age—Population Division, Department of Economic and Social Affairs (DESA), United Nations.
2. The old age dependency ratio is the number of persons who are 65 years and over per 100 persons 15–64 years of age—Population Division, DESA, United Nations.
3. Healthy life expectancy at birth is defined as the average number of years that a newborn could expect to live, if he or she was subject to the age-specific mortality rates of a given period—Population Division, DESA, United Nations.
4. The difference is calculated in percentage points (p.p.) for the relevant indicators of AAI.

References

ActivAge Consortium. (2008). Overcoming the barriers and seizing the opportunities for active ageing policies in Europe. *International Social Science Journal, 58*, 617–631.

Parlamento Italiano, Legge costituzionale n. 3 del 18 ottobre. (2001). *Modifiche al titolo V della parte seconda della Costituzione.* Retrieved October 13, 2016, from http://www.parlamento.it/parlam/leggi/01003lc.htm

Walker, A., & Maltby, T. (2012). Active ageing: A strategic policy solution to demographic ageing in the European Union. *International Journal of Social Welfare, 21*(S1), 117–130.

World Health Organisation (WHO). (2002). *Active ageing: A policy framework.* Retrieved October 10, 2016, from http://www.who.int/ageing/publications/active_ageing/en

Zaidi, A., Gasior, K., Hofmarcher, M. M., Lelkes, O., Marin, B., Rodrigues, R., Schmidt, A., Vanhuysse, P., & Zolyomi, E. (2013). *Active Ageing Index 2012: Concept, methodology and final results.* EC/UNECE, Active Ageing Index Project, UNECE Grant ECE/GC/2012/003. Vienna: European Centre for Social Welfare Policy and Research. Retrieved from http://www.euro.centre.org/data/aai/1253897823_70974.pdf

Luciana Quattrociocchi is a senior researcher at the Italian National Institute of Statistics. She deals with analysis on the family (transformations, behaviors, values) and social groups: youth, women, the elderly, and immigrants. She devotes her attention to the development of indicators, analysis, and dissemination of results on the living conditions of social subjects to meet the needs of various stakeholders. She is a member of the United Nations Economic Commission for Europe (UNECE) "Active Aging Index" Task Force Expert Group.

Daria Squillante has PhD in Demography, Political Sciences, and in Psychological Sciences and Techniques graduate, and is a researcher at the Italian National Institute of Statistics (ISTAT), focusing her involvement in Press Office and Communication strategies and production. Expert in social research (ageing, justice, ICT, aspects of daily life), her research interest has been also concentrated in managing many public and private projects aimed at promoting analysis and interventions on gender issues.

Mauro Tibaldi is researcher at the Italian National Institute of Statistics (ISTAT). His work concerns analysis of trends and evolving dynamics of the aggregates of the labor market and in-depth studies on specific issues, such as women's status, young people, foreign presence, and the ageing of the labor force. He has been responsible for the implementation of the national version of European Union Labour Force Survey (EU LFS) ad hoc module "Transition from work to retirement" (E.C. No.249/2011) of year 2012.

Part III

Comparative Analysis of Active Ageing

10

Differences in Active Ageing Index in Eastern European Countries: A Comparison of Poland and the Czech Republic

Kasia Karpinska

10.1 Introduction

The Active Ageing Index (AAI) was developed primarily as a measure to compare active aging phenomena between different European countries and to identify strengths and weaknesses in a country (Zaidi et al. 2013). Moreover, the longitudinal character of the index (indices for 2010, 2012 and 2014 are now available) allows policy makers and researchers to evaluate the differences between active aging outcomes over time. The general idea behind the development of the AAI was that, equipped with adequate information, policy makers can follow the examples of other countries and apply well-targeted policy responses to increase the potential for active aging (Zaidi et al. 2013). A thorough evaluation of the policies that underlie active aging outcomes is thus necessary to both evaluate the current situation and identify the most desired directions for future policy changes. This chapter aims to examine welfare state policies that affected active aging outcomes in Poland and the Czech Republic, as measured in the 2012 AAI.

K. Karpinska (✉)
Erasmus University Rotterdam, Rotterdam, Netherlands

A. Zaidi (eds.), *Building Evidence for Active Ageing Policies*,
https://doi.org/10.1007/978-981-10-6017-5_10

The Czech Republic and Poland (together with Hungary and Slovenia) are often grouped together under the label of Central and Eastern European countries that share a similar historical legacy and social background (Aidukate 2011). Their simultaneous transition from the centrally planned to market economy, and successful accession to the EU structures in 2004, further reinforced this classification (Perek-Bials et al. 2006). In his empirical analysis of welfare regimes, Fenger (2007) indeed showed that the Czech Republic, Poland and Hungary share many features, such as governmental programs (defined as values of expenditures and incomes), social situation (i.e., levels of inequality, participation in the labor market but also life expectancy) and political participation. Fenger (2007) has shown that the so-called 'post-communist European type' welfare regime is distinct not only from the traditional western European regimes but also from the former USSR countries and developing welfare states (such as Romania, Moldova, and Georgia). This welfare regime is characterized by its conservative approach to governmental expenditures and relatively higher well-being levels in comparison to other post-communist countries.

Yet, despite those similarities in social policies, active aging outcomes vary significantly between the two countries: the Czech Republic ranks 13th in the overall Active Ageing Index for 2012, while Poland occupies the bottom positions, in both overall and domain-specific indices. Although the Czech Republic can intensify efforts to improve the active potential of older people, it precedes Poland and other Eastern European countries (all of which occupy bottom positions in the 2012 AAI, with a notable exception of Estonia). The question arises as to what the most relevant differences between the Czech Republic and Poland are in terms of the active potential for older people and how those differences can be attributed to welfare state characteristics and existing policies.

A shared past and similarities in social developments have led many researchers to believe that their social and economic problems are common and require similar solutions and yet, as Perek-Bialas et al. (2006) have shown, there are many differences in approaches toward active aging. Piecing together—wherever possible—information on policies and cultural differences between the two countries will shed more light on the current outcomes for both countries.

When using the AAI for policy analysis, one can either decide to look at the general ranking of the countries or at the values of specific indices. The aggregate measure is suitable for comparison between countries and identifying priorities at the domain level while looking at specific indicators is helpful to identify and design appropriate policy measures. To identify differences in the active aging outcomes, the origin of the most striking differences at the indicator level was evaluated.

The current work focuses predominantly on three aspects—employment, participation in care and health services. This focus is determined by three considerations. First, those three elements represent the most relevant differences between the countries and can lead to the most relevant policy recommendations, especially for Poland, as a country that needs to improve the most. Second, some aspects of active aging are not easily amenable to policies and require a life course approach toward active aging—a principle stating that experiences in earlier life influence how individuals age. It is therefore difficult to identify specific policies that might affect, for instance, the share of healthy years of remaining life or mental health. This is regrettable, as those indicators are important predictors of participation in employment and caring and volunteering. Third, responsibilities for some aspects of active aging, such as the use of ICT (Information and Communications Technology), are likely to rest with regional- or local-level authorities rather national government (Karpinska and Dykstra 2014), which was our focus. Therefore, identifying policies that might have affected those outcomes is unfeasible.

In our attempt to understand the reasons for the AAI ranking of those countries, we focus on the policies, cultural and historical processes as well as general welfare state characteristics that occurred in the period before the year 2010 (when the data used for calculation of the 2012 AAI was collected). Although the rankings for 2014 are already available, the comparison over time is beyond the scope of this chapter. In the following sections, we present background information on both countries, describe briefly the main features of the AAI, analyze the scores in the 2012 AAI and describe the policies that in our view underlie those differences. Lastly, we briefly discuss our findings and identify possible policy interventions.

10.2 Background Information

10.2.1 Demographic Situation in Poland and the Czech Republic

In addition to a common past and shared experiences, the Czech Republic and Poland also share challenges of demographic decline; in both countries, the percentage of older people is increasing while the fertility rates remain low. The total fertility rate dropped in the Czech Republic from 2.11 in 1980 to 1.46 in 2013, while in Poland, in the same period it dropped from 2.26 to 1.29 (Sobotka et al. 2015). Although Poland is almost four times larger in terms of population than the Czech Republic (38 million and 10.5 million, respectively), both countries face the threat of population decline, which is projected to be more severe in Poland than in the Czech Republic (Eurostat 2011). Migration does play an important role here; while the Czech Republic is characterized by low levels of out-migration, Poland experienced a dramatic outflow of predominantly young citizens after its EU accession in 2004, and emigration surpasses immigration (Strelkowski et al. 2013). At the moment, the median age in Poland and the Czech Republic is still relatively low (37.7 and 40 years, respectively). Yet, the aging of the population is speeding up and by 2060 both countries are expected to belong to the world's oldest populations: Poland with a median age of 51 and the Czech Republic with 48. The share of the population aged 65+ will continue to rise in both countries, as will the proportion of the population aged 80+ and the old age depen-

Table 10.1 Older population in the Czech Republic, Poland, Germany, Italy and EU28 countries

	Percentage aged 65+			Percentage aged 80+			Old age dependency ratio (%)		
	2005	2010	2060	2005	2010	2060	2005	2010	2060
Czech Republic	14.1	15.2	28.3	3.0	3.6	11.4	19.8	21.7	50.3
Poland	13.1	13.5	32.9	2.5	3.3	12.0	18.8	19.0	60.9
Germany	16.6	20.7	32.3	4.3	5.1	13.4	28.8	31.4	59.2
Italy	19.5	20.4	30.0	4.9	5.8	13.2	29.4	31.2	53.1
EU28	16.6	17.4	28.4	4.0	4.6	11.8	24.6	25.9	52.6

Source: Eurostat: demo_pjanind; proj_13ndbims. Accessed on January 16, 2016

dency ratio (see Table 10.1)—all measures are expected to align and even exceed the EU28 averages by 2060. The changes are even more striking when comparing Poland to Germany or Italy, other European countries that experience low fertility rates but high inward migration. Although those countries also age, the speed of changes is less sharp.

10.2.2 Active Aging Policies in Poland and the Czech Republic

The demographic changes that are occurring in both countries require adequate policy responses. As shown by Perek-Bialas et al. (2006), the issues related to active aging appeared only in the mid-2000s on the policy agenda in Poland. Prior to this, issues such as youth unemployment were considered more important. Still, in 2005, when the study was concluded, Poland did not yet have a comprehensive active aging strategy in place. The reasons for this were 'the lack of institutional mechanisms, discriminatory socio-cultural perceptions and the unfavorable economic climate' (Perek-Bialas et al. 2006, p. 568). Although the Czech Republic attempted to develop active aging strategies from the late 1980s, those failed to produce the envisioned results or were not implemented (Perek-Bialas et al. 2006). Similarly to Poland, the lack of cooperation between different administrative bodies, financial restrictions and the negative connotation of the term 'senior' have limited the possibilities for implementing active aging policies effectively. The authors concluded that a small number of existing measures were adopted as a result of the pressure from the European authorities and, in some cases, seem to have emerged accidentally or as a response to short-term needs. Given that both countries were unsuccessful in implementing active aging policies, a question arises as to how between-country differences arose.

10.3 Active Ageing Index

Before we answer this question, we briefly present the main features of the AAI. The AAI is a composite index that measures the contribution older people make to a society (Zaidi et al. 2013). As such, it does not measure their well-being or welfare but their untapped potential to participate in society. With the use of 22 indicators, the Active Ageing Index captures the multidimensionality of active aging. The indicators are grouped into four domains: *Employment; Participation in society; Independent, healthy and secure living;* and *Capacity and enabling environment for active aging.* Table 10.2 summarizes the indicators for each domain. The first three domains depict the actual *experience* of active aging, while the last domain relates to the *capacity* to actively age. For each domain, the arithmetic weighted average of the indicators is calculated.[1]

The outcome of the Active Ageing Index is a value that ranges between 0 and 100, with higher outcomes pointing toward greater contributions of older people to society and better enabling conditions. The maximum of 100 is not likely to be achieved as it implies the 'fullest active aging' (Zaidi et al. 2013) and forms a theoretical possibility, rather than an attainable goal. The results of the AAI are available for 2012 and since recently, also for 2014, for all 28 countries of the European Union.

10.4 Results

10.4.1 Difference in the AAI in the Czech Republic and Poland

The Czech Republic and Poland, despite their similarities in social and demographic structure, rank very differently in the overall and domain-specific indices; the Czech Republic ranks 13th in the 2012 AAI with its score of 34.0, while Poland occupies a bottom position, with a score of 27.2.[2] In comparison, the frontrunner in the Index, Sweden, obtained the score of 44. Table 10.2 summarizes the domain scores and rankings and scores for the individual indicators for the Czech Republic and Poland. The maximum and mean values obtained in the AAI are included for reference (countries not named). In general, the results for all domains and individual indicators are higher for the Czech Republic than for Poland, which is rather intuitive, given the difference in the overall rankings of those countries. The results of the Czech Republic are fairly

Table 10.2 AAI results for the Czech Republic and Poland

	The Czech Republic		Poland		Mean score	Highest score
	Score	Ranking	Score	Ranking	AAI 2012	AAI 2012
Overall AAI 2012	34.3	13	27.3	28		44.0
1st domain: Employment	**26.4**	**14**	**19.8**	**25**		**41.0**
1.1 Employment rate 55-59	67.1		45.8		60.7	80.7
1.2 Employment rate 60-64	25.2		19.1		30.4	61.0
1.3 Employment rate 65-69	9.5		9.4		11.2	24.5
1.4 Employment rate 70-74	3.6		5.0		6.1	21.4
2nd domain: Participation in society	**19.4**	**12**	**12.2**	**27**		**25.2**
2.1 Voluntary activities	12.9		4.8		14.9	32.7
2.2 Care to children, grandchildren	37.2		22.5		32.4	53.7
2.3 Care to older adults	14.8		13.3		12.8	17.1
2.4 Political participation	12.3		7.1		12.1	26.5
3rd domain: Independent and secure living	**73.8**	**11**	**67.5**	**21**		**79.0**
3.1 Physical exercise	5.4		7.0		11.0	28.9
3.2 No unmet needs of health and dental care	94.8		81.6		91.5	99.0
3.3 Independent living arrangements	86.2		70.0		84.3	99.3
3.4 Relative median income	82.0		93.0		84.4	100.0
3.5 No poverty risk	98.7		93.2		92.5	98.7
3.6 No material deprivation	95.7		83.5		90.0	99.9
3.7 Physical safety	89.9		95.1		79.0	95.1
3.8 Lifelong learning	4.2		0.6		4.7	22.3
4th domain: Capacity and enabling environment for active aging	**54.4**	**14**	**46.7**	**22**		**69.5**
4.1 RLE achievement at age 55	50.6		50.0		53.4	59.2
4.2 Share of healthy life years in the RLE at age 55	57.6		49.8		53.4	77.1
4.3 Mental well-being	61.0		49.6		63.9	87.2
4.4 Use of ICT	31.0		22.0		38.3	75.0
4.5 Social connectedness	47.5		30.8		51.5	75.6
4.6 Educational attainment	83.4		69.2		56.6	85.7

Source: AAI, 2013

consistent across different domains, while for Poland domains ranking ranged from 22 to 27. Clearly, the overall ranking of the country plummeted due to the second domain, *Participation in society*, for which Poland ranked the lowest.

10.4.2 Economic Performance

The first, general observation related to those outcomes pertains to countries' economic performance. As shown in the analysis of the AAI, there is a positive relationship between per capita GDP and the AAI outcomes: countries characterized by a better economic performance (and, consequently, better living conditions) obtained, in general, higher AAI results.

Indeed, the Czech Republic higher GDP per capita is also related to higher AAI results in this country in comparison to Poland. However, this correlation does not denote causal relationship, and high levels of active aging potential can lead to an increase in GDP by, for instance, increased tax revenues related to higher labor market participation of older citizens. Moreover, the analysis of AAI shows that economic prosperity does not always translate into similar AAI outcomes (Zaidi et al. 2013). This stresses the need for the analysis of welfare state characteristics and policies adopted by countries.

10.4.3 Employment

When looking specifically into the *Employment* domain, the most striking differences refer to employment rates of those aged 55–59. In Poland, the employment in this age category is estimated at 45.9%, while 67.1% of the Czechs in this age category is employed. The participation of the 60–64 age group is still higher in the Czech Republic, although the difference is smaller. The scores for the two highest age categories are comparable.

The employment rates of older people aged 55–59 differ greatly between the countries and are thus essential for countries' different position in the domains. In both countries, this is the age category in closest proximity to the official retirement age at the time of data collection. In

Poland, the statutory retirement age in 2010 was 60 for women and 65 for men, while in the Czech Republic it was 61 for men and 58.7 for women (OECD 2011a). The average effective exit age in the Czech Republic matched the statutory retirement age, while in Poland both men and women leave the labor force before they are eligible for their statutory pension. Consequently, early exit from the workforce has a great impact on the AAI results in Poland.

The labor market participation patterns differ between the countries. Although in the mid-2000s the early exit in both countries was perceived as substantial, the OECD (2006, 2011a) shows that the employment rates in the Czech Republic have always been relatively high and, in contrast to Poland, they did not drop dramatically in recent decades. In fact, participation rates for Czech men aged 50–64 remained roughly the same between the 1970s, under the communist regime striving for employment of all citizens, and the early 2000s. The participation rates of Czech women increased in the same period (OECD 2006) to above 50%. For Polish women in the same age category and in the same period, participation rates decreased from roughly 60% to 40%, while Polish men experienced a drop from almost 90% to approximately 60% (OECD 2006).

There are several reasons for those differences. For Poland, the low participation rates are a result of difficulties during the initial stages of economic transition; Poland's national debt was higher than the Czech Republic's and Poland also experienced a more severe recession. During the 1990s, pre-retirement benefits were introduced, further decreasing the economic activity of older persons. Reduction of entitlements for employees and the relaxation of the labor code in the 2000s—which occurred as a result of economic downturn—decreased the participation rates in Poland even further (ActiveAge 2005). Incentives to remain employed until the mandatory retirement age were scarce (European Commission 2008).

The Czech Republic, although it also faced difficulties at the initial stages of the economic transformation, began adopting policies to increase labor market participation of older workers earlier than Poland. The official retirement age has been increasing gradually ever since the late 1990s (in Poland, this measure was undertaken only recently; OECD 2013). Moreover, following the OECD recommendations, the Czech

Republic recently implemented policies to improve the labor market prospects of older workers. Policies aimed at making retirement financially less attractive have been introduced (the minimum insurance period required for a pension entitlement is gradually being increased) and restrictions on the concurrent receipt of an old-age pension and income from work have been abolished. Additionally, the Assistance in Material Need Act ensures that those with low incomes obtain material support if employed or actively seeking employment.

Gender differences in the labor market exit can also account for differences in labor market participation in both countries. Studies have repeatedly shown that men tend to exit the labor market mostly via retirement, unemployment or disability paths, while women are more likely to exit the labor market via 'other inactivity' paths (OECD 2011a). Van Bavel and de Winter (2013) showed that grandparenting speeds up retirement and women are especially likely to leave the workforce before the official retirement age to fulfill their roles as grandparents. The inactivity exit path is often related to care responsibilities for family members in a situation when publicly provided alternatives for care are not available (OECD 2011a). Interestingly, older women in Poland are more likely to take this exit path than older women in the Czech Republic (OECD 2011a). This coincides with the fact that Poland has a lower provision of care services than the Czech Republic (see the next section) and suggests that policies regarding care responsibilities might have also affected labor market participation, particularly of older women.

10.4.4 Participation in Society

Poland's low ranking in the 2012 AAI is mostly due to low participation in care responsibilities. While the observed scores for care for older are comparable (14.8 and 13.3, respectively), Poland scores very low in terms of care for children and grandchildren (22.5 vs. 37.2 in the Czech Republic). The highest score for this indicator was 53.7 in Italy.

The low score of Poland is rather surprising. Although in both countries the socialist regime produced strong dependency of the young family on their parents to provide a place to live and financial and instrumental support, their care regimes differ. According to the distinction made by Saraceno and Keck (2010), Poland can be characterized in terms of *familialism by default*. This term describes countries that offer neither financial support nor publicly provided alternatives for care, imposing care responsibilities on family members. The Czech Republic, on the contrary, is described in terms of *supported familialism* where care for the young is concerned (i.e., the state provides financial means to support family members in fulfilling their care responsibilities). Following this distinction, one would assume Polish seniors would be more involved in the caring responsibilities for the young, predominantly to support younger women's footing in the labor market.

This assumption seems corroborated by Poland's higher rate of persons aged 55–69 that are inactive in the labor force on the ground of caring responsibilities or other family-related issues as compared to the Czech Republic. Using the MITILINKS database and SHARE data, Bordone et al. (2012) showed that the care gap for children under two (a number of weeks not covered by compensated leave or available child care facilities) is the highest in Poland and in Mediterranean countries, suggesting high care burden on families. For the Czech Republic, this gap was much smaller. Robila (2012) reported an identical rate of children under three in formal care in both countries, while the uptake of formal care for children aged 3–6 was significantly higher in the Czech Republic (69%) than in Poland (31%). This does not necessarily reflect meeting the demand for care—in general, the care provision in both countries was low as the financial contributions were directed toward parental leaves rather than to formal care provisions. While Polish parents use care facilities predominantly as a full-time provision (30 hours or more), Czech children attend those facilities part-time (Mills et al. 2014). Due to a declining number of nurseries, an increasing number of children under three attend kindergartens, which are intended for children aged 3–6 (25.5% in 2005; Platenga and Remery 2009). This arrangement, although allowed Czech parents to participate in the labor market, required organizing additional care, as preschool facilities are open only part-time. The main difficulty in using daycare facilities in Poland compared to the Czech Republic was related to high costs (availability and accessibility were rated similarly in both countries) (Mills et al. 2014). The more frequently voiced concerns about the costs of care facilities seem to, again, stress the need of working Polish parents to rely on grandparental care.

When looking at the employment rates for mothers of children under six, Poland surpasses the Czech Republic (58.9% and 34.7%, respectively; Robila 2012). Despite the lower number of working mothers in the Czech Republic, the care levels reported in the AAI were higher than in Poland, where grandparental care appears needed the most. Bordone et al. (2012) showed that while Polish grandparents are more likely to offer daily care than Czech grandparents, the probability of at least weekly care is similar (and lower in absolute terms). This suggests that patterns of care can differ between countries; while Czech grandparents are more likely to step in to fill care gaps, Polish grandparents are involved in care less frequently but the volume of care is possibly greater. Feasibly, the Czech grandparents offer incidental support for non-working mothers or take care for grandchildren after they leave their (pre)school. In Poland, a pattern bearing resemblance to Southern regimes seems to apply—mothers either do not work (and do not require help from grandparents) or involve in meaningful employment and need full-time care from grandparents, as the level of formal care provisions is lower.

This explanation, however, does not solve the puzzle of low rates of grandparental support in Poland as reported in the AAI. The difference may be attributable to the question asked in different surveys. The data used for constructing the AAI inquired into general care responsibilities, while the SHARE data (where care incidences reported by Polish grandparents were much higher than those described in AAI) inquired into care offered to specific grandchildren. This might have made respondents to think more carefully about the care incidences and report it more accurately than when referring to general care provided.

10.4.5 Independent, Healthy and Secure Living

The Czech Republic also ranks higher than Poland in this domain. The difference that stands out the most refers to the unmet needs of medical and dental care. Almost 95% of the Czech seniors report their medical needs being met, while this is the case only for 81% of Polish seniors. This number is lower than the EU27 average.

Both the Czech Republic's and Poland's health systems have been subject to many transformations since the early 1990s. In the Czech Republic, the centralized system has been replaced by private health care that offers the possibility of choosing health care providers without restrictions. Andel (2014) suggests that the system provides adequate health care for older adults of all levels of functional status, despite the initial staffing and financial problems. Similarly, the Polish health care system has also undergone a major transition to a privately administered health insurance system, resulting in a low quality of health care provisions (Aspalter et al. 2009). The information collected during the ActiveAge project (ActiveAge 2005) shows that, indeed, the situation in Poland in the mid-2000s was worse compared to the Czech Republic's. Poland had the lowest share of GDP devoted to health care among the countries under the study, and also one of the lowest numbers of medical personnel per hundred thousand inhabitants. Waiting times in public health system are often long, leading to higher levels of health-related inactivity in the labor market (OECD 2015), which shows also the direct link between the health status and labor market participation of older workers. Better health condition is necessary to prolong working lives. In the Czech Republic, health conditions of older person were shown to be slightly better (OECD 2015), which partly accounts for higher employment rates.

With respect to *living independently* indicators, Poles also score lower than the Czech seniors. This can be linked to differences in publicly provided alternatives for care arrangements—in both countries, the majority of care (80%) was performed by informal carers (OECD 2011b). In Poland, services that would support independent living are still scarce and, consequently, many Polish seniors co-reside with their family members. This is less the case in the Czech Republic.

10.5 Discussion and Policy Implications

The current study shows how—despite presumed similarities of demographic and economic situation—the countries differ in their active aging outcomes. The Czech Republic obtained higher results in virtually all indicators of the AAI. Yet, this outcome is not the result of a comprehensive active aging policy. In both countries, some of the AAI outcomes arose as a result of more general social policies and different transition trajectories. The disparity in active aging outcomes between the neighbors can be thus partly attributed to their economic situation and differences in policies applied.

Clearly, when looking into those differences, one can derive more specific policy recommendations for both Poland and the Czech Republic. In case of Poland, the most required changes refer to the labor market participation, especially of older women. Promoting employment until the statutory retirement age and investments in lifelong learning or training are among the measures that can help achieve this goal. In the Czech Republic, further improvement in this domain would rely on promoting labor market participation beyond the mandatory retirement age, with similar measures as in Poland.

The current study suggests that labor market participation of older people in Poland and in the Czech Republic is also related to care responsibilities and good health. Although active aging does not require that individuals involve in all aspects of active aging equally, the preferred level of activity can be largely determined by the provision of services available. Involvement in care and, to some extent, exit from the labor market for older women can be linked to lack of or insufficient provision of care arrangements. Expansion of institutionalized care facilities (for both old and young) would give older people more agency over their activity and would eventually help increase the labor market participation of older women, especially in Poland. Moreover, the labor market participation of older people is strongly related to their health situation. Improvements in health care (i.e., shortening waiting times to health specialist, more focus on (occupational) health prevention) can help increase the labor market participation of older people further.

The presented recommendations in other domains are linked to the labor market; however, they should also be seen as independent goals. Informal care is a valuable, non-monetary contribution to family and society, while participation in employment is relevant, for both the macro-level of a country (in terms of tax revenues and sustainability of welfare systems) and individuals (employment leads to higher disposable incomes and higher future pension). Improvement in those domains removes barriers to active aging and offers older people more possibilities to determine the preferred level and focus of activity (either in care or employment).

In analyzing the differences between the countries, we aimed at piecing together policies for the indicators that affected the ranking of the countries the most. Some of the aspects of active aging, however, are not easily amenable to policy measures and require a life course approach (i.e., remaining healthy life expectancy). These aspects are crucial preconditions of employment, participation in voluntary activities or care. Responsibilities for those aspects of active aging are most likely to rest with regional- or local-level authorities and not at the national level, which was our focus. Analysis of active aging outcome at the regional level is indeed very interesting, but not suitable for cross-country comparison. Consequently, information on policies regarding some of those indicators was not provided in the current study.

Recently, both countries have been intensifying their efforts to implement active aging strategies, and analysis of the future AAIs will help establish how those strategies shaped active aging outcomes of older people. Moreover, the data for the 2012 AAI was collected at the onset of the economic crisis. Poland's economic standing is fairly good in comparison to other European countries. Whether Poland will compensate for the economic disadvantages of the early transformation and what consequences this will have to increasing active potential for older people in this country are interesting questions for future inquiry.

Notes

1. For details regarding the source of data for each of the indicators and weighting, see Zaidi et al. (2013).
2. In the 2014 AAI, the Czech Republic moved to 12th position (with the slightly increased score of 34.6%) while Poland ranked 27th, with the slightly increased score of 28.2%, preceding only Greece.

References

ActiveAge. (2005). *Overcoming the barriers and seizing the opportunities for active ageing policies in Europe*. Final Report. Retrieved from http://cordis.europa.eu/documents/documentlibrary/100124401EN6.pdf

Aidukate, J. (2011). Welfare reforms and socio-economic trends in the 10 new EU member states of Central and Eastern Europe. *Communist and Post-Communist Studies, 44*, 211–219.

Andel, R. (2014). Ageing in the Czech Republic. *Gerontologist, 54*(6), 893–900.

Aspalter, C., Jinsoo, K., & Sojeung, P. (2009). Analysing the welfare state in Poland, the Czech Republic, Hungary and Slovenia: An ideal-typical perspective. *Social Policy and Administration, 43*(2), 170–185.

Bordone, V., Arpino, B., & Aassve, A. (2012). Policy perspectives of grandparenting in Europe. *Working Paper No. 51*. Available from www.dondena.unibocconi.it/wp51

European Commission. (2008). *Promoting longer working lives through pension reforms. Early exits form the labour market*. Luxembourg: Office for Official Publications of the European Communities.

Fenger, H. J. M. (2007). Welfare regimes in Central and Eastern Europe: Incorporating post-communist countries in a welfare regime typology. *Contemporary Issues and Ideas in Social Sciences, 3*(2), 1–30.

Karpinska, K., & Dykstra, P. (2014). *The Active Ageing Index and its extension to the regional level: Discussion Paper.* Brussels: European Commission.

OECD. (2006). *Living longer, working longer*. Paris: OECD Publishing.

OECD. (2011a). *Pensions at a Glance 2011. Retirement-income systems in OECD and G20 countries*. Paris: OECD Publishing.

OECD. (2011b). *Help wanted? Providing and paying for long term help*. Paris: OECD Publishing.

OECD. (2013). *Pensions at a Glance 2013. Retirement-income systems in OECD and G20 countries*. Paris: OECD Publishing.

Perek-Bials, J., Ruzik, A., & Vidovicova, R. (2006). Active ageing policies in the Czech Republic and Poland. *International Social Science Journal, 58*(190), 559–570.

Platenga, J., & Remery, C. (2009). The provision of childcare services: A comparative review of 30 European countries. *European Commission's Expert group on Gender and Employment Issues.* Retrieved from http://ec.europa.eu/social/main.jsp?langId=en&catId=89&newsId=545&

Robila, M. (2012). Family policies in Eastern Europe: A focus on parental leave. *Journal of Child and Family Studies, 21*, 32–41.

Saraceno, C., & Keck, W. (2010). Can we identify intergenerational policy regimes in Europe? *European Societies, 12*(5), 675–696.

Sobotka, T., Zeman, K., Potančoková, M., Eder, J., Brzozowska, Z., Beaujouan, E., & Matysiak, A. (2015). *Fertility Datasheet 2015.* Vienna Institute of

Demography/Wittgenstein Centre for Demography and Global Human Capital (IIASA, VID/ÖAW, WU).

Strielkowski, W., Filipec, P., Stefanik, M., & Kowalska, K. (2013). Outward labour migration in the Czech Republic, Poland and Slovakia after the EU Enlargement in 2004. *Czech Economic Review, 7*, 42–54.

Van Bavel, J., & De Winter, T. (2013). Becoming a grandparent and early retirement in Europe. *European Sociological Review, 29*, 1295–1308.

Zaidi, A., Gasior, K., Hofmarcher, M. M., Lelkes, O., Marin, B., Rodrigues, R., Schmidt, A., Vanhuysse, P., & Zolyomi, E. (2013). *Active Ageing Index 2012: Concept, methodology and final results*. EC/UNECE, Active Ageing Index Project, UNECE Grant ECE/GC/2012/003. Vienna: European Centre for Social Welfare Policy and Research. Retrieved from http://www.euro.centre.org/data/aai/1253897823_70974.pdf

Kasia Karpinska is a postdoctoral researcher at the Department of Public Administration and Sociology at Erasmus University Rotterdam. Her work includes labor market position of older workers and ageism in the workplace. Currently, she also works on issues related to family relationships of migrants.

11

Ageing in a Multicultural Europe: Perspectives and Challenges

Eralba Cela and Mariateresa Ciommi

11.1 Introduction

Population ageing is one of the key features of European demographic trends and one of the main challenges of western European countries. Fertility decline below the replacement level alongside increasing life expectancy have transformed Europe into one of the oldest regions in the world, with a median age of 42 years in 2015 (UN 2015).

Migration is the second major phenomenon affecting population structure and challenging the policies of European societies. Europe is the most important destination of international migration (UN 2013); here migrants are older compared to other world regions and the number of older migrants will continue to increase, as not only the first generation is ageing, but by 2020 migrants' children born in Europe will approach retirement age as well (White 2006), introducing a further dimension to the ageing challenge, namely, that of an ageing population with different cultural and ethnic backgrounds.

E. Cela (✉) • M. Ciommi
Università Politecnica delle Marche, Ancona (AN), Italy

A. Zaidi (eds.), *Building Evidence for Active Ageing Policies*,
https://doi.org/10.1007/978-981-10-6017-5_11

Ageing is a demanding process, requiring changes and adaptations; migration is a complex life event that may bring both stress and opportunities. Their intersection could cause a double risk of being vulnerable in terms of poor economic and health status and limited opportunities for social outdoor activities. Empirical research shows, that on average older migrants have more physical and mental health problems, compared to native peers; moreover, their lower socio-economic conditions reflected in a low level of education, poverty, poor housing conditions or embeddedness in deprived neighbourhoods increase their vulnerability and social exclusion (Fokkema and Naderi 2013). In addition, acculturation to the host societies could also have negative effects on migrants' health, especially when poor socio-economic conditions are coupled with lifestyle changes related, for example, to Westernization of dietary habits which has been proved to lead to increased risk of 'Western diseases' such as obesity, diabetes and cardiovascular disease, which were almost absent in the migrants' origin communities (Darmon and Khlat 2001). Other migrant-specific factors like language and cultural barriers, homesickness, discrimination and stigmatization in the destination countries restrict ageing migrants' participation in social outdoor activities, resulting in relatively small and homogeneous social networks and increasing the risk of social isolation and loneliness (King et al. 2014).

Besides the above risk-enhancing factors, there are, however, other factors that may protect older migrants from vulnerability. The most important one is the embeddedness within the family and community, both locally and transnationally; indeed multigenerational co-residence and intergenerational support and obligations are more prevalent in migrant families (Fokkema and Naderi 2013; King et al. 2014), although there is the risk that filial obligations might become less available in the future due to changed conditions in terms of time and willingness (Ajrouch 2005; Fokkema et al. 2016). Moreover, innovations in information and communication technologies and low-cost transportations have facilitated mobility, enabling strong ties with people left behind in the countries of origin and fuelling older migrants' active participation in the administration of (moral, emotional and material) care both locally and across borders (Baldassar and Merla 2013).

Demographic change and the resulting population ageing have been placed high in the political agenda especially because of the fiscal sustainability concerns, related to the increase of public finance expenditure, in particular for social protection and pension systems, health and long-term care (EC 2014). Nevertheless, although ageing represents a priority challenge for Europe, no special focus has been given to the ageing migrant population.

2012 has been designated by the European Union as the Year for Active Ageing and Solidarity between Generations (EY2012), in order to promote a culture of active ageing in the European countries. The idea behind the concept of 'active ageing' is to address the ageing challenge by improving conditions that enable the elderly to enjoy a better quality of life; a more active role in society is expected to increase the number of healthy-life years and thus reduce the burden on public health and social services. It is highly questionable, however, whether the common conceptualization of successful and active ageing, developed in the Western context for 'native' older adults, is fully effective for older migrants due to their cultural specificities and their higher risk of being vulnerable in different respects. Understanding of successful ageing that is simultaneously shaped by two cultures is poorly developed (Torres 2001).

The aim of this chapter is to stimulate the debate on the ageing-migration nexus in relation to active ageing. For this purpose, we apply the Active Ageing Index (hereafter AAI), developed in the framework of the EY2012, to a specific target of the ageing population, that is, ageing migrants in seven European countries.

The remainder of the chapter is structured as follows. The next section presents methodological aspects and the description of the data used in the analysis. The third part moves on to the results, discussing the four domains and the AAI. The final part contains conclusions, policy implications of our findings and future research directions.

11.2 Methodology

In order to analyse the AAI for the migrant population, we use micro data from the first wave (2004/2005) of the Survey of Health, Ageing and Retirement in Europe (hereafter SHARE).[1] We focus on migrants living in seven EU countries (out of the 11 countries of the survey), namely, Austria, Belgium, Germany, France, the Netherlands, Switzerland and Sweden. More in detail, in each country, we compare migrants to natives by computing a modified version of the AAI. The choice of these countries is justified by a minimum threshold on the percentage of immigrant resident respondents in the sample that we fixed at a 5% level in order to avoid analysing countries where the immigrant sample is too small (Table 11.1). In SHARE, households were selected following the criteria according to which at least one member was born in 1954 or before (individuals aged 50 years or over), based on nationally representative samples. Since AAI focuses on people aged 55 years or over, we reduced the sample, excluding individuals that do not fulfil this requirement. SHARE provides information on 18 out of the 22 indicators of the AAI (see Table 11.2 for more details). Following Zaidi et al. (2013), the construction of our AAI accounts for two aggregation step procedures: computation of an aggregated index for each domain and the collapse of the four domains into the overall AAI.

There are several differences between our formulation of the AAI and the official one. Firstly, our AAI accounts for 18 instead of 22 indicators. Secondly, we do not take into consideration gender gap in the different indicators because of the small size of the sample in each EU country. Thirdly, we use a modified system of weights in order to take into account the weight lost by the dropped variables (Table 11.3). More in detail, as for the AAI, in the two aggregation steps we adopt implicit weights reflecting the magnitudes of each indicator; the weights associated to the dropped variables were redistributed proportionally to the weighs already assigned to the indicators within each domain. Individual calibrated weights developed by SHARE were also considered in the computation of the indicators.[2]

11.3 Results

In this section, for each domain, we present, separately for natives and migrants, actual scores of each indicator as well as their relative contribution to the overall domain scores. At the end, the four domains are aggregated in the Active Ageing Index. The analysed European countries are thus ranked according to the score achieved in the four domains and in the overall index. Some specificities of the migrant population both in the single domains and in the overall AAI are highlighted.

11.3.1 First Domain: Employment

The first domain measures the contribution of older people to society through paid activities. We compute the employment rate, both for natives and migrants, for three of the four age groups, namely, 55–59, 60–64 and 65–69.[3]

Figure 11.1 shows the actual scores for each country as well as the magnitude of their gap between both the overall scores and each indicator.

Sweden and Switzerland are the best performing countries for both migrants and natives. Nevertheless, in the case of migrant population, these countries still have a high potential for further improvements compared to the situation of the native population.

Although the results in Fig. 11.1 show that some countries perform better than others, this does not mean that these countries perform better also in the single indicators. For example, although Sweden ranks first in

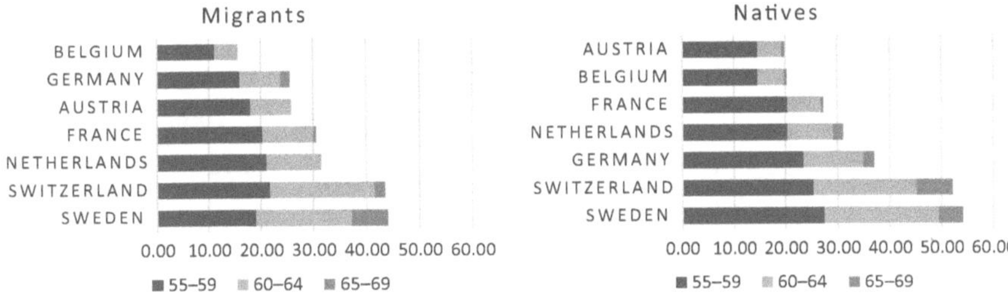

Fig. 11.1 Ranking of European countries according the first domain for migrants and natives

the domain for migrant population, it leads only in the third indicator (employment rate for the age group 65–69), whereas Switzerland performs best in the first two ones.

In Fig. 11.2 we show the relative impact of each indicator on the composition of the first domain for each country. The age group 55–59 gives the major contribution to the domain; the maximum value is achieved in Sweden (27.52) and the minimum regards Belgium (11.12) for natives and migrants, respectively. In the age group 60–64, Belgium and Sweden have the minimum and maximum values (4.57 for migrants and 21.95 for natives). The relative contribution of the indicator for the last age group (65–69) is inexistent for migrants in Austria, Belgium and the Netherlands; it is, however, low for all the other countries with the only exception for migrants in Sweden (6.82) and natives in Switzerland (7.01).

Analysing the relative contribution of the three age groups for the migrant population, it is possible to differentiate three different group countries: countries where the first two age groups (55–59 and 60–64) contribute to around 70% and 30% each (Austria, Belgium, the Netherlands and France); a second group represented by Sweden and Switzerland where the first two classes contribute almost equally (around 40%) and the last group represented by Germany where the three age groups contribute in a different way.

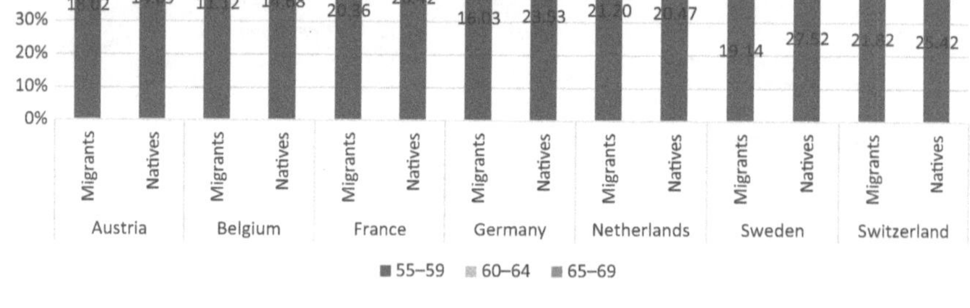

Fig. 11.2 Relative contribution of indicators to the first domain for migrants and natives

The different situation between migrants and natives is not surprising and can be read through the lens of empirical evidence which shows that older migrants are characterized by a higher occupational morbidity and mortality compared to natives, as they usually work in unsafe environments where the risk for health and safety is high (Ahonen et al. 2007; Schenker 2011). Thus, extending the working career in later life, especially for those whose working lives have been characterized by hard manual labour and who have more health problems, might not be considered as the desired outcome. Nevertheless, although older migrants might wish to retire, the combination of employment patterns and lower number of years of employment, and thus less contribution to pension schemes, might expose them to lack of entitlements for full pension rights and thus to risk of poverty in old age. And this might be the main reason for continuing to be active in the labour market in older age. This aspect requires however further detailed analysis for each country, as it is also context dependent.

11.3.2 Second Domain: Participation in Society

The second domain accounts for the contribution of old people in unpaid but still productive activities. It is the outcome of four individual indicators: voluntary activities, care to children and grandchildren, care to older adults and political participation.

The top performing countries in this domain are Sweden and Belgium for migrants and Belgium and the Netherlands for natives, whereas France and Austria are ranked in the last positions for both migrants and natives.

Natives have a higher score than migrants in almost all the countries, with the only exception represented by Sweden where migrants score slightly better. But whereas Sweden is the best performing country for migrants, it is not high ranking for natives. On the contrary, Belgium, although being second best performing country for migrants, has high potential for further improvements compared to natives, as the overall index for the migrant population is much lower (16.9 migrants in Sweden versus 21.53 natives in Belgium).

When looking at the relative contribution of each indicator in the overall domain score, we find significant difference between countries.

The most important indicator contributing to the overall domain is 'care to older adults' for both migrants and natives, partly due to the higher weight associated to this indicator. The maximum and minimum values are achieved in the Netherlands and France for natives and migrants, respectively (12.57 and 6.65). On the contrary, the least contributing indicator for all countries and for both groups is related to 'voluntary activities'.

If we compare the two high-ranking countries (Fig. 11.3), they show higher difference between migrants and natives in the 'care to (grand) children' indicator, being much more important for natives in Belgium and for migrants in Sweden (Fig. 11.4).

In France, Germany and Sweden, the indicator 'care to (grand)children' has a higher score for migrants than for natives. A possible interpretation of the different picture between the two groups might be related to migrants' stronger sense of intergenerational obligations and accordingly a stronger engagement in childcare (de Valk and Schans 2008; Fuligni and Pedersen 2002). Care provision depends also on public care services in each country. The case of Sweden is interesting in this respect, as here State provisions for childcare are higher compared to other EU countries (Zaidi et al. 2013, p. 29), so the differences between migrants and natives might be related to the different types of family engagement. In migrant

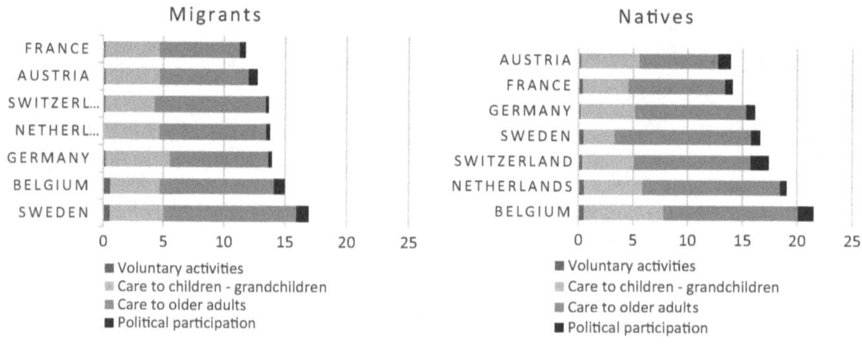

Fig. 11.3 Ranking of European countries according to the second domain for migrants and natives

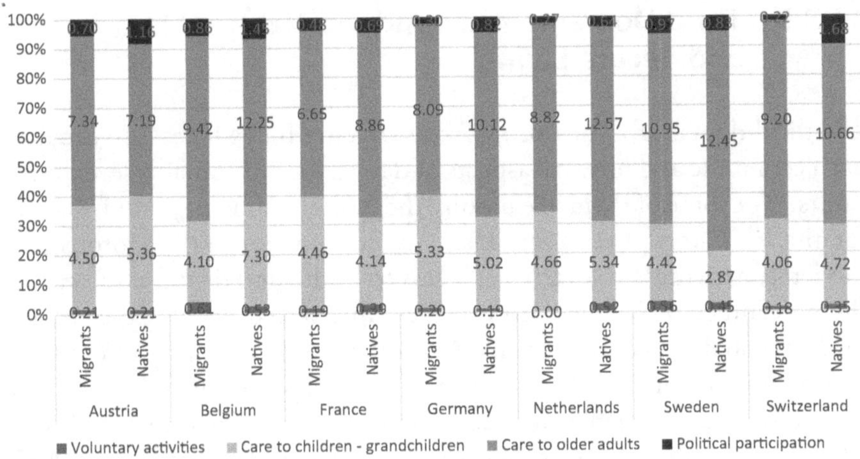

Fig. 11.4 Relative contribution of indicators to the second domain for migrants and natives

families, the cultural norm of being 'grannies' reinforces older migrants' sense of fulfilment as caregivers and is considered as a key activity (King et al. 2014).

Even though the worst ranking countries for migrants, France and Austria, present similar results for 'voluntary activities' and 'care to children and grandchildren' indicators, Austria is placed better because of the higher contribution of the 'care to older adults' indicator.

An interesting difference between countries and among groups is related to the 'political participation' indicator. Although it is the second least contributing indicator for both migrants and natives, the latter perform better in all countries with the only exception of Sweden. A reasonable explanation might be related to the fact that migrants do not have strong political participation first and foremost because they do not have voting rights in many countries (Groenendijk 2008). Moreover, spending a lot of time in caring duties (for children, grandchildren and/or other adults) means that older migrants might have less or no time at all to allocate to other outdoor activities, or might give less importance to those activities (Cela and Fokkema 2016).

11.3.3 Third Domain: Independent, Healthy and Secure Living

The third domain for measuring Active Ageing Index takes into account both economic and physical aspects of daily lives and contains seven indicators.[4] For some of them, we modify the AAI methodology. For instance, for the indicator 'physical exercise', we account for both vigorous and moderate activities. The variable 'access to health and dental care' reports the percentage of people who can face unexpected health expenses for medical and dental examinations. As for the financial security dimension, we only differ from the AAI in the 'no severe deprivation' indicator since we use only one variable (instead of the nine indicators used in AAI),[5] whereas the indicator 'no poverty risk' has been computed, as in AAI, accounting for the percentage of people whose income is above 50% of the national median equalized disposable income after social transfers (Table 11.2).

As Fig. 11.5 shows, the top performing countries in this domain are Austria and Switzerland for migrants and Sweden and Austria for natives, whereas in the last positions are ranked Belgium and Germany for migrants and Germany and France for natives. Natives and migrants

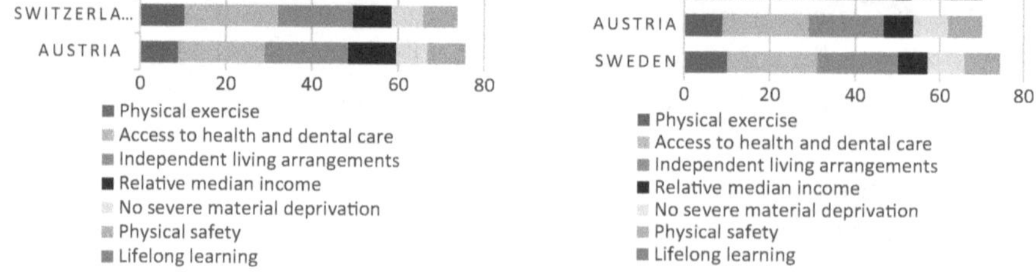

Fig. 11.5 Ranking of European countries according the third domain for migrants and natives

have similar scores in all the countries, with the only exception of Belgium where natives perform much better. The absence of striking differences between migrants and natives might indicate that the level of migrant's integration with respect to this domain is similar to that of the native population.

The top position for Austria and Sweden for both migrants and natives reflects their high performance in all the single indicators in the overall domain. If we look at the relative contribution of each indicator to the overall domain, we observe that the ability to access health care and independent living arrangements are the indicators that mainly contribute to the overall domain (both around 20% on average). The 'relative median income' variable accounts for the economic aspect of older people living. Migrants in Belgium achieve the worst performance compared both to natives in Belgium and to migrants and natives in other countries.

Although there are no striking differences between migrants and natives in this domain, it is worth pointing out some specifications of older migrants in relation to some indicators. For example, regarding the 'independent living arrangements' indicator, we think it is worth stressing that, as shown by previous research findings, migrants are more likely to co-reside with their adult children, compared to native peers, for several reasons, ranging from low financial independence to the high cultural endorsement of multigenerational households, which enable hands-on intergenerational support (Burr and Mutchler 1999; King et al. 2014). Co-residence, however, for older migrants, differently from the AAI approach, does not necessarily mean loss of independence; on the contrary, being embedded within the family might represent one of the main protective factors against social isolation in later life (Cela and Fokkema 2016). What really becomes a strong barrier to independence and engagement in outdoor activities in later life is represented by a poor financial situation (King et al. 2014) (Fig. 11.6).

Furthermore, being active through physical exercise, although an important way to cope with health decline in later life, might not be feasible for older migrants, due to different circumstances like lack of time, lack of personal and economic resources (e.g. language barriers, money to buy a bus ticket) or because they live in deprived areas with limited services (King et al. 2014).

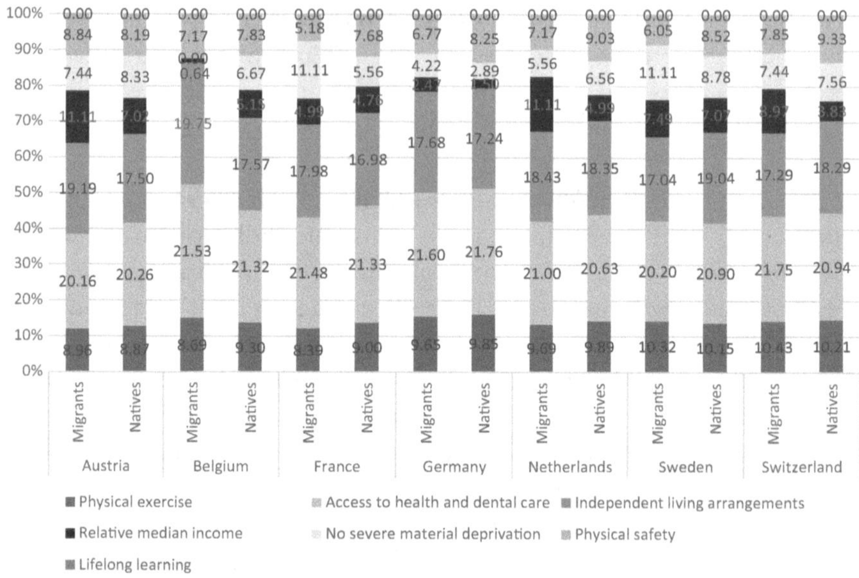

Fig. 11.6 Relative contribution of indicators to the third domain for migrants and natives

11.3.4 Fourth Domain: Capacity and Enabling Environment for Active Ageing

The fourth AAI domain accounts for six indicators that represent some pre-conditions of active ageing and other important factors that may contribute to being active in later life (Fig. 11.7).

Because of data unavailability, we reduced the indicators to four, namely, 'remaining life expectancy achievement at age 55', 'mental well-being', 'social connectedness' and 'educational attainment'. 'Life expectancy at age 60 years' (hereafter RLE60) was computed using data from the World Health Organization[6] for 2000 and it is the same for both natives and migrants.[7]

A further difference with the AAI methodology is the computation of the 'mental well-being' indicator that we define as the aggregation of five items (Table 11.2). Finally, to account for the percentage of old people with an upper secondary or tertiary educational attainment, since there is no harmonized definition of the education level among the different

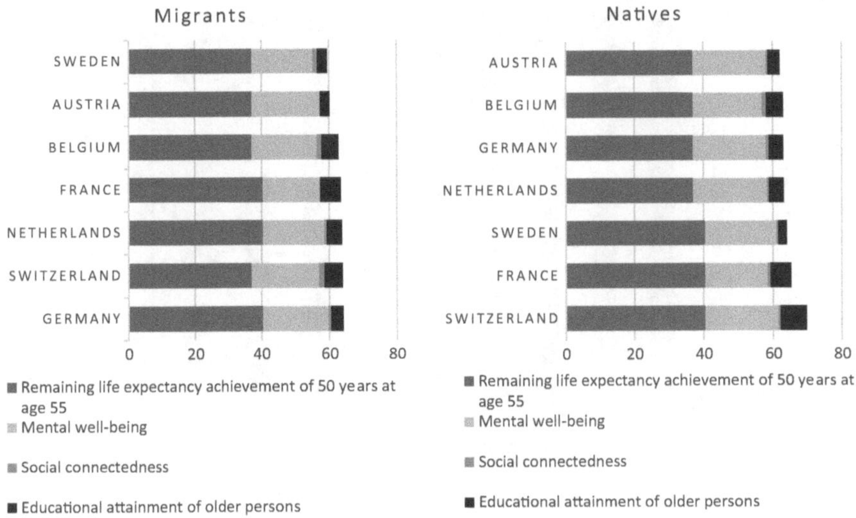

Fig. 11.7 Ranking of European countries according the forth domain for migrants and natives

countries, we use the International Standard Classification of Education (ISCED) code[8] computing the number of years of schooling instead of the level achieved.

The best performing countries are Germany and Switzerland for migrants and Switzerland and France for natives, whereas Austria ranks among the last positions for both groups. Germany in this domain is the best performing country for the migrant population, differently from the other domains where this country stands at the bottom of the country ranking (Fig. 11.8).

The contribution of 'social connectedness indicator' on the overall domain is almost null for both migrants and natives.

If we look at the single indicators, the most important one contributing to the overall domain is 'remaining life expectancy' for both migrants and natives, partly due to the higher weight associated to it. The maximum and minimum values are achieved in France and Switzerland for migrants and natives, respectively (73% and 57%, respectively, as displayed in the vertical axis). By contrast, the less contributing indicator for all countries and for both groups is related to 'social connectedness', with the lowest value reached by migrants in France (0.03%).

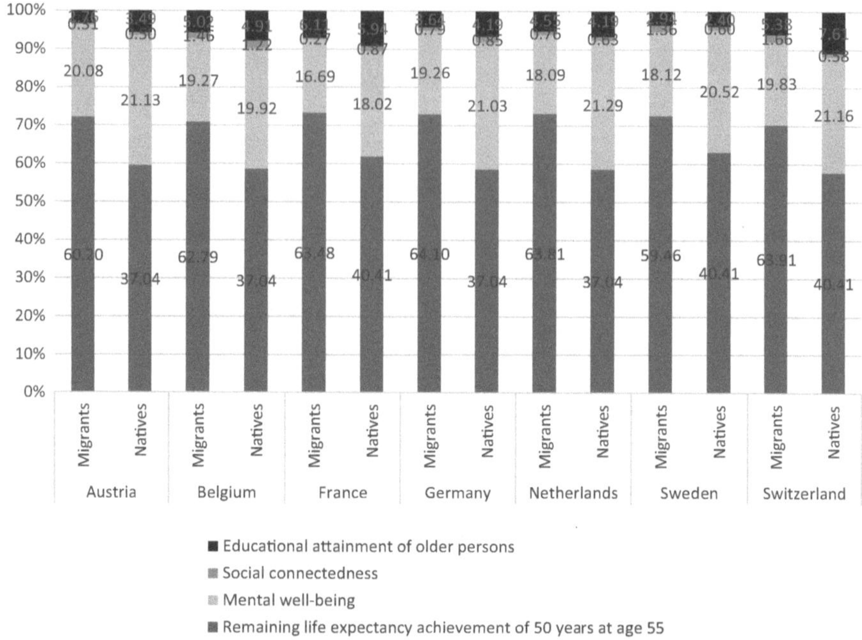

Fig. 11.8 Relative contribution of indicators to the forth domain for migrants and natives

11.3.5 Overall Index

The overall index is the result of the aggregation of the four domains, namely, 'employment rate', 'participation in society', 'independent healthy and secure living' and 'capacity and enabling for active ageing'.

Looking at the best performing countries for both migrants and natives (Fig. 11.9), it is worth noting that there is still room for further improvements; even Switzerland and Sweden, the front runners for both groups, have still a high potential for further improvement as they fall short by about 55% and 60%, respectively, from the most desired status possible. It is the migrant group, however, that falls behind natives in all the seven countries.

Looking at the contribution of each domain to the overall index (Fig. 11.10), the relative contribution of Domain 1 and Domain 3 is the highest for natives in Sweden, while for the same group in Switzerland,

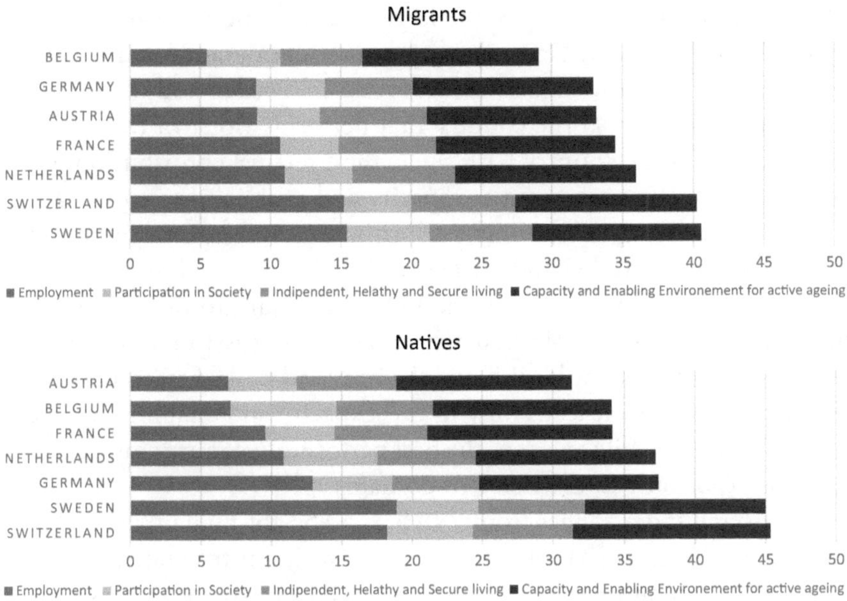

Fig. 11.9 Ranking of countries by the overall AAI for migrants and natives

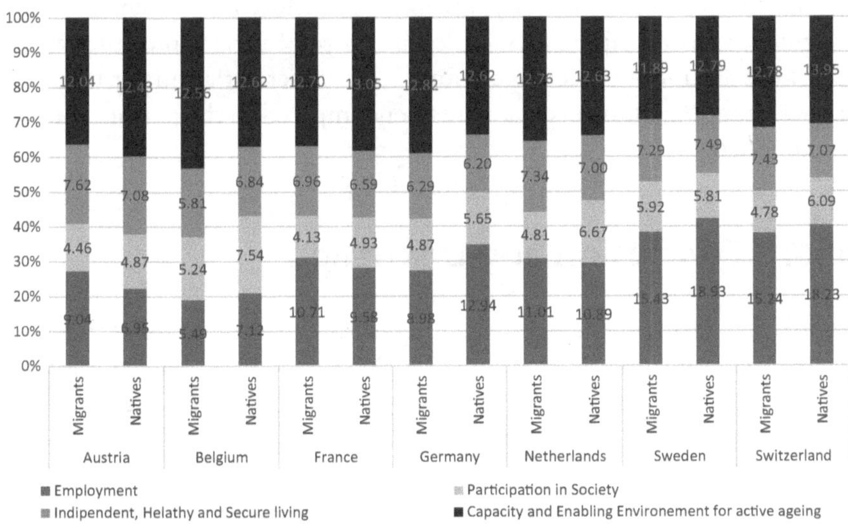

Fig. 11.10 Relative contribution of indicators to overall AAI for migrants and natives

Domain 2 and Domain 4 give the largest contribution; Belgium has the highest contribution of Domain 2. In the case of migrant population, the first Domain gives the highest contribution in Sweden and the lowest in Belgium, which is also the country that performs worst in this domain, while Austria is the country with the highest relative contribution in the third domain. For both groups, 'capacity and enabling for active ageing' is the domain that gives the higher contribution to the overall index, with the exception of Sweden and Switzerland.

The result on relative contribution of the domains to the overall index does not necessarily imply that the same ranking is found also within each domain: for example, although Switzerland ranks first in the overall domain for natives, it leads only in the 'capacity and enabling for active ageing' domain, while Sweden, the best performing country for natives in the AAI, leads only in the first two domains.

The overall index ranges from 29.10 to 40.53 for migrants in Belgium and Sweden, and from 31.34 to 45.34 for natives in Austria and Switzerland, with a variation range of 11.43 for migrants and 14.01 for natives.

If we drop out Austria, which is the least active country for natives, the range of variation of the overall AAI for natives (11.24) becomes more similar to the value of the AAI for migrants. Moreover, when dropping Belgium from the migrants' sample, the range of variation of the overall AAI loses around 4 percentage points. The higher variation for migrants suggests that Belgium acts as an outsider compared to the performance of the other countries.

11.4 Conclusion and Discussion

Population ageing is the distinguishing feature of demographic trends in developed countries. Over the last decades, ageing has involved migrants as well, with the result of an increasing complexity of the challenge for Western societies in terms of the ageing process.

In this chapter, we computed a modified version of the Active Ageing Index, for a specific target of the ageing population, namely, ageing migrants. Our aim was to stimulate the debate on the necessity to link ageing and migration concepts. In doing so, we computed the AAI for migrants and natives, using SHARE dataset in seven European countries.

Although we are aware of the limits of our analysis due to the small sample sizes of foreign-born population in SHARE, this dataset is, to the best of our knowledge, the only available instrument to account for over 50 migrant populations in Europe that allows cross-country comparisons.

Our results show that even the best performing countries for migrants still have high potential for further improvements compared to the situation of native population in the different domains and in the overall index. There are some significant differences between the two groups especially in the first two domains; regarding the first domain, they concern different working conditions and health, that might affect the desire to extending working career in later life. Migrants' participation in society (second domain) is strongly unbalanced towards private instead of public activities, as caring duties towards younger and older generations are much more important than voluntary activities and political participation and strongly endorsed in migrant families.

Our results show that similar values in one domain and in each single indicator in different countries do not necessarily imply the same level of being active in that domain or indicator, because different countries might attribute different values to them. We think this might be crucial when comparing different population groups. Indeed, when comparing migrants versus natives within the same country they show different patterns of being active, which could reflect social and cultural specificities of different groups of population. Accordingly, the choice of the weights for both indicators and domains should reflect these different propensities of being active. In addition, it would be interesting to monitor over time the performance of the same country. For this purpose, it might be useful to develop a country-specific AAI that takes into account specific factors with a system of weights that reflects the above differences. It could also be interesting to investigate the relationship between AAI and life satisfaction in order to assess if being active is positively valued by individuals themselves, as well as compute the AAI for different age groups. The latter might support policymakers to targeting different groups of older people with specific measures taking into account both socially defined needs and obligations and cultural awareness.

The neglected issue of ageing migrants in Europe has generated a lack of reliable and comparable data over time and across countries on the situation and characteristics of ageing migrants. While there are several databases on ageing, few of them are appropriate for analysing older migrants because of non-representative samples. Given the increasing number of ageing migrants in Europe, it is important to provide further research on the interplay between migration and ageing. For this purpose, we think it is necessary to make them 'visible' in statistics through both cross-country statistics that enable comparisons between countries and more systematic longitudinal data, which also capture naturalized migrants, allowing assessment of the long-term integration process within the destination countries.

Acknowledgements This chapter uses data from SHARE Waves 1 (DOIs: 10.6103/SHARE.w1.260). The SHARE data collection has been primarily funded by the European Commission through the FP5 (QLK6-CT-2001-00360), FP6 (SHARE-I3: RII-CT-2006-062193, COMPARE: CIT5-CT-2005-028857, SHARELIFE: CIT4-CT-2006-028812) and FP7 (SHARE-PREP: N°211909, SHARE-LEAP: N°227822, SHARE M4: N°261982). Additional funding from the German Ministry of Education and Research, the U.S. National Institute on Aging (U01_AG09740-13S2, P01_AG005842, P01_AG08291, P30_AG12815, R21_AG025169, Y1-AG-4553-01, IAG_BSR06-11, OGHA_04-064) and from various national funding sources is gratefully acknowledged (see www.share-project.org).

Appendix

Table 11.1 SHARE data, wave 1

Country	No. of observations	No. of immigrants	% immigrants
Austria	1594	139	8.72
Belgium	3827	270	7.06
Switzerland	1004	169	16.83
Germany	3008	558	18.55
Denmark	1707	65	3.81
Spain	2396	55	2.30
France	3193	503	15.75
Greece	2898	69	2.38
Italy	2559	38	1.48
Netherlands	2979	186	6.24
Sweden	3053	259	8.48
Total	28,218	2311	8.19

Table 11.2 AAI domains and correspondent variables in SHARE

AAI variable[a]		SHARE	Question	Notes
1.1	Employment rate for the age group 55–59	ep005_	Current job situation	
1.2	Employment rate for the age group 60–64			
1.3	Employment rate for the age group 65–69			
1.4	Employment rate for the age group 70–74			The variable has not been used although present in the dataset because of several missing data.
2.1	Voluntary activities	ac002d1	Done voluntary or charity work	
2.2	Care to children, grandchildren	sp014_	Look after grandchildren	
2.3	Care to older adults	sp008_	Did you give help to others outside the household	
2.4	Political participation	ac002d7	Taken part in a political or community-related organization	

(continued)

Table 11.2 (continued)

AAI variable[a]		SHARE	Question	Notes
3.1	Physical exercise	br015_	Sports or activities that are vigorous	
		br016_	Activities requiring a moderate level of energy	
3.2	Access to health and dental care	hc002_	How often seen or talked to medical doctor last 12 months	
		hc010_	Seen a dentist/ dental hygienist	
		hc040_	Forgo any types of care because of costs	
3.3	Independent living arrangements	ch007_	Where does children live	
3.4	Relative median income	ep041e1	Taken home from work before tax, (main) job	Computed as the ratio between the median equalized disposable income of people aged above 65 and the median equalized disposable income of those who are aged less than 65 years old.

(*continued*)

Table 11.2 (continued)

AAI variable[a]		SHARE	Question	Notes
3.5	No poverty risk	ep041e1		To compute the 50% of the national median equalized disposable income after social transfers, we use the "ilc_di04" variable collected by Eurostat (http://appsso.eurostat.ec.europa.eu/nui/show.do?dataset=ilc_di04&lang=en) in 2004 and 2005, depending on the year in which SHARE has been conducted. For countries for which we do not have data available for the specific year (i.e. Switzerland, Netherlands and Germany in 2004) we use the available years to construct the trend (year by year) and obtain the missing value.
3.6	No severe material deprivation	co007_	Is household able to make ends meet?	
3.7	Physical safety			No data available in SHARE
3.8	Lifelong learning	ac002d4	Attended an educational or training course	
4.1	Remaining life expectancy achievement of 50 years at age 55	No data available in SHARE		Life expectancy at age 60 http://apps.who.int/gho/data/node.main.688?lang=en

(continued)

Table 11.2 (continued)

AAI variable[a]		SHARE	Question	Notes
4.2	Share of healthy life years in the remaining life expectancy at age 55			No data available in SHARE
4.3	Mental well-being	mh002_	Sad or depressed last month	
		mh010_	Irritability	
		mh013_	Fatigue	
		mh007_	Trouble sleeping	
		mh008_	Less or same interest in things	
4.4	Use of ICT			No data available in SHARE

(continued)

Table 11.2 (continued)

AAI variable[a]		SHARE	Question	Notes
4.5	Social connectedness	ac002d1	Done voluntary or charity work	We define as active an individual that perform at least once a week two of these activities.
		ac002d2	Cared for a sick or disabled adult	
		ac002d3	Provided help to family, friends or neighbours	
		ac002d5	Gone to a sport, social or other kind of club	
		ac002d6	Taken part in a religious organization (church, synagogue, mosque etc.)	
		ac002d7	Taken part in a political or community-related organization	
4.6	Educational attainment of older persons	ac002d4	Attended an educational or training course	

[a]For a detailed description of the variables used in the official AAI see Zaidi et al. (2013)

Table 11.3 Weights

Domain	Weight of the domain in the overall AAI	Variable	Weight of the variable
1. Employment	7/20	Employment rate for the age group 55–59	1/3
		Employment rate for the age group 60–64	1/3
		Employment rate for the age group 64–69	1/3
2. Participation in society	7/20	Voluntary activities	1/4
		Care to children, grandchildren	1/4
		Care to older adults	3/10
		Political participation	1/5
3. Independent healthy and secure living	1/10	Physical exercise	1/9
		Access to health and dental care	2/9
		Independent living arrangements	2/9
		Relative median income	1/9
		No severe material deprivation	1/9
		Physical safety	1/9
		Lifelong learning	1/9
4. Capacity and enabling environment for active aging	2/10	Remaining life expectancy achievement of 50 years at age 55	33/70
		Mental well-being	17/70
		Social connectedness	1/10
		Educational attainment of older persons	13/70

Notes

1. We focus on the first wave since countries in this wave present the highest percentage of non-native respondents. In the appendix (Table 11.2), we report an exhaustive description of the variables used in our analysis as coded in the SHARE Manual.
2. In other words, for a given indicator, the value achieved for a single unit of analysis was multiplied by SHARE's sampling weights, that is the number of people that each individual represents.

3. We don't account for the last age group 70–74 years old, because of the absence of observations for some countries.
4. The results for migrants in Belgium are biased for the missing value in the indicator 'no severe material deprivation'.
5. This is mainly due to the lack of such information in wave 1 of SHARE that does not collect information on individuals or households that cannot afford certain basic needs such as the possession of a television, car, telephone, washing machine and so on. Accordingly, material deprivation is computed only taking into account the ability of a household to make ends meet.
6. Data are available at: http://apps.who.int/gho/data/node.main.688
7. Values are standardized using the max-min method.
8. See http://www.uis.unesco.org/Education/Pages/international-standard-classification-of-education.aspx for a definition of ISCED code and its comparability across countries.

References

Ahonen, E. Q., Benavides, F. G., & Benach, J. (2007). Immigrant populations, work and health—A systematic literature review. *Scandinavian Journal of Work Environment & Health, 33*(2), 96–104.

Ajrouch, K. J. (2005). Arab-American immigrant elders' views about social support. *Ageing & Society, 25*(5), 655–673.

Baldassar, L., & Merla, L. (Eds.). (2013). *Transnational families, migration and the circulation of care*. London: Routledge.

Burr, J. A., & Mutchler, J. E. (1999). Race and ethnic variation in norms of filial responsibility among older persons. *Journal of Marriage and the Family, 61*(3), 674–687.

Cela, E., & Fokkema, T. (2016). How to cope with loneliness in a foreign country? A qualitative study among Albanian and Moroccan older migrants in Italy. *Ageing & Society*. https://doi.org/10.1017/S0144686X16000209.

Darmon, N., & Khlat, M. (2001). An overview of the health status of migrants in France, in relation to their dietary practices. *Public Health Nutrition, 4*(2), 163–172.

De Valk, H., & Schans, D. (2008). 'They ought to do this for their parents': Perceptions of filial obligations among immigrant and Dutch older people. *Ageing & Society, 28*(1), 49–66.

European Commission. (2014). *Population ageing in Europe. Facts, implications and policies* [Online]. Retrieved from https://ec.europa.eu/research/social-sciences/pdf/policy_reviews/kina26426enc.pdf

Fokkema, T., & Naderi, R. (2013). Differences in late-life loneliness: A comparison between Turkish and native-born older adults in Germany. *European Journal of Ageing, 10*(4), 289–300.

Fokkema, T., Cela, E., & Witter, Y. (2016). Pendular migration of the older first generations in Europe: Misconceptions and nuances. In C. Schweppe & V. Horn (Eds.), *Transnational aging: Current insights and future challenges* (pp. 141–162). New York: Routledge.

Fuligni, A. J., & Pedersen, S. (2002). Family obligation and the transition to young adulthood. *Developmental Psychology, 38*(5), 856–868.

Groenendijk, K. (2008). *Local voting rights for non nationals in Europe: What we know and what we need to learn* [Online]. Washington, DC: Migration Policy Institute. Retrieved from http://www.migrationpolicy.org/research/local-voting-rights-non-nationals-europe-what-we-know-and-what-we-need-learn

King, R., Cela, E., Fokkema, T., & Vullneatri, J. (2014). The migration and wellbeing of the zero generation: Transgenerational care, grandparenting and loneliness amongst Albanian older people. *Population, Space and Place, 20*(8), 728–738.

Schenker, M. (2011). *Migration and occupational health: Understanding the risks* [Online]. Migration Information Source. Retrieved from http://www.migrationpolicy.org/article/migration-and-occupational-health-understanding-risks

Torres, S. (2001). Understandings of successful aging in the context of migration: The case of Iranian immigrants to Sweden. *Ageing & Society, 21*(3), 333–355.

United Nations. (2013). *The number of international migrants worldwide reaches 232 million*. Population Facts No. 2013/2 [Online]. Retrieved from http://esa.un.org/unmigration/documents/the_number_of_international_migrants.pdf.

United Nations. (2015). The 2015 revision, key findings and advance tables. *Working Paper No. ESA/P/WP.241* [Online]. Retrieved from http://esa.un.org/unpd/wpp/publications/files/key_findings_wpp_2015.pdf

White, P. (2006). Migrant populations approaching old age: Prospects in Europe. *Journal of Ethnic and Migration Studies, 32*(8), 1283–1300.

Zaidi, A., Gasior, K., Hofmarcher, M. M., Lelkes, O., Marin, B., Rodrigues, R., Schmidt, A., Vanhuysse, P., & Zolyomi, E. (2013). *Active Ageing Index 2012: Concept, methodology and final results*. EC/UNECE, Active Ageing Index

Project, UNECE Grant ECE/GC/2012/003. Vienna: European Centre for Social Welfare Policy and Research. Retrieved from http://www.euro.centre.org/data/aai/1253897823_70974.pdf

Eralba Cela is a Postdoctoral Researcher in Demography at the Department of Economics and Social Sciences, Università Politecnica delle Marche, Italy. Her research interests focus on migration, remittances, transnational families and intergenerational relations, ageing and well-being. She has served as an expert on migration for organizations such as ILO-International Labour Organisation and Centre for Economic and Social Studies (Albania).

Mariateresa Ciommi is a Postdoctoral Researcher in Economic Statistics at the Department of Economics and Social Sciences, Università Politecnica delle Marche, Italy. Her area of research interest includes the measurement of diversity, inequality, poverty and welfare economics. Her current research is directed on the analysis of multidimensional well-being with a focus on the construction of composite indicators, from both theoretical and empirical perspectives.

12

Estimating Weights for the Active Ageing Index (AAI) from Stated Preferences: Proposal for a Discrete Choice Experiment (DCE)

Christian Ernst Heinrich Boehler, Timea Mariann Helter, Ibrahim-Kholilul Rohman, and Fabienne Abadie

12.1 Introduction

Demographic change is a major societal challenge for EU Member States, which may increase demand for health and care services and socioeconomic inequalities (WHO 2002; EC 2012). In this context, *Active and*

The views expressed in this chapter are those of the authors and may not in any circumstances be regarded as stating an official position of the European Commission. The authors would like to thank Csaba Kucsera, Rene van Bavel and Ioannis Maghiros for their contributions and take full responsibility for all remaining errors and shortcomings of this chapter.

C.E.H. Boehler (✉)
European Centre for Social Welfare Policy and Research, Vienna, Austria

T.M. Helter
Main Association of Austrian Social Security Institutions, Vienna, Austria

I.-K. Rohman
Chalmers University of Technology, Göteborg, Sweden

F. Abadie
DG JRC-IPTS, European Commission, Sevilla, Spain

A. Zaidi (eds.), *Building Evidence for Active Ageing Policies*,
https://doi.org/10.1007/978-981-10-6017-5_12

Healthy Ageing (AHA) has been identified as a EU policy priority, and an opportunity to strengthen the sustainability of health and care systems and improve the health of older people, allowing them to participate in employment and society (EC 2006). Many initiatives on Active and Healthy Ageing are underway across Europe, most prominently the *European Innovation Partnership on Active and Healthy Ageing* (EIP on AHA) (EC 2011), and several concepts are being developed to assess active ageing policies and interventions. For instance, the monitoring framework for the EIP on AHA seeks to extrapolate outcomes from individual interventions in terms of *health-related quality of life* and *health and care expenditure* in a bottom-up approach (Boehler and Abadie 2015; Boehler et al. 2015). Another concept, which incorporates information from macro-level indicators drawn from comparative international datasets, is the *Active Ageing Index* (AAI) (Zaidi et al. 2013; Zaidi and Stanton 2015).

The AAI is a composite indicator reflecting the multidimensional character of active ageing (Zaidi et al. 2013). It combines information on 22 indicators drawn from the EU-SILC (Survey of Income and Living Conditions), the EU-LFS (Labour Force Survey) and the European Quality of Life Survey (EQLS). These are grouped in four domains related to healthy and independent living and more active participation in employment and society (Zaidi et al. 2013). The AAI indicators and domains are linearly aggregated using constant weights elicited from a group of experts. The AAI-Expert Group (2014, p. 4) recently 'expressed their concerns about the arbitrary weighting of indicators and domains'.

This chapter discusses the opportunity to further AAI methods regarding the weighting issue and suggests an approach for basing complementary weight sets on stated preferences. We propose using Discrete Choice Experiments (DCEs) and outline how this could be implemented across different population subgroups and geographic contexts. DCEs 'force respondents to make trade-offs among different choice sets, unlike other methods such as ranking or rating' (Wong et al. 2014, p. 2), which may provide more credible information on the relative importance of each indicator in the Index. The chapter addresses the relevance and feasibility of DCEs to define alternative weights for the AAI, but does not report actual DCE results.

The next section reviews the existing weighting system and introduces DCEs as a promising method to estimate complementary, preference-based AAI weights. Section 12.3 outlines the design of a potential DCE to estimate such preference-based weights. In the discussion, we consider the benefits of increasing the flexibility of the AAI as a tool for policy evaluation, by providing complementary weight sets suitable for subgroup, scenario and sensitivity analyses of AAI results. We further discuss the validity of stated preferences, especially when respondents are not active ageing policy experts (e.g. general population). In this context, we stress the importance of an underlying theoretical construct for empirically estimating weights and discuss factors influencing the resources required for conducting the proposed DCE.

12.2 Estimating Preference-based Weights for the AAI

Composite indicator weights constitute value judgements and many statistical or participatory methods allow their estimation (OECD/JRC 2005). They should be derived based on an underlying theoretical framework, and both 'correlation and compensability issues among indicators need to be considered and either be corrected for or treated as features of the phenomenon that need to be retained in the analysis' (OECD/JRC 2005, p. 15). The current AAI weight set is particularly useful as expert opinion is essential to ensure adequate consideration of existing policy priorities, theoretical factors or budget constraints (OECD/JRC 2005). Further, transparency and comparability are key features of composite indicators (OECD/JRC 2005), which make constant weights particularly appealing for comparative country analysis, a key objective of the AAI (Zaidi et al. 2013; Zaidi and Stanton 2015). However, expert-based weights may induce bias and not reflect actual preferences of the target population, whilst constant weights may sometimes provide adverse incentives for policymakers (OECD/JRC 2005; Sharpe and Andrews 2012). Such concerns were recently raised about the AAI (AAI-Expert Group 2014).

For instance, the current AAI weighting system does not consider opportunity costs and marginal substitution rates between indicators. However, according to consumer theory, the marginal utility from increasing the score of an indicator should be higher if its current score is low (Fig. 12.1) (OECD/JRC 2005). Hence, if a particular AAI indicator has a low value, compensation through other indicators should be more difficult (OECD/JRC 2005). Moreover, if the optimal score of an indicator is below 100%, an increase beyond the optimum should be counterbalanced through negative weights (Fig. 12.1). Constant AAI weights may therefore lead policymakers to maximise scores for indicators and domains which are comparatively easier to increase. Further, AAI results based on constant weights ignore distributional aspects and may underestimate the burden associated with low scores for a particular indicator in disadvantaged populations, whilst overestimating it in others. Preference-based weights could therefore help provide greater incentives to address problems in underserved populations (Fig. 12.1). Finally, expert-based constant weights have no theoretical foundation, and may not reflect what people actually *want* in different social, cultural or geographic contexts. The importance of citizens' voice in public policy and planning being increasingly recognised (Cogan et al. 1986), citizens' participation should be intrinsic to active ageing policies, and the AAI could

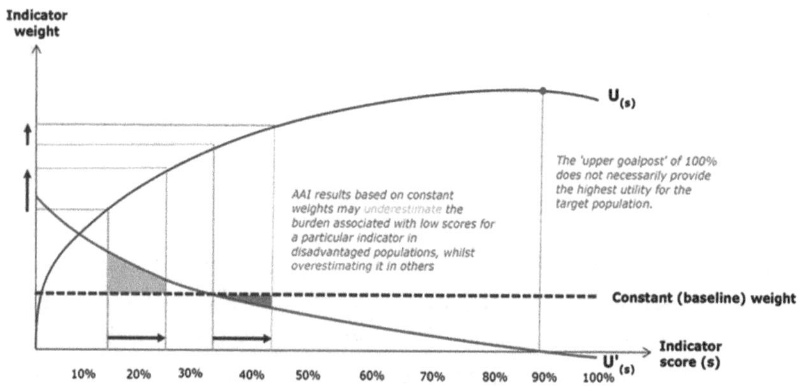

The population utility for an indicator or domain of the AAI is represented by U(s). The utility gain from increasing the indicator score by the same amount would be higher at lower absolute indicator scores, as indicated by the blue arrows. Likewise, compensation through other indicators would be more difficult if the absolute indicator score is low so that the positive impact on the overall index is higher when improving on policy areas which are particularly underserved. The weight for the indicator would be estimated from the derivative of the utility function U(s), the marginal utility function U'(s). The score of an indicator that provides the highest utility does not necessarily match the upper goalpost of 100%, so that increasing the indicator score beyond the maximum should even be weighted negatively.

Source: own drawing

Fig. 12.1 Preference-based weights reflecting diminishing marginal utility of increasing indicator scores

gain from reflecting public preferences. Decision makers should then ensure that public participation leads to better planning and prioritisation, safeguarding citizens' interest.

The key issue is how to empirically estimate preference-based AAI weights. DCEs seem promising as a method to improve the evaluation of active and healthy ageing policies and interventions. The concept is based upon Random Utility Theory and the assumptions that (Lancaster 1966, p. 134):

- 'The good, per se, does not give utility to the consumer; it possesses characteristics, and these characteristics give rise to utility': The AAI components (characteristics) contribute to the utility realised by members of the target population of active ageing policies and interventions (the good).
- 'In general, a good will possess more than one characteristic, and many characteristics will be shared by more than one good': The AAI comprises several component-indicators, and different active ageing policies and interventions may affect different subsets of components.
- 'Goods in combination may possess characteristics different from those pertaining to the goods separately': The utility of the target population derived from policies impacting several AAI components may differ from the sum of the utilities influenced by policies and interventions targeting each component separately.

In DCEs, respondents are given hypothetical scenarios representing competing alternatives (here active ageing policies or interventions) whose attributes (potential determinants of active and healthy ageing) vary (Lancsar and Louviere 2008). Their purpose is to estimate a utility function of the type (Amaya-Amaya et al. 2008):

$$U_{in} = ASC_i + \beta_1 X_{i1} + \beta_2 X_{i2} + \ldots + \beta_k X_{ik} + \varepsilon_{in}$$

where 'ASC_i' is the 'alternative specific constant', 'k' (=1, 2, ... K) represents the number of attributes with coefficients 'β_k' to be estimated across alternatives 'i' and respondents 'n', and 'ε' represents unmeasured random variation in respondents' preferences for different alternatives

(Amaya-Amaya et al. 2008). The information obtained could help estimate AAI weights, especially (Ryan and Farrar 2000):

- The *direction of the effect* of a change in attribute levels, for example, whether respondents prefer an increase in the use of ICT over a reduction
- The *relative importance* of one attribute over another, for example, whether improving access to health services is valued higher than greater political participation
- The *trade-off or marginal substitution rate* between attributes, that is, to what extent people would accept decreasing one attribute (e.g. employment) to increase another one (e.g. financial (in)security) at different attribute levels

DCEs can potentially reveal how attributes influence respondents' choice or decision (Scott 2002), and what utilities they attach to different levels of achievement for each attribute (Lancsar and Louviere 2008; Mangham et al. 2009). DCEs are therefore promising for estimating AAI weights as the value of active ageing to an individual can be defined by a set of attributes reflecting the multidimensional character of active ageing and the composite character of the Index.

12.3 Designing a DCE to Estimate Preference-based AAI Weights

Several authors described the steps for developing DCEs (Fig. 12.2), and unanimously highlighted that decisions taken in each stage may have downstream implications for subsequent stages of the experiment (e.g. Amaya-Amaya et al. 2008; Viney et al. 2002; Bridges et al. 2011; Johnson et al. 2013). In this section, we discuss the key elements for a DCE to estimate preference-based AAI weights.

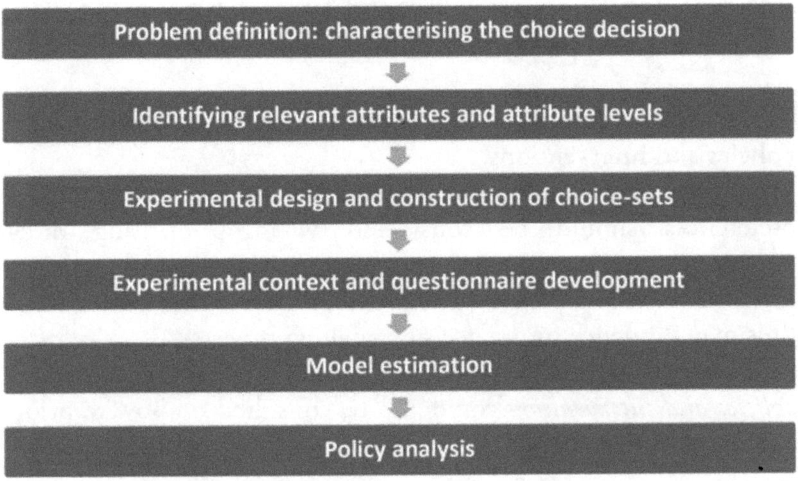

Fig. 12.2 Key stages for developing a Discrete Choice Experiment (adapted from Amaya-Amaya et al. 2008)

12.3.1 Problem Definition: Characterising the Choice Decision

Developing a DCE starts with considering explicitly how individuals make decisions about the problem in question and which dimensions driving the choice process to include as attributes (Amaya-Amaya et al. 2008; Bridges et al. 2011). The 22 indicators composing the AAI pre-determine our list of potential attributes for a DCE as we assume that they provide a meaningful, measurable and comprehensive list of active ageing determinants.

Another issue regarding the choice decision is to 'identify the possible choosers' and potential sources of variability across individuals, such as gender or socioeconomic characteristics, which could become explanatories for behavioural differences (Amaya-Amaya et al. 2008). Regarding potential choosers, several sources of opinions could help ascertain which outcomes of active ageing policies and interventions matter to people (e.g. Dolan 1999; Shumway et al. 2003; Ubel et al. 2003; Watson et al. 2012). For instance, priorities could be elicited from:

- The *general public* whose tax contributions provide funding for such policies and interventions.
- The *ageing population* being recipients of active ageing policies, their preferences should be considered when prioritising different outcomes.
- *Policy planners and policymakers* are responsible for policy setting and thus may influence future decision-making processes.
- As with the current weighing methodology, *experts on active ageing policies and interventions* could also be consulted on how to prioritise outcomes in various policy decisions.

12.3.2 Identifying Attributes and Attribute Levels

Various methods have been used or suggested for identifying attributes for DCEs. Helter and Boehler (2016) distinguish four general stages of attribute development: raw data collection, data reduction, removing inappropriate attributes and wording.

Though the list of potential attributes derives from the 22 indicators composing the index, further refinement may be required. Whilst all attributes influencing an individual's decision should be included (Coast et al. 2012), DCEs should generally have a maximum of six or seven attributes to reduce the cognitive burden on respondents (e.g. Ryan and Gerard 2003). Further, attributes should be salient, plausible and capable of being traded (Ryan 1999); they should not be too close to the latent construct investigated or dominant for a decision; and they should not be intrinsic to a person's personality and experimentally 'manipulable' by the intervention (Coast et al. 2012).

To reduce the number of attributes, some AAI indicators like employment rates for different age cohorts could be merged, and differential weights assigned according to the relative importance different respondents (e.g. from different age cohorts) place on employment. Further, 'relative mean income', 'poverty risk' and 'severe material deprivation' could merge into 'financial security' as their correlation is probably high and their capability of being traded against each other doubtful. Attributes could also be divided into separate *blocks* and different choice sets presented to different respondents (Viney et al. 2002; Street et al. 2008), as discussed below.

Equally important for the proposed DCE are the attribute levels (Lancsar and Louviere 2008). Levels may be categorical (e.g. degree of physical safety) or continuous (e.g. relative median income), and attributes can have different numbers of levels (WHO 2012; Huber and Zwerina 1996). Indeed, attribute levels are particularly important as the current AAI implies a linear and flat relationship between indicator scores and active ageing outcomes, as Fig. 12.1 shows (Zaidi et al. 2013; Zaidi and Stanton 2015). However, whilst a greater number of levels increases the ability to detect non-linear marginal utility relationships, so does complexity as possible attribute-level combinations increase exponentially, which impacts the experimental design of the DCE (Huber and Zwerina 1996).

12.3.3 Experimental Design and Construction of Choice Sets

This is about combining attributes and levels into different alternatives (e.g. potential outcomes of competing policy interventions) and choice sets (of at least two competing alternatives) to be presented to respondents (Ryan and Farrar 2000). Figure 12.3 shows an exemplary choice question for indicators of the second AAI domain.

Johnson et al. (2013, p. 5) state that 'the experimental-design step consists of defining a systematic plan that determines the content of the choice questions to generate the variation in the attribute levels required to elicit a choice response'. The aim is to select, among all possible attribute-level combinations (or 'full factorial design'), a subset providing the best balance between obtaining reliable parameter estimates and limitations to the number of potential alternatives (Mangham et al. 2009). A full factorial design would provide respondents with too many choice sets and pose an unmanageable cognitive burden (Street et al. 2008; Louviere

Below are the outcomes of two competing policy interventions to improve participation in society. Which alternative would you prefer depending on the current score of each indicator?

Attributes (indicators)	Policy intervention A	Policy intervention B
Voluntary activities	10% up	unchanged
Care to children, grandchildren	5% up	5% up
Care to older adults	unchanged	unchanged
Political participation	unchanged	10% down

Note that this example has been constructed for demonstrative purposes only. Different level definitions and customisation to different populations or geographic contexts (e.g. by linking attribute levels to current indicator scores within each geographic setting), is possible for different attributes.

Fig. 12.3 Potential choice question for participation in society

et al. 2000). Amaya-Amaya et al. (2008, p. 19) suggest that 'the number of combinations chosen should be equal or greater than the number of parameters the analyst is likely to estimate from the data collected'. Further, the number and types of parameters to be estimated depend on the model specification, the number of attributes and levels, and the functional form of the attributes (Johnson et al. 2013).

The AAI structure with indicators grouped in four domains of active and healthy ageing suits an experimental design which assigns respondents randomly to different blocks of choice questions, as depicted in Fig. 12.4 (Viney et al. 2002; Street et al. 2008): choice sets in each block would comprise attributes from each AAI domain and across domains, respectively. The weights '$w(s,x)_{i,j}$' for each indicator (i) in block (j) depend on the marginal rate of substitution at different indicator scores (s) and a vector of covariates (x) denoting relevant population characteristics (e.g. age or gender). A blocked design may reduce the complexity of the choice task and the cognitive burden on respondents whilst providing a full preference-based weight set for the Index.

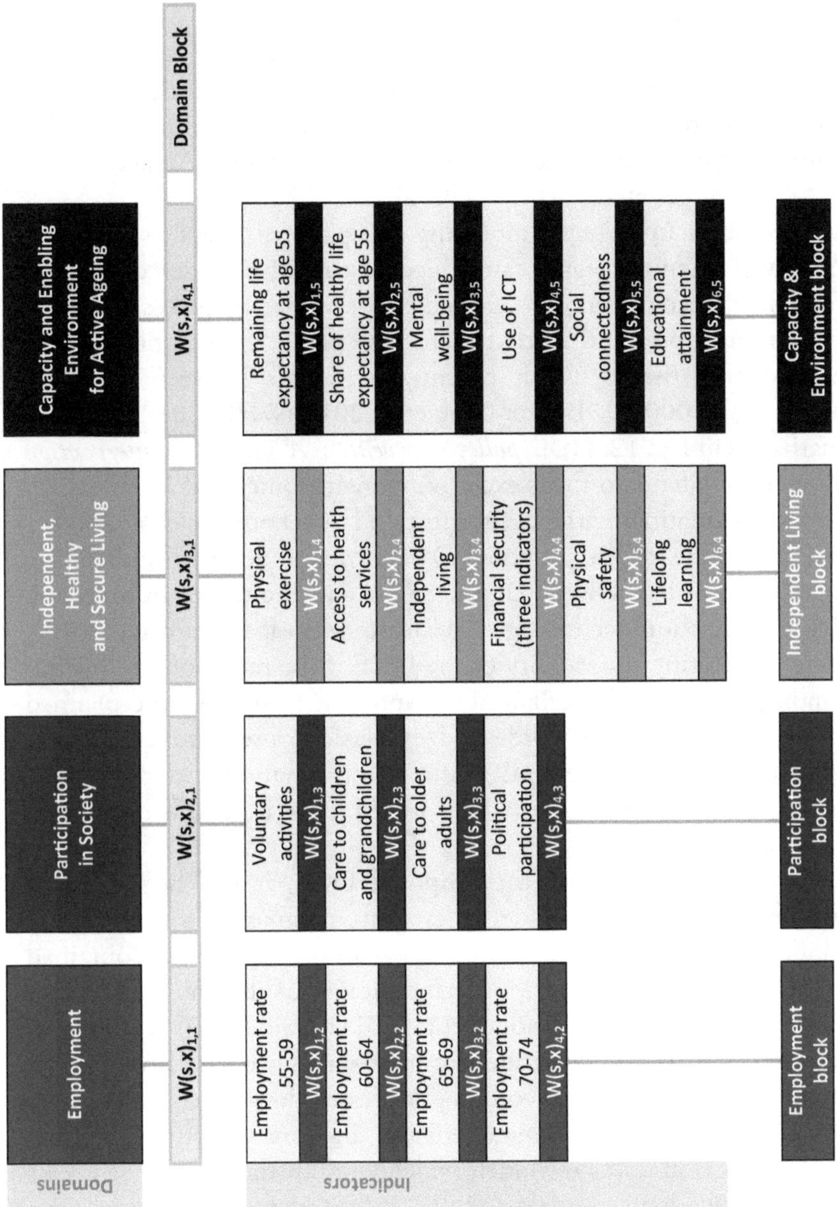

Fig. 12.4 Blocked DCE-design to estimate preference-based AAI weights. Source: Own drawing adapted from Active Ageing Index Project: http://www1.unece.org/stat/platform/display/AAI/Active+Ageing+Index+Home

12.3.4 Questionnaire Development and Model Estimation

The labelling of alternatives, the administration mode and the quantitative modelling methods to estimate indicator weights from the choice responses are some additional issues for consideration (e.g. Amaya-Amaya et al. 2008; Bridges et al. 2011; WHO 2012).

Labelled alternatives (e.g. 'improving occupational safety' vs. 'improving quality of part-time work') can reduce the cognitive burden on respondents as the choice presented is clearer. However, the emphasis may shift away from attributes and levels towards the policy interventions themselves, reducing the focus on potential trade-offs between attributes (Louviere et al. 2000; De Bekker Grob et al. 2010; WHO 2012). Generic alternatives as in Fig. 12.3 (i.e. '*policy intervention A*' vs. '*policy intervention B*') allow respondents to focus exclusively on attributes and levels, which matters for estimating marginal substitution rates (Louviere et al. 2000).

Also important for the proposed DCE is the mode of administration (Bridges et al. 2011). Digital competencies and the use of information and communication technologies decrease with age (Eurostat 2015), potentially affecting the validity of the DCE if the mode of administration induces selection bias. It is also important to pre-test the planned DCE to ensure that the words and expressions used are understood unequivocally by everyone, and to obtain information about potential correlations between attributes which would make trade-offs impossible (Coast et al. 2012).

Finally, when conducting the proposed DCE, it will be crucial to determine an appropriate model to estimate parameters for measuring the random utility between outcome-alternatives from the data obtained (WHO 2012). The cross-sectional panel design of the choice data and within-subject correlation (Bridges et al. 2011; Train 2009) require particular attention. Further, preferences may vary systematically between population subgroups, which could be addressed through covariates (e.g. gender and socioeconomic characteristics) and interactions with attributes (Bridges et al. 2011). Models widely used in the context of DCEs are multinomial logit models (MNL) (McFadden 1974) which can account for both the panel structure of the data and potential preference variation (Bridges et al. 2011; Train 2009). A whole body of literature exists around model estimation for DCEs and available options to relax the assumptions of the MNL (Train 2009).

12.4 Discussion and Conclusion

The aim of the AAI is to produce high-quality, independent evidence to evaluate and compare outcomes from a diverse range of policies targeting active and healthy ageing. We believe that the AAI can achieve this but hope to contribute to its further development by proposing DCEs for estimating complementary weight sets. Indeed, we do not advocate replacing the existing expert-based and constant weights as expert opinion is essential for policy evaluation, and constant weights ensure transparency and comparability, particularly for cross-country analysis. Our intention is rather to provide additional options for conducting policy-relevant analyses through the AAI, thereby improving the evidence base upon which to evaluate active ageing policies. A DCE could provide alternative weights from stated preferences which also recognise potential trade-offs between the indicators and domains of the Index. This could deepen our understanding of the potential impact of alternative active ageing policies in various contexts and further enhance the acceptance of the Index through citizens' participation.

Participatory methods for eliciting composite indicator weights help incorporating the views of various stakeholders, such as the general population, the target cohort of active ageing policies, policymakers, or active and healthy ageing experts. The methodological literature for composite indicator development suggests a variety of participatory methods for weight development, including among others budget allocation process, analytic hierarchy process and conjoint analysis (CA) (OECD/JRC 2005). DCEs are often called a variant of CA as both methods confront individuals with choices between alternative goods and services that vary along several attributes and levels (Louviere et al. 2010). However, DCEs are, unlike traditional CA, thoroughly rooted in Random Utility Theory (Lancaster 1966), which assumes that *utility* is a latent construct in a person's mind that cannot be observed by the researcher (Louviere et al. 2010).

A potential objection to our approach relates to whether survey-based estimates can provide relevant and meaningful information for estimating AAI weights, especially when elicited from lay people rather than experts of active ageing policies. Survey-based preference estimates may raise scepticism when results 'seem inconsistent with economic intuition' and may be 'interpreted as evidence of (a) the hypothetical nature of the question, (b) strategic behaviour, or (c) preferences which are either ill-defined or inconsistent with economic theory' (Carson and Groves 2007, p. 182). In other words, respondents do not always answer a question truthfully, and neither do they always understand the question posed correctly (Carson and Groves 2007). This reinforces the importance of an underlying theoretical construct. Random Utility Theory assumes that utilities consist of a systematic (explainable) component and a random (unexplainable) component which comprises all unidentified factors that impact individual choices (Louviere et al. 2010). It further assumes that individuals are 'imperfect measurement devices', hence, the random component may reflect 'variability and differences in choices associated with individuals and not choice options per se' (Louviere et al. 2010, p. 62.) We can therefore predict the probability that an individual chooses one alternative over another, but never the exact alternative an individual *will* choose (Louviere et al. 2010). We can investigate different types of information or strategic incentives used in DCEs and their impact on respective responses so as to assess changes in consumer behaviour in response to changes in the choice context (Louviere et al. 2010). We can also externally validate trade-offs made in DCEs through comparison with trade-offs observed elsewhere (such as real market behaviour) (Louviere et al. 2010; Carson and Groves 2007). Hence, DCEs may not only allow assessing the variability in choices through covariates representing differences in individual choosers (which would matter for subgroup analysis of AAI weights), but also potential causes of bias in choice responses (Louviere et al. 2010).

Finally, it is also important to consider key elements of the study design likely to drive the cost of implementing the proposed DCE, such as the size of the sample required for estimating preference-based AAI weights that allow for both marginal substitution rates between indicators and disaggregation of subgroups and geographic contexts. It is difficult to estimate ad hoc the sample size required as it depends on several issues discussed here: such as the question format, the complexity of the choice tasks, the desired degree of precision, heterogeneity in the target population and the need for subgroup analysis (Bridges et al. 2011; Louviere et al. 2000). Methods are available for sample size estimation (Louviere et al. 2000) and the overall population size actually plays a relatively small role in these calculations (Amaya-Amaya et al. 2008). The mean sample size for health-related DCEs published between 2005 and 2008 was 259 (Marshall et al. 2010), and Johnson et al. (2013) state that 'sample sizes in the range of 1000 to 2000 will produce small confidence intervals' (p. 6). Further, to demonstrate the value of preference-based AAI weights in the context of cross-country analysis, the experiment would have to be conducted in different geographic domains, with obvious implications for the overall number of respondents required and additional resources needed for customising questionnaires to different contexts (e.g. translations, data collection in different countries). Finally, the mode of administration is likely to increase the resources required, as a self-administered, computer-based format without face-to-face contact may be inadequate for some groups of respondents. This may also require more extensive pre-testing of the questionnaire, which in turn affects costs.

In conclusion, we believe that our proposed DCE can enhance the AAI as an effective tool for policymaking. Preference-based weights account for potential trade-offs and marginal substitution rates between indicators, and our proposed method allows including the views of different groups of stakeholders in evaluating policy outcomes. We think that our approach can complement the current AAI methodology, enhance its flexibility and improve our understanding of what people actually want in different social, cultural or geographic contexts, thus helping to define more targeted active and healthy ageing policies.

References

AAI-Expert Group. (2014). *Report of the third meeting of the AAI Expert Group*. Brussels, Belgium. Retrieved from http://www1.unece.org/stat/platform/download/attachments/76287849/Report%20of%20the%20Third%20meeting_Final.pdf?version=1&modificationDate=1395409157483&api=v2

Amaya-Amaya, M., Gerard, K., & Ryan, M. (2008). Discrete choice experiments in a nutshell. In M. Ryan, K. Gerard, & M. Amaya-Amaya (Eds.), *Using discrete choice experiments to value health and healthcare*. Dordrecht: Springer.

Boehler, C., & Abadie, F. (2015). *Monitoring and Assessment Framework for the European Innovation Partnership on Active and Healthy Ageing (MAFEIP)*. Conceptual description of the monitoring framework. European Commission, Joint Research Centre, Seville, Spain. ISBN:978-92-79-50574-4. Retrieved from http://publications.jrc.ec.europa.eu/repository/handle/JRC96205

Boehler, C., de Graaf, G., Steuten, L., Yang, Y., & Abadie, F. (2015, September). Development of a web-based tool for the assessment of health and economic outcomes of the European Innovation Partnership on Active and Healthy Ageing (EIP on AHA). *BMC Medical Informatics and Decision Making, 15*(Suppl. 3), S4. Retrieved from http://www.biomedcentral.com/1472-6947/15/S3/S4

Bridges, J. F., Hauber, A. B., Marshall, D., Lloyd, A., Prosser, L. A., Regier, D. A., Johnson, F. R., & Mauskopf, J. (2011). Conjoint analysis applications in health-a checklist: Report of the ISPOR good research practices for conjoint analysis task force. *Value in Health, 14*(4), 403–413. Retrieved from http://www.ispor.org/taskforces/documents/ISPOR-CA-in-Health-TF-Report-Checklist.pdf

Carson, R. T., & Groves, T. (2007). Incentive and informational properties of preference questions. *Environmental and Resource Economics, 37*, 181–200.

Coast, J., Al-Janabi, H., Sutton, E. J., Horrocks, S. A., Vosper, A. J., Swancutt, D. R., & Flynn, T. N. (2012). Using qualitative methods for attribute development for discrete choice experiments: Issues and recommendations. *Health Economics, 21*(6), 730–741.

Cogan, A., Sharpe, S., & Hertzberg, J. (1986). *"Citizen Participation." The practice of state and regional planning*. Chicago.

De Bekker-Grob, E. W., Hol, L., Donkers, B., van Dam, L., Habbema, J. D., van Leerdam, M. E., Kuipers, E. J., Essink-Bot, M. L., & Steyerberg, E. W. (2010). Labeled versus unlabeled discrete choice experiments in health economics: An application to colorectal cancer screening. *Value in Health, 13*(2), 315–323.

Dolan, P. (1999). Whose preferences count? *Medical Decision Making, 19*(4), 482–486.

European Commission. (2006). *The demographic future of Europe—From challenge to opportunity*. COM/2006/0571. Retrieved from http://eur-lex.europa.eu/legal-content/EN/ALL/;ELX_SESSIONID=X5cbJ7hYDFQjnJHrLT7LMD1nPSNcrVk1ZmvhnnXgkVLvX0Ln11Qw!-82020822?uri=CELEX:52006DC0571

European Commission. (2011). *Strategic implementation plan for the European innovation partnership on active and healthy ageing*. Steering Group Working Document. Retrieved from http://ec.europa.eu/research/innovation-union/pdf/active-healthy-ageing/steering-group/implementation_plan.pdf#view=fit&pagemode=none

European Commission. (2012). *The 2012 Ageing Report, Economic and budgetary projections for the 27 EU Member States (2010–2060)*. European Economy 2/2012. Retrieved from http://ec.europa.eu/economy_finance/publications/european_economy/2012/pdf/ee-2012-2_en.pdf

Eurostat. (2015). *Information society statistics*. Retrieved January 21, 2015, from http://ec.europa.eu/eurostat/web/information-society/data/database

Helter, T. M., & Boehler, C. (2016). Developing attributes for discrete choice experiments in health—A systematic literature review and case study of alcohol misuse interventions. *Journal for Substance Use, 21*(6), 662–668.

Huber, J., & Zwerina, K. (1996). The importance of utility balance in efficient choice designs. *Journal of Marketing Research, 33*, 307–317.

Johnson, F. R., Lancsar, E., Marshall, D., Kilambi, V., Mühlbacher, A., Regier, D. A., Bresnahan, B. W., Kanninen, B., & Bridges, J. F. (2013). Constructing experimental designs for discrete-choice experiments: Report of the ISPOR conjoint analysis experimental design good research practices task force. *Value in Health, 16*, 3–13. Retrieved from http://www.ispor.org/ValueInHealth/ShowValueInHealth.aspx?issue=3551E3F3-83A8-4895-9415-5272AF29A9DF.

Lancaster, K. J. (1966). A new approach to consumer theory. *Journal of Political Economy, 74*(2), 132–157. Retrieved from http://www.dklevine.com/archive/refs41385.pdf.

Lancsar, E., & Louviere, J. (2008). Conducting discrete choice experiments to inform healthcare decision making: A user's guide. *PharmacoEconomics, 26*(8), 661–677.

Louviere, J., Hensher, D. A., & Swait, J. D. (2000). *Stated choice methods: Analysis and applications*. Cambridge: Cambridge University Press.

Louviere, J. J., Flynn, T. N., & Carson, R. T. (2010). Discrete choice experiments are not conjoint analysis. *Journal of Choice Modelling, 3*(3), 57–72.

Mangham, L. J., Hanson, K., & McPake, B. (2009). How to do (or not to do)….Designing a discrete choice experiment for application in a low-income country. *Health Policy and Planning, 24*, 151–158.

Marshall, D., Bridges, J. F. P., Hauber, B., Cameron, R., Donnalley, L., Fyie, K., & Johnson, F. R. (2010). Conjoint analysis applications in health—How are studies being designed and reported? *The Patient: Patient-Centered Outcomes Research, 3*(4), 249–256.

McFadden, D. (1974). Conditional logit analysis of qualitative choice behavior. In P. Zarembka (Ed.), *Frontiers of econometrics*. New York: Academic Press.

OECD/JRC. (2005). *Handbook of constructing composite indicators: Methodology and user guide*. OECD Statistics Working Paper. Retrieved from http://www.oecd.org/std/42495745.pdf

Ryan, M. (1999). Using conjoint analysis to take account of patient preferences and go beyond health outcomes: An application to in vitro fertilisation. *Social Science and Medicine, 48*(4), 535–546.

Ryan, M., & Farrar, S. (2000). Using conjoint analysis to elicit preferences for healthcare. *BMJ, 320*, 1530–1533.

Ryan, M., & Gerard, K. (2003). Using discrete choice experiments to value health care programmes: Current practice and future research reflections. *Applied Health Economics and Health Policy, 2*(1), 55–64.

Scott, A. (2002). Identifying and analysing dominant preferences in discrete choice experiments: An application in healthcare. *Journal of Economic Psychology, 23*(3), 383–398.

Sharpe, A., & Andrews, B. (2012). *An assessment of weighting methodologies for composite indicators: The case of the index of economic well-being*. CSLS Research Report 2012–10. Retrieved from http://www.csls.ca/reports/csls2012-10.pdf

Shumway, M., Saunders, T., Shern, D., Pines, E., Downs, A., Burbine, T., & Beller, J. (2003). Preferences for schizophrenia treatment outcomes among public policy makers, consumers, families, and providers. *Psychiatric Services, 54*(8), 1124–1128.

Street, D., Burgess, L., Viney, R., & Louviere, J. (2008). Designing discrete choice experiments for healthcare. In M. Ryan, K. Gerard, & M. Amaya-Amaya (Eds.), *Using discrete choice experiments to value health and healthcare*. Dordrecht: Springer.

Train, K. (2009). *Discrete choice methods with simulation* (2nd ed.). Cambridge: Cambridge University Press.

Ubel, P. A., Loewenstein, G., & Jepson, C. (2003). Whose quality of life? A commentary exploring discrepancies between health state evaluations of patients and the general public. *Quality of Life Research, 12*(6), 599–607.

Viney, R., Lancsar, E., & Louviere, J. (2002). Discrete choice experiments to measure consumer preferences for health and healthcare. *Expert Review of Pharmacoeconomics Outcomes Research, 2*(4), 319–326.

Watson, V., Carnon, A., Ryan, M., & Cox, D. (2012). Involving the public in priority setting: A case study using discrete choice experiments. *Journal of Public Health, 34*(2), 253–260. Retrieved from http://jpubhealth.oxfordjournals.org/content/34/2/253.full.pdf+html.

WHO. (2002). *Active ageing: A policy framework.* WHO/NMH/NPH/02.8. Retreievd from http://whqlibdoc.who.int/hq/2002/WHO_NMH_NPH_02.8.pdf?ua=1

WHO. (2012). *How to conduct a discrete choice experiment for health workforce recruitment and retention in remote and rural areas: A user guide with case studies.* Retrieved from http://www.who.int/hrh/resources/dceguide/en/

Wong, S. F., Norman, R., Dunning, T. L., Ashley, D. M., & Lorgelly, P. K. (2014). A protocol for a discrete choice experiment: Understanding preferences of patients with cancer towards their cancer care across metropolitan and rural regions in Australia. *BMJ Open, 4*, e006661. Retrieved from http://bmjopen.bmj.com/content/4/10/e006661.full.pdf+html

Zaidi, A., Gasior, K., Hofmarcher, M. M., Lelkes, O., Marin, B., Rodrigues, R., Schmidt, A., Vanhuysse, P., & Zolyomi, E. (2013). *Active Ageing Index 2012. Concept, methodology, and final results.* Research Memorandum/Methodology Report. European Centre Vienna, March 2013. Retrieved from www.euro.centre.org/data/aai/1253897823_70974.pdf

Zaidi, A., & Stanton, D. (2015). *Active ageing index 2014: Analytical report.* Report produced at the Centre for Research on ageing, University of Southampton, under contract with UNECE (Geneva), co-funded by European Commission, Brussels. Retrieved from http://www.southampton.ac.uk/assets/sharepoint/groupsite/Administration/SitePublisher-document-store/Documents/aai_report.pdf

Christian Ernst Heinrich Boehler is a Health Economist at the European Centre for Social Welfare Policy and Research in Vienna (Austria). He obtained his PhD at Brunel Universities' Health Economics Research Group (HERG) in London (UK), and has years of experience in the conduct of economic evaluation of innovative health and care technologies. Previously, he worked as a scientific officer at the European Commission's Joint Research Centre (DG-JRC IPTS) in Seville (Spain), where he was in charge of developing a Monitoring and Assessment Framework for the European Innovation Partnership on Active and Healthy Ageing (MAFEIP-project).

Timea Mariann Helter is an economist with a wide range of international work experience and education related to the health sector and is currently working for the Main Association of Austrian Social Security Institutions. She holds Masters degrees in economics, health informatics and health research. Timea is also experienced within the area of international health outcomes research through her positions at British, Hungarian and Austrian consultancy firms and the national Health Technology Assessment office in Hungary.

Ibrahim-Kholilul Rohman is a PhD in Technology Management of Economics from Chalmers University of Technology. Ibrahim is a Research Fellow at the United Nations University Operating Unit on Policy-Driven Electronic Governance (UNU-EGOV). His research analyses the role of eGov to combat the shadow economy in developing countries. Previously, he worked at the Information Society Unit, JRC IPTS prior to joining UNU. He was in charge in the project called Prospective Insights on R&D in ICT (PREDICT).

Fabienne Abadie, Scientific Officer at JRC Seville since 2006, managed a research team working on research projects in the field of eHealth until 2015. She led the Strategic Intelligence Monitor on Personal Health Systems (SIMPHS), a multi-annual project on remote patient monitoring and telehealth developments and the Monitoring and Assessment Framework for the European Innovation Partnership on Active and Healthy Ageing (MAFEIP) project whose aim was to assess the impact of European Innovation Partnership on Active and Healthy Ageing (EIP on AHA) initiatives.

Part IV

Methodological Improvements in Measuring Active Ageing

13

A Cross-country and Cohort Analysis of Active Ageing Differences Among the Elderly in Europe

Javier Olivera

13.1 Introduction

The Active Ageing Index (AAI) is a composite index computed in each EU-28 country that captures the contributions of the elderly, through activity, engagement and independent living, which in turn can be considered prerequisites for well-being. The index is aimed at measuring the active and healthy ageing experienced by the old individuals of a given country and period, and therefore it can compare the quality of ageing across countries and monitor its evolution over time. In this way, the AAI can be a useful tool to detect areas of active ageing where the elderly are lacking and promote an adequate policy response. The index was generated in the context of the 2012 European Year for Active Ageing and Solidarity between Generations and is rooted in the concept of active ageing defined by the World Health Organization (2002): "Active ageing

J. Olivera (✉)
Luxembourg Institute of Socio-Economic Research, Luxembourg, Luxembourg

Pontificia Universidad Catolica del Peru, Lima, Peru

A. Zaidi (eds.), *Building Evidence for Active Ageing Policies*,
https://doi.org/10.1007/978-981-10-6017-5_13

is the process of optimizing opportunities for health, participation and security in order to enhance quality of life as people age". More details on the methodology and results of the index are in Zaidi et al. (2013).

The AAI can be regarded as part of a broader family of composite indices aimed at measuring multidimensional well-being in the society, such as the well-known Human Development Index. Although the AAI shares interesting properties with other related indices, this still faces some of the common problems found in the elaboration of these measures. Among these limitations are the choice of appropriate weights, indicators and dimensions, and the inability to assess individual heterogeneity in each indicator and dimension because of aggregation and the use of different databases.

However, given the clear methods and nationally representative data sources employed to produce the AAI, the index can potentially be computed for distinctive groups of older individuals. Therefore, a key contribution of this chapter is the replication of the index for cohorts formed by age groups, sex and country and the study of active ageing differences across these groups. In the case of the elderly population, differences among cohorts can be substantial. Some of the AAI outcomes, past experiences and expectations of the 55–59-year-old persons can be very different from those aged 75+. Think, for example, on differences in schooling policies and pension participation and statutory rights among birth cohorts because of changes in social policies. Furthermore, it is likely that differences among birth cohorts can become more pronounced when the cohort is also disaggregated by sex. In addition, life expectancy is larger for younger cohorts and for females, and hence, active ageing indicators should be rightly assessed according to different groups and composition of old people.[1]

In this chapter, the AAI is fully replicated following the official methodology, but not without some adjustments, and is computed for cohorts formed by sex, five age groups (55–59, 60–64, 65–69, 70–74, 75+) and 28 EU countries. This means that the total number of cohorts with AAI outcomes is 280 (= 2 × 5 × 28). Another contribution is the study of the predictors of AAI outcomes with the implementation of a cross-country analysis. This analysis is made with regressions of the AAI outcomes on cohort identifying variables and relevant macro variables at the level of the country. The purpose is to uncover what drives differences across cohorts and countries in Europe. This is an important goal as the analysis can be informative for policy making and expand our understanding on active ageing in a comparative perspective. Indeed, this study finds important differences among cohorts in Europe. In general, females are behind males in active ageing and present a gap that grows in older cohorts. The regression analysis indicates that, in general, wealth, equity and pension settings of the country are important predictors for better active ageing. Regarding welfare state regimes in Europe, it is found that the Social-Democratic regime (Nordic countries)—with its strong redistributive policies—is the the most favourable for active ageing.

The chapter is organized as follows. The next section presents a description of the AAI and its decomposition by cohorts. Section 13.3 presents and discusses the results of the cohort analysis and the study of the predictors of the AAI. Finally, Sect. 13.4 provides a conclusion.

13.2 The Active Ageing Index

The AAI includes 22 indicators grouped in 4 domains: (i) employment; (ii) participation in society; (iii) independent, healthy and secure living; and (iv) capacity and enabling environment for active ageing. Table 13.1 shows these domains, indicators, weights and data sources.

The precise definition and corresponding survey questions of each indicator used in the official methodology is available in Zaidi et al. (2013) and the website of the AAI.[2] Note that 9 out of 22 indicators are computed for the group of individuals aged 55+, but there are other indicators that correspond to other reference groups. For example, the indicators of financial security (3.4, 3.5 and 3.6) are measured for the population aged 65+, while the indicator of independent living (3.3) corresponds to individuals aged 75+. Other indicators are also capped at age 74. The heterogeneity in the age reference groups is largely because the different age groups capture better the perspective relevant for each of those indicators. In any case, the replication of the AAI performed in this study considers, first, a homogenous group of individuals aged 55+, and then five different age groups (55–59, 60–64, 65–64, 70–74 and 75+). The reproduction of the AAI has been done with the same data sources as in

Table 13.1 Composition of Active Ageing Index

Domain	Indicator		Age group	Weight indicator	Weight domain	Data source	Adjustments
1. Employment	1.1	Employment rate 55–59	55–59	0.25	0.35	LFS, 2012	SILC-rev1, 2012 used
	1.2	Employment rate 60–64	60–64	0.25		LFS, 2012	Idem
	1.3	Employment rate 65–69	65–69	0.25		LFS, 2012	Idem
	1.4	Employment rate 70–74	70–74	0.25		LFS, 2012	Idem
2. Participation in society	2.1	Voluntary activities	55+	0.25	0.35	EQLS, 2012	
	2.2	Care to older children, grandchildren	55+	0.25		EQLS, 2012	
	2.3	Care to older adults	55+	0.3		EQLS, 2012	
	2.4	Political participation	55+	0.2		EQLS, 2012	
3. Independent, healthy and secure living	3.1	Physical exercise	55+	0.10	0.10	EQLS, 2012	
	3.2	Access to health and dental care	55+	0.20		SILC-rev1, 2012	
	3.3	Independent living arrangements	75+	0.20		SILC-rev1, 2012	
	3.4	Relative median income of 65+ relative to those aged below 65	65+	0.10		SILC-rev1, 2012	Relative median income of each cohort relative to those aged 25–54
	3.5	No poverty risk for older persons	65+	0.10		SILC-rev1, 2012	Done for each cohort
	3.6	No severe material deprivation rate	65+	0.10		SILC-rev1, 2012	Done for each cohort
	3.7	Physical safety	55+	0.10		ESS, 2012	
	3.8	Lifelong learning	55–74	0.10		LFS, 2012	Eurostat [trng_lfs_01]

(*continued*)

Table 13.1 (continued)

Domain	Indicator		Age group	Weight indicator	Weight domain	Data source	Adjustments
4. Capacity and enabling environment for active ageing	4.1	Remaining life expectancy at age 55	55	0.33	0.20	EHLEIS, 2010	Eurostat 2012 [demo_mlexpec]
	4.2	Share of healthy life expectancy at age 65	55	0.23		EHLEIS, 2010	Eurostat 2012 [hlth_hlye]
	4.3	Mental well-being	55+	0.17		EQLS, 2012	
	4.4	Use of information and communications technology (ICT)	55–74	0.07		Eurostat, ICT Survey, 2012	
	4.5	Social connectedness	55+	0.13		ESS, 2012	
	4.6	Educational attainment	55–74	0.07		LFS, 2012	SILC-rev1, 2012 used

Acronyms: LFS, European Union Labour Force Survey; SILC, European Union Statistics on Income and Living Conditions; EQLS, European Quality of Life Survey; ESS, European Social Survey; EHLEIS, European Health and Life Expectancy Information System; ICT Survey, Community Survey on ICT Usage in Households and by Individuals

the official methodology, with the exception of the indicators from the employment domain and educational attainment (item 4.6) which use the SILC-rev1[3] instead of LFS data. Other adjustments are reported in the last column of Table 13.1. Due to all these adjustments, one should not expect identical results between the official AAI and the performed replication, although both indices should be highly correlated. Table 13.2 reports the official results (version December 2014) computed for 2012 and the simulated results.

The simulated index is lower than the official one in each country, although the correlation is very high at 0.970. The average gross value of

Table 13.2 Computation results of the Active Ageing Index

	AAI (official)		AAI		Diff	
Country	Index	Rank	Index	Rank	Index	Rank
Sweden	0.448	1	0.419	1	0.029	0
Denmark	0.405	2	0.387	3	0.018	−1
Netherlands	0.399	3	0.363	4	0.037	−1
Finland	0.396	4	0.360	5	0.036	−1
United Kingdom	0.392	5	0.388	2	0.004	3
Ireland	0.390	6	0.360	6	0.030	0
Germany	0.359	7	0.329	9	0.030	−2
Luxembourg	0.358	8	0.335	8	0.023	0
France	0.357	9	0.337	7	0.020	2
Austria	0.352	10	0.328	11	0.023	−1
Estonia	0.348	11	0.313	16	0.035	−5
Czech Republic	0.346	12	0.321	14	0.025	−2
Cyprus	0.344	13	0.329	10	0.016	3
Italy	0.340	14	0.326	13	0.015	1
Belgium	0.338	15	0.326	12	0.012	3
Portugal	0.336	16	0.296	17	0.040	−1
Spain	0.328	17	0.313	15	0.015	2
Malta	0.318	18	0.282	21	0.037	−3
Lithuania	0.317	19	0.288	20	0.029	−1
Latvia	0.316	20	0.292	19	0.024	1
Croatia	0.313	21	0.279	22	0.034	−1
Bulgaria	0.300	22	0.292	18	0.008	4
Slovenia	0.299	23	0.275	24	0.024	−1
Romania	0.297	24	0.246	28	0.051	−4
Hungary	0.286	25	0.276	23	0.010	2
Slovakia	0.285	26	0.274	25	0.011	1
Poland	0.282	27	0.260	27	0.022	0
Greece	0.277	28	0.260	26	0.017	2

the official index is 0.340, and that of the replication is 0.316, that is, 7% lower. When inspecting within each domain, the main difference is observed in the employment domain which shows a drop of 17% in the simulated index with respect to the official value.[4] The inclusion of the group of persons aged 75+ in the simulated index is the main reason of this fall because they have a very low participation in employment. Sweden is the country with the highest performance in active ageing in both official and simulated indices, but the country with worst performance differs in both indices. Romania is at the bottom in the simulated index, while Greece was placed at the bottom with the official index. Regarding the ranking, in 15 out of 28 countries, the difference in ranks is 1 at most. In six countries, the difference in ranks is 2; and in three countries the difference in ranks is 3. Estonia is the country that presents the largest difference. The official figures rank Estonia in place 11 with an index of 0.348, while the simulated figures rank this country 16 with an index of 0.313. The Spearman correlation of the ranks is very high as well at 0.965. In general, the replication of results is satisfactory and very close to what is observed with the official figures. From this point in this chapter, any reference to the index will only correspond to the simulated results.

The active ageing observed in males is higher than that in females for each country, except in Finland, Estonia and Latvia (see Fig. 13.1). On average, the AAI is 0.303 for females and 0.331 for males, although there are important differences in some countries. Figure 13.2 is useful to observe the intensity of the gender gap in the quality of ageing in each country. The vertical axis shows the ratio of AAI of males to females, which indicates how much is the difference between the quality of ageing between males and females. Females of any country placed over the unity line are in worse situation with respect to males. For example, in Cyprus and Malta, the active ageing of males is 25% and 22% larger than that of females. In Luxembourg, Romania, Italy and Czech Republic, males also report a high AAI, which is 15–20% larger than the AAI of females. Although there is a negative relationship between the AAI male to female ratio (the gender gap) and the overall AAI, this is not a clear-cut relationship (correlation is −0.24) as one can observe countries simultaneously with high performance in the overall AAI and high gender gaps. For example, this is the case for Luxembourg, Austria and Italy.

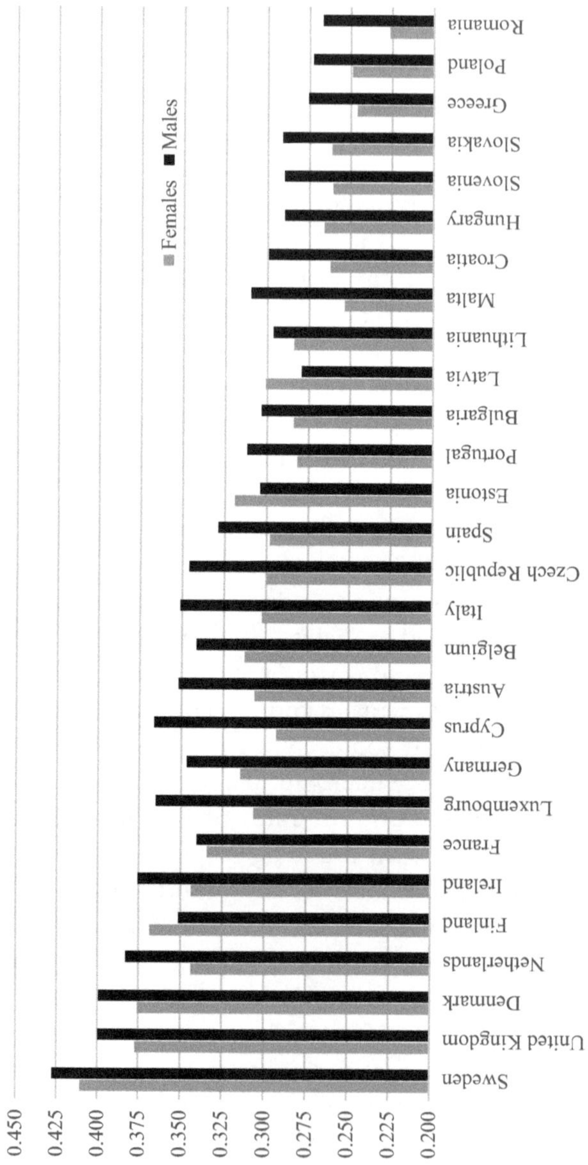

Fig. 13.1 The Active Ageing Index by sex. Note: the countries are placed according to their ranking in the overall AAI

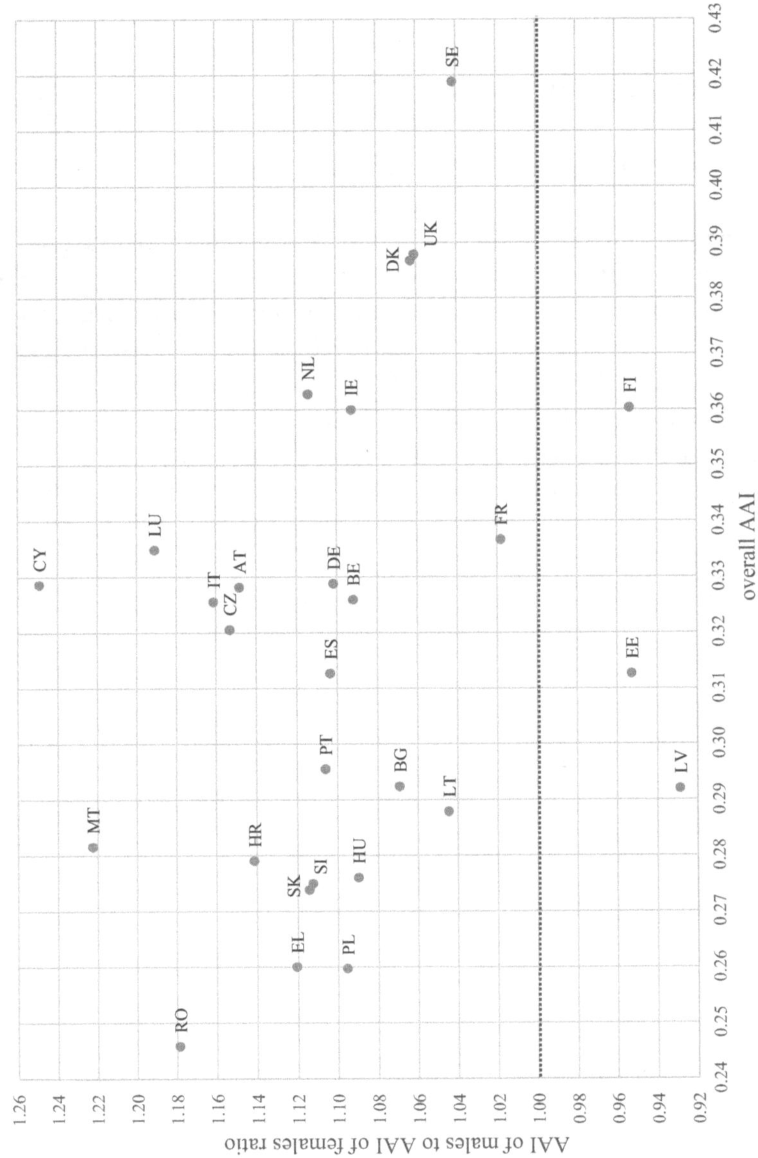

Fig. 13.2 Gender gap in the Active Ageing Index

There are also important differences by sex in each AAI domain. The results can be consulted in Tables 13.10 and 13.11 in the Appendix.

13.3 Analysis of Cohorts

13.3.1 Disentangling the AAI

One of the first observed results when the AAI is broken by age group is a large heterogeneity in ageing quality experienced by each group in each country. Younger cohorts are always better in every country, although the size of the inter-cohort difference greatly differs among countries (see Fig. 13.3). The gross average AAI for the age groups 55–59, 60–64, 65–64, 70–74 and 75+ are 0.479, 0.371, 0.274, 0.232 and 0.19, respectively. A country that performs well in the AAI in a given age group can obtain a low AAI in another group. The ranking of countries for the oldest group (75+) is correlated at 0.75 with the ranking of countries for the youngest group (55–59). For example, Ireland is sixth with the overall indicator (and fourth for the groups older than 65) but it is placed only

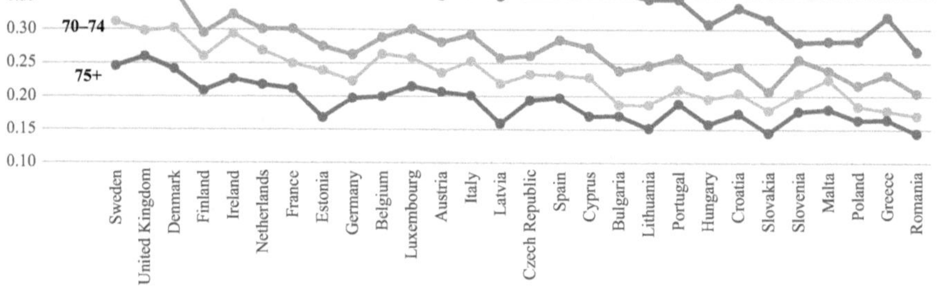

Fig. 13.3 The Active Ageing Index by age group. Note: the countries are placed according to their ranking in the overall AAI

14th for the 55–59 age group. An opposite case is Cyprus as the 55–59 age group ranks 9th, while the 75+ group ranks only 20th. Other countries with a high variation in their rankings per age group are Estonia, Lithuania, Slovenia, Germany and Austria. Table 13.3 reports the complete ranking.

In some countries there are important gender gaps in the quality of ageing, which also differs by age group. The vertical axis of Fig. 13.4 shows the AAI of males and the horizontal axis reports the AAI of females. Each point corresponds to the same country and age group. For comparison reasons only three age groups are depicted (55–59, 65–69 and 75+). Hence, the points located above the dashed diagonal line indicate that males are better off than females for a given country and age group. It seems that in general, females are worse off than males in the oldest group. At least, one can observe five countries in the youngest age group where females are better off (Finland, Lithuania, Estonia, Bulgaria and Latvia). Males and females aged 55–59 from the United Kingdom and Ireland also show a very similar AAI. But, in the oldest group, only Estonia shows that females are better off than males.

A way to observe the effect of each variable that identifies the cohort is to employ Ordinal Least Squares (OLS) and regress the cohort identifying variables on the AAI outcome. Recall that the sample is formed by 280 cohorts produced from 28 countries, 2 sexes and 5 age groups. The regressions use robust standard errors clustered at the country level and the following specification:

$$AAI_{i,j,c} = \alpha + \beta_1 \text{sex}_i + \beta_2 \text{age}_j + \beta_3 \text{country}_c + \varepsilon_{i,j,c} \qquad (13.1)$$

The subscripts i, j and c refer to sex (1 = female, 0 = male), age group and country, respectively. The Active Ageing Index of a given cohort is $AAI_{i,j,c}$ and the rest of the variables are dummies. The error term $\varepsilon_{i,j,c}$ is assumed to be normally distributed. An alternative model specification is given by Eq. 13.2. In this case, an interaction term between age group and sex is added. The goal is to be able to observe differences in age groups that are sex specific. Given the preliminary graphical results one should expect larger gender gaps in older age groups.

Table 13.3 The Active Ageing Index by age group

	AAI						Ranking					
Country	55–59	60–64	65–69	70–74	75+	Total	55–59	60–64	65–69	70–74	75+	Total
Sweden	0.597	0.552	0.369	0.312	0.244	0.419	1	1	2	1	2	1
United Kingdom	0.529	0.445	0.381	0.298	0.259	0.388	7	3	1	3	1	2
Denmark	0.550	0.466	0.360	0.302	0.240	0.387	3	2	3	2	3	3
Finland	0.550	0.445	0.295	0.260	0.208	0.360	4	4	8	7	8	5
Ireland	0.484	0.431	0.322	0.294	0.226	0.360	14	6	4	4	4	6
Netherlands	0.555	0.444	0.300	0.269	0.217	0.363	2	5	7	5	5	4
France	0.536	0.364	0.301	0.250	0.211	0.337	6	14	5	10	7	7
Estonia	0.467	0.391	0.275	0.239	0.167	0.313	16	9	13	11	21	16
Germany	0.506	0.425	0.262	0.223	0.197	0.329	8	8	15	17	12	9
Belgium	0.489	0.382	0.288	0.263	0.199	0.326	13	10	10	6	11	12
Luxembourg	0.499	0.372	0.301	0.258	0.215	0.335	11	12	6	8	6	8
Austria	0.540	0.351	0.281	0.235	0.206	0.328	5	17	12	12	9	11
Italy	0.491	0.361	0.292	0.253	0.201	0.326	12	15	9	9	10	13
Latvia	0.444	0.350	0.257	0.219	0.158	0.292	22	18	17	18	24	19
Czech Republic	0.501	0.366	0.260	0.233	0.193	0.321	10	13	16	13	14	14
Spain	0.451	0.373	0.284	0.232	0.197	0.313	19	11	11	14	13	15
Cyprus	0.504	0.426	0.273	0.228	0.169	0.329	9	7	14	15	20	10
Bulgaria	0.466	0.357	0.238	0.187	0.170	0.292	17	16	23	23	19	18
Lithuania	0.474	0.345	0.246	0.187	0.151	0.288	15	19	20	24	26	20
Portugal	0.442	0.345	0.257	0.209	0.188	0.296	23	20	18	19	15	17
Hungary	0.450	0.307	0.231	0.195	0.157	0.276	20	24	24	22	25	23
Croatia	0.401	0.332	0.243	0.204	0.173	0.279	26	21	21	20	18	22
Slovakia	0.460	0.314	0.206	0.178	0.144	0.274	18	23	27	26	27	25
Slovenia	0.428	0.281	0.255	0.204	0.176	0.275	24	27	19	21	17	24
Malta	0.445	0.282	0.238	0.224	0.180	0.282	21	26	22	16	16	21
Poland	0.404	0.283	0.216	0.184	0.163	0.260	25	25	26	25	23	27
Greece	0.375	0.318	0.231	0.178	0.164	0.260	28	22	25	27	22	26
Romania	0.390	0.267	0.204	0.171	0.144	0.246	27	28	28	28	28	28

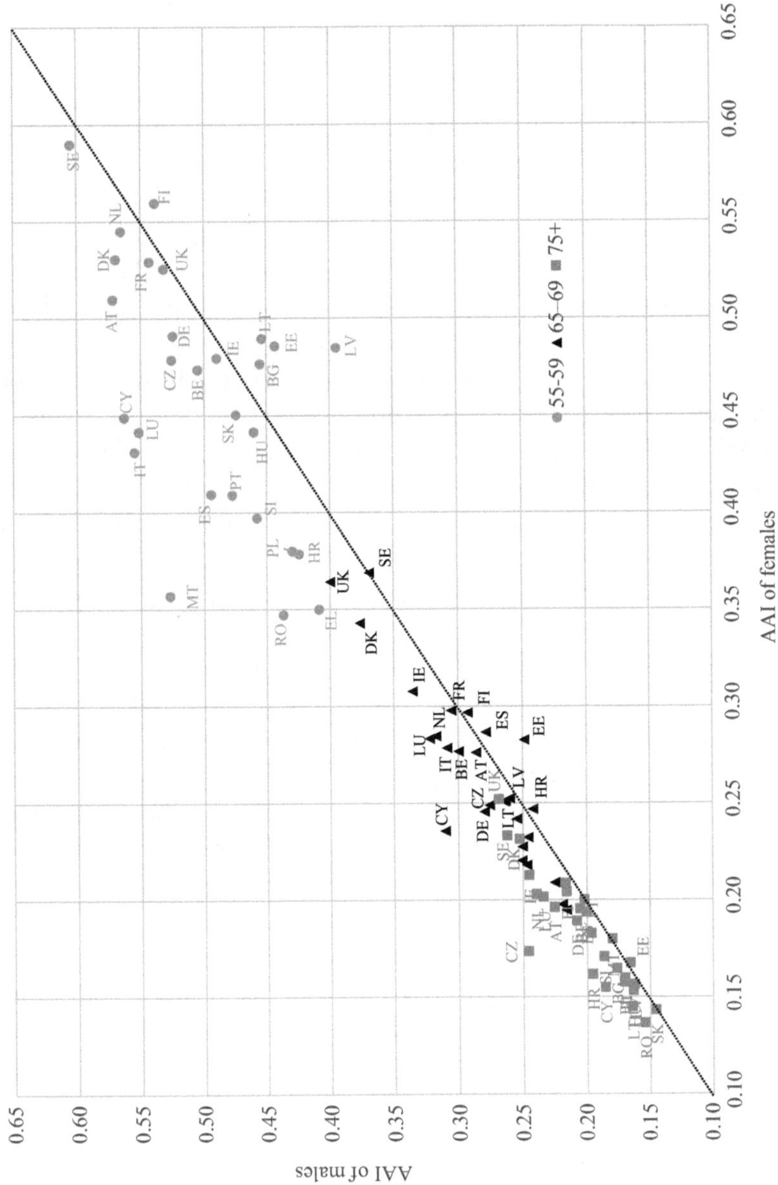

Fig. 13.4 Gender gap by age group in the Active Ageing Index

$$AAI_{i,j,c} = \alpha + \beta_1 \text{sex}_i + \beta_2 \text{age}_j + \beta_3 \left(\text{sex}_i \times \text{age}_j \right) + \beta_4 \text{country}_c + \varepsilon_{i,j,c} \tag{13.2}$$

Table 13.4 reports the OLS results for the overall AAI and also for each domain. The first column corresponds to Eq. 13.1 and columns 2–6 correspond to Eq. 13.2. In model 1, being female is penalized with a drop of 0.026 (2.6 in a 0–100 scale, or 8.2% of the average AAI). In addition, the decreasing and significant coefficients of age groups indicate that younger age groups are better off than old age groups. For example, the cohort 55–59 has an AAI that is larger than that of the 75+ by 0.288 points in the AAI (this difference is 91% of the average AAI). The introduction of interaction terms between sex and age groups indicates that the sex penalty (against females) in the overall AAI increases with age. This can be more clearly observed in Table 13.5, which reports the predicted values of the AAI outcomes by sex and age group. Note that only the coefficients of the interactions of sex and groups 55–59 and 60–64 are statistically different from zero, confirming that sex has specific effects in these groups with respect to the reference group (individuals aged 75+).

The employment domain (third column in Table 13.4 and second column in Table 13.5) shows a clear and even more pronounced penalty in detriment of females. This result is in line with the lower participation of females observed in the labour market. Although female participation rates have been increasing during the last years, the birth cohorts analysed with the AAI data are old (born in 1957 or earlier) and show, in general, a much lower participation rate. The domain of participation in society also shows a clear negative relationship with age, although the specific effects of sex are different for each age group. Table 13.5 reports that females of age groups 55–59 and 60–64 show a higher participation in society than males of the same age, but females also show a lower participation in society than males in the age groups 70–74 and 75+. It seems that after retirement, males catch up with females in the activities measured in the domain of social participation (voluntary activities, care to older children and grandchildren, care to older adults and political participation) which is in line with a higher rate of male labour participation before retirement and more disposable time after retirement. Although

Table 13.4 Ordinal Least Square regressions for the Active Ageing Index

	(1)	(2)	(3)	(4)	(5)	(6)
Variable	AAI	AAI	AAI.1 (employment)	AAI.2 (participation)	AAI.3 (independent)	AAI.4 (capacity)
Female	−0.0260***	−0.0184***	−0.0122***	−0.0230***	−0.0375***	−0.0117*
	(0.0045)	(0.0031)	(0.0039)	(0.0082)	(0.0045)	(0.0059)
Age 75+ (reference)						
Age 70–74	0.0403***	0.0385***	0.0282***	0.0529***	0.0015	0.0496***
	(0.0034)	(0.0047)	(0.0048)	(0.0108)	(0.0046)	(0.0050)
Age 65–69	0.0820***	0.0823***	0.1072***	0.0823***	−0.0187***	0.0891***
	(0.0039)	(0.0047)	(0.0121)	(0.0100)	(0.0049)	(0.0046)
Age 60–64	0.1795***	0.1895***	0.4080***	0.0963***	−0.1153***	0.1228***
	(0.0093)	(0.0086)	(0.0254)	(0.0097)	(0.0041)	(0.0055)
Age 55–59	0.2875***	0.2979***	0.6935***	0.1104***	−0.1202***	0.1429***
	(0.0074)	(0.0079)	(0.0180)	(0.0112)	(0.0049)	(0.0055)
Age 70–74 × female		0.0036	−0.0106**	0.0158	0.0168***	0.0003
		(0.0053)	(0.0045)	(0.0125)	(0.0055)	(0.0050)
Age 65–69 × female		−0.0006	−0.0377***	0.0227*	0.0365***	0.0050
		(0.0042)	(0.0082)	(0.0117)	(0.0049)	(0.0055)
Age 60–64 × female		−0.0200***	−0.1276***	0.0489***	0.0675***	0.0041
		(0.0072)	(0.0211)	(0.0090)	(0.0063)	(0.0052)
Age 55–59 × female		−0.0208*	−0.1291***	0.0523***	0.0346***	0.0130**
		(0.0114)	(0.0278)	(0.0126)	(0.0055)	(0.0061)

(continued)

Table 13.4 (continued)

	(1)	(2)	(3)	(4)	(5)	(6)
Variable	AAI	AAI	AAI.1 (employment)	AAI.2 (participation)	AAI.3 (independent)	AAI.4 (capacity)
Constant	0.2052***	0.2014***	0.0218**	0.1077***	0.7087***	0.4260***
	(0.0052)	(0.0042)	(0.0105)	(0.0068)	(0.0034)	(0.0042)
N	280	280	280	280	280	280
R^2	0.9463	0.9484	0.9306	0.8008	0.9604	0.9528

The top row indicates the dependent variable used in each model equation. Robust and clustered (by country) standard errors are provided in parenthesis. $^{*}p < 0.1$, $^{**}p < 0.05$, $^{***}p < 0.01$. Each model includes dummies for countries

Table 13.5 Predicted Active Ageing Index per sex and age group (based on models 2–6)

Age group	AAI	AAI.1 (employment)	AAI.2 (participation)	AAI.3 (independent)	AAI.4 (capacity)
			Female		
Age 75+	0.183	0.010	0.085	0.671	0.414
Age 70–74	0.225	0.027	0.153	0.690	0.464
Age 65–69	0.265	0.080	0.190	0.689	0.508
Age 60–64	0.353	0.291	0.230	0.623	0.541
Age 55–59	0.460	0.574	0.247	0.586	0.570
			Male		
Age 75+	0.201	0.022	0.108	0.709	0.426
Age 70–74	0.240	0.050	0.161	0.710	0.476
Age 65–69	0.284	0.129	0.190	0.690	0.515
Age 60–64	0.391	0.430	0.204	0.593	0.549
Age 55–59	0.499	0.715	0.218	0.589	0.569
			Female/male ratio		
Age 75+	0.909	0.440	0.786	0.947	0.973
Age 70–74	0.938	0.544	0.955	0.971	0.976
Age 65–69	0.933	0.619	0.998	0.999	0.987
Age 60–64	0.902	0.676	1.127	1.051	0.986
Age 55–59	0.921	0.803	1.134	0.995	1.002

The predicted values are computed with the results of models 2–6 from Table 13.4

one cannot discard that this can also be explained by selective survival of males. Contrary to the previous domains, the domain of independent, healthy and secure living shows a positive relationship with age.[5] The direction of this relationship is confirmed in each sex as well, and it seems that being a female is more penalized in older groups (see fourth column of Table 13.5). Regarding the domain of capacity and enabling environment for active ageing, the older the group the lower the score in this domain. The coefficients of the interaction effects are only statistically significant for the age group 55–59, so that it is not possible to establish specific effects by sex on age groups.

13.3.2 Explaining Country Differences

It is expected that some specific effects of country variables will affect the well-being of the elderly because the age groups are embedded in a particular country. These are contextual variables with potential effects on the extent of active ageing, which can also show important country variation. For example, the gross average of GDP per capita (in purchasing power standard prices) exhibits average living standards which is 23,414 euros in EU-28 in 2012, but one can find vast differences across countries, such as Luxembourg with 67,100 euros per inhabitant or Bulgaria with 12,100 euros. Generosity of pensions and other variables related with pensions can also account for the ability of the elderly to have financial independence. In a more general picture, it is possible that the type of welfare state in the country is an important determinant of the active ageing. In order to account for the effect of contextual variables on the AAI, the following specification is used in further OLS regression models:

$$AAI_{i,j,c} = \alpha + \beta_1 \text{sex}_i + \beta_2 \text{age}_j + \beta_3 X_c + \varepsilon_{i,j,c} \qquad (13.3)$$

X_c is a set of macro variables at the country level that are potentially related with the AAI. These variables are the GDP per capita (in logs, annual, purchasing power standard prices), social protection expenditures in old age (as percentage of GDP), Gini index of equivalised disposable income (ranges from 0 to 100, where 0 indicates full equality and 100 indicates maximum inequality), pension per capita (in logs, annual, purchasing power standard prices), statutory average retirement age, percentage of population in eligible age covered by pensions, age dependency ratio (percentage of population aged 0–14 or 65+ over population aged 15–64), and the percentage of females in the population aged 55+.

The variables correspond to 2012 and are drawn from Eurostar, except the retirement age and pension coverage which are drawn from the International Labour Organization. Table 13.12 in the Appendix shows the values of these variables per country.

Models 1–5 of Table 13.6 report the OLS estimates of equations that include one macro variable or a group of related macro variables at once, while model 6 shows the results of a specification that includes all macro variables together. Country dummies cannot be introduced, otherwise the effect of the macro variable cannot be identified. The results are not completely unexpected and indicate that the elderly are able to age with more quality in countries that are richer, more egalitarian and expend more in social protection for old age. It is worth to mention that in a recent review of social policy in developed countries, Marx et al. (2015) observe the significant role of pensions to reduce poverty and income inequality, so that higher social protection expenditure in old age (mostly composed by pensions) should improve the well-being of the elderly. The results for model 4 from Table 13.6 confirm the significant effects of pensions in determining the active ageing experiences of the elderly. Higher average pensions, pension coverage and statutory retirement age in a country are associated with a larger AAI. There are a number of explanations why a higher legal retirement age has a positive effect on the AAI. In general, pension systems with higher retirement ages can offer better pensions, which is the case in the EU-28 countries (the correlation between pension per capita and retirement age is 0.514), and better pension means better resources to fulfil more quality in ageing. In addition, a late retirement age means that the individual is active for more time in the labour market and hence can present a better performance in the domain of employment and financial independence.

Two variables related to population composition are included in model 5 of Table 13.6. These are the age dependency ratio and the percentage of females in the population aged 55+. The age dependency ratio indicates the relative weight of "economic inactive" individuals (0–14 years and 65+ group) with respect to active individuals (aged 15–64) and is interpreted as a measure of the effort needed by the active population to support children and the elderly. A straightforward explanation for obtaining a positive relationship between the

Table 13.6 Ordinal Least Square regressions for the Active Ageing Index

	(1)	(2)	(3)	(4)	(5)	(6)
Variable	AAI	AAI	AAI	AAI	AAI	AAI
Female	−0.0260***	−0.0260***	−0.0260***	−0.0260***	−0.0260***	−0.0260***
	(0.0043)	(0.0043)	(0.0043)	(0.0043)	(0.0043)	(0.0044)
Age 75+ (reference)						
Age 70–74	0.0403***	0.0403***	0.0403***	0.0403***	0.0403***	0.0403***
	(0.0032)	(0.0032)	(0.0032)	(0.0032)	(0.0032)	(0.0033)
Age 65–69	0.0820***	0.0820***	0.0820***	0.0820***	0.0820***	0.0820***
	(0.0037)	(0.0037)	(0.0037)	(0.0037)	(0.0037)	(0.0038)
Age 60–64	0.1795***	0.1795***	0.1795***	0.1795***	0.1795***	0.1795***
	(0.0089)	(0.0089)	(0.0089)	(0.0089)	(0.0089)	(0.0090)
Age 55–59	0.2875***	0.2875***	0.2875***	0.2875***	0.2875***	0.2875***
	(0.0070)	(0.0070)	(0.0070)	(0.0070)	(0.0070)	(0.0071)
Log of GDP pc	0.0860***					0.0443**
	(0.0262)					(0.0162)
Gini index		−0.0042*				−0.0010
		(0.0024)				(0.0017)
Old age social protection exp.			0.0063*			−0.0015
			(0.0034)			(0.0020)
Log of pension pc				0.0432***		
				(0.0108)		
Retirement age				0.0086**		0.0034
				(0.0037)		(0.0034)
Pension coverage				0.0010**		0.0011**
				(0.0005)		(0.0004)

(*continued*)

Table 13.6 (continued)

	(1)	(2)	(3)	(4)	(5)	(6)
Variable	AAI	AAI	AAI	AAI	AAI	AAI
Age dependency ratio					0.0053***	0.0043***
					(0.0013)	(0.0014)
Share of 55+ females					−0.0063***	−0.0038*
					(0.0017)	(0.0021)
Constant	−0.6597**	0.3293***	0.1426***	−0.7709***	0.2981**	−0.5032*
	(0.2618)	(0.0719)	(0.0335)	(0.2237)	(0.1160)	(0.2592)
N	280	280	280	280	280	280
R^2	0.8753	0.8244	0.8309	0.8827	0.8803	0.9130

The top row indicates the dependent variable used in each model equation. Robust and clustered (by country) standard errors are provided in parenthesis. $*p < 0.1$, $**p < 0.05$, $***p < 0.01$

dependency ratio and the AAI is that countries with higher dependency ratios have more elderly and higher life expectancy and hence their populations can be more active and healthier. The percentage of females in old age is negatively related with the AAI, which is in line with previous results on the gender gap in the AAI in favour of males. It is worrying to observe a systematic female disadvantage in the AAI. This disadvantage is, perhaps, a consequence of less labour market participation and differential social protection. For example, the average of pension coverage is 98.3% for males and 83.5% for females within the population of eligible age; and labour market participation is 73.0% for males and 59.5% for females in the group aged 55–59. The last model of Table 13.6 includes all predictors at once, except the log of pension per capita which is highly correlated with GDP per capita (correlation is 0.86).[6]

Table 13.7 reports the results of OLS regressions for each domain. The share of individuals in eligible age that receive pensions is the only statistically significant macro variable to explain the employment domain. Given the debate of increasing income inequality in the world and its negative effects on social outcomes, it is interesting to observe that income inequality is significantly and negatively related with social participation and independence. This result is in line with the findings of Lancee and Van de Werfhorst (2012) who uncover a negative relationship between income inequality and civic and social life participation in a sample of 24 European countries. Thus, income inequality matters for the active ageing outcomes and well-being in old age. The results show that the share of elderly females is negatively and significantly associated only with the independence and capacity domains. Finally, the variables that affect more domains are the log of GDP per capita and the age dependency ratio (all domains, except employment).

Given the multiple effects of distinctive macro variables on the well-being of the elderly, it is worth to use a more aggregate variable that somewhat summarizes the features of a country that are relevant for the elderly. It is perhaps the set of variables related to benefits, social policy,

Table 13.7 Ordinal Least Square regressions for each domain of the Active Ageing Index

	(1)	(2)	(3)	(4)
Variable	AAI.1 (employment)	AAI.2 (participation)	AAI.3 (independent)	AAI.4 (capacity)
Female	−0.0732***	0.0049	−0.0064***	−0.0072
	(0.0088)	(0.0050)	(0.0022)	(0.0049)
Age 75+ (reference)				
Age 70–74	0.0229***	0.0608***	0.0098***	0.0498***
	(0.0040)	(0.0071)	(0.0030)	(0.0034)
Age 65–69	0.0883***	0.0936***	−0.0005	0.0916***
	(0.0094)	(0.0060)	(0.0036)	(0.0033)
Age 60–64	0.3442***	0.1207***	−0.0816***	0.1249***
	(0.0257)	(0.0071)	(0.0035)	(0.0046)
Age 55–59	0.6289***	0.1366***	−0.1029***	0.1494***
	(0.0194)	(0.0076)	(0.0044)	(0.0052)
Log of GDP pc	0.0159	0.0556***	0.0484***	0.0723***
	(0.0348)	(0.0107)	(0.0137)	(0.0226)
Gini index	0.0031	−0.0024**	−0.0070***	−0.0030
	(0.0037)	(0.0009)	(0.0013)	(0.0020)
Old age social protection exp.	−0.0023	0.0006	−0.0009	−0.0041
	(0.0031)	(0.0027)	(0.0020)	(0.0036)
Retirement age	0.0083	−0.0028	0.0019	0.0062*
	(0.0088)	(0.0022)	(0.0027)	(0.0035)
Pension coverage	0.0032***	−0.0009**	0.0005	0.0011
	(0.0008)	(0.0004)	(0.0004)	(0.0007)
Age dependency ratio	0.0033	0.0035***	0.0043***	0.0076***
	(0.0032)	(0.0009)	(0.0010)	(0.0015)
Share of 55+ females	−0.0034	−0.0018	−0.0040**	−0.0081**
	(0.0047)	(0.0017)	(0.0017)	(0.0038)
Constant	−0.9667	−0.2178	0.2726	−0.5794
	(0.5729)	(0.1938)	(0.2171)	(0.4477)
N	280	280	280	280
R^2	0.8897	0.7108	0.8732	0.8439

The top row indicates the dependent variable used in each model equation. Robust and clustered (by country) standard errors are provided in parenthesis. $^*p < 0.1$, $^{**}p < 0.05$, $^{***}p < 0.01$

generosity, health, pensions, tax structure, social trust, equality and so on that matter for active ageing. This set can be reduced and operationalized with a classification of welfare states in Europe. There is a long tradition in studying and classifying welfare states in the political and social policy literature. On the basis of the revision of Esping-Andersen (1990), Fenger (2007), Kammer et al. (2012) and Sapir (2006), Table 13.8 proposes a classification of welfare regimes in EU-28.

Table 13.9 reports the OLS estimates after including dummy variables for each welfare regimen, the Post-Communism regimen being the reference group. It seems clear that a welfare state like that in the Nordic countries—with high level of transfers, tax collection and equity—is the most favourable for active ageing. It is surprising to find the liberal regime (United Kingdom and Ireland) as the second best regime to develop good active ageing. This regime is characterized by means-tested benefits, low universal allowances and more income inequality. However, the results show that this regime favours, importantly, more employment and social participation in old age. The conservative type of welfare regimen is the third best regimen for active ageing. Then, it follows the Southern type and the former USSR countries (Baltic countries), the Post-Communist block being the least favourable for active ageing.

Table 13.8 Welfare state regimen classification in EU-28

Social-Democratic	Southern	Liberal (Anglo-Saxon)	Conservative	Former USSR (Baltic)	Post-communist
Denmark	Greece	United Kingdom	Austria	Estonia	Bulgaria
Finland	Italy	Ireland	France	Latvia	Croatia
Sweden	Spain		Germany	Lithuania	Czech Republic
	Portugal		Luxembourg		Hungary
	Malta		Belgium		Poland
	Cyprus		Netherlands		Slovakia
					Romania
					Slovenia

Table 13.9 Ordinal Least Square regressions for the AAI and the welfare state

	(1)	(2)	(3)	(4)	(5)
Variable	AAI	AAI.1 (employment)	AAI.2 (participation)	AAI.3 (independent)	AAI.4 (capacity)
Female	−0.0260***	−0.0732***	0.0049	−0.0064***	−0.0072
	(0.0043)	(0.0087)	(0.0050)	(0.0022)	(0.0049)
Age 75+ (reference)					
Age 70–74	0.0403***	0.0229***	0.0608***	0.0098***	0.0498***
	(0.0032)	(0.0039)	(0.0070)	(0.0030)	(0.0034)
Age 65–69	0.0820***	0.0883***	0.0936***	−0.0005	0.0916***
	(0.0037)	(0.0094)	(0.0059)	(0.0036)	(0.0033)
Age 60–64	0.1795***	0.3442***	0.1207***	−0.0816***	0.1249***
	(0.0089)	(0.0256)	(0.0071)	(0.0035)	(0.0046)
Age 55–59	0.2875***	0.6289***	0.1366***	−0.1029***	0.1494***
	(0.0071)	(0.0193)	(0.0076)	(0.0044)	(0.0051)
Type: Social-Democratic	0.1129***	0.1332***	0.0618***	0.1168***	0.1649***
	(0.0177)	(0.0312)	(0.0113)	(0.0130)	(0.0292)
Type: liberal (Anglo-Saxon)	0.0968***	0.1214**	0.0700***	0.0675***	0.1154***
	(0.0137)	(0.0465)	(0.0191)	(0.0134)	(0.0160)
Type: conservative	0.0616***	0.0328	0.0586***	0.0738***	0.1110***
	(0.0095)	(0.0206)	(0.0147)	(0.0144)	(0.0184)
Type: Southern	0.0239*	0.0196	0.0342*	0.0021	0.0243
	(0.0131)	(0.0223)	(0.0173)	(0.0153)	(0.0220)

(continued)

Table 13.9 (continued)

	(1)	(2)	(3)	(4)	(5)
Variable	AAI	AAI.1 (employment)	AAI.2 (participation)	AAI.3 (independent)	AAI.4 (capacity)
Type: former USSR (Baltic)	0.0197*	0.0885***	−0.0094	−0.0272	−0.0265
	(0.0102)	(0.0214)	(0.0113)	(0.0227)	(0.0179)
Type: Post-communist (reference)					
Constant	0.1657***	0.0086	0.0633***	0.6625***	0.3717***
	(0.0091)	(0.0151)	(0.0109)	(0.0134)	(0.0150)
N	280	280	280	280	280
R^2	0.9168	0.8950	0.6492	0.8243	0.8166

The top row indicates the dependent variable used in each model equation. Robust and clustered (by country) standard errors are provided in parenthesis. *$p < 0.1$, **$p < 0.05$, ***$p < 0.01$

13.4 Conclusions

This study reports significant differences in the Active Ageing Index among cohorts of the elderly in Europe. Therefore, it is important to consider subgroups of individuals when a composite index of active ageing is computed. This practice can contribute to detect areas of active ageing where some groups of elderly are lacking and promote an appropriate policy response. Several regression models indicate that, in general, wealth, income equality and favourable pension characteristics of the country are important predictors for a better active ageing. Furthermore, it is worrying to observe a systematic female disadvantage in active ageing among the elderly, which is, perhaps, a consequence of lower labour market participation and differential social protection. The review of antipoverty policies in rich economies (mostly Europe) by Marx et al. (2015) reveals that the very old females are the ones at more risk of poverty because they have a higher life expectancy, less time expended in the labour market, fewer social security contributions and more probability of living alone. This study also performs an evaluation of welfare regimes regarding their effects on prompting favourable active ageing. The results show that the Social-Democratic regime (Nordic countries), with its strong redistributive policies, is the most favourable for active ageing. In the other side of the ranking, the set of policies and characteristics of Post-Communist countries are the least favourable for active ageing. Interestingly, the Liberal regime (United Kingdom and Ireland) is associated with better outcomes in employment and social participation in old age than the conservative western European countries. Analysing the effects of welfare regimes on active ageing can be an important task for future research and improve our understanding of the relationship between policies and outcomes in old age.

Appendix

Table 13.10 The Active Ageing Index by domain and sex

	Women						Men						Total					
Country	Emp	Soc	Liv	Cap	AAI	Rank	Emp	Soc	Liv	Cap	AAI	Rank	Emp	Soc	Liv	Cap	AAI	Rank
Austria	0.158	0.188	0.704	0.573	0.306	12	0.244	0.224	0.695	0.593	0.352	8	0.201	0.204	0.701	0.582	0.328	11
Belgium	0.166	0.189	0.667	0.607	0.312	10	0.220	0.215	0.688	0.601	0.341	13	0.193	0.200	0.676	0.604	0.326	12
Bulgaria	0.218	0.128	0.573	0.526	0.284	18	0.264	0.125	0.612	0.530	0.303	19	0.239	0.127	0.590	0.526	0.292	18
Croatia	0.096	0.172	0.635	0.524	0.262	22	0.191	0.175	0.635	0.538	0.299	20	0.141	0.173	0.635	0.528	0.279	22
Cyprus	0.203	0.188	0.625	0.469	0.293	17	0.379	0.180	0.638	0.533	0.366	6	0.289	0.185	0.631	0.498	0.329	10
Czech Republic	0.180	0.173	0.651	0.555	0.300	15	0.295	0.197	0.649	0.543	0.346	12	0.233	0.183	0.651	0.548	0.321	14
Denmark	0.308	0.178	0.753	0.651	0.376	3	0.354	0.210	0.736	0.640	0.399	3	0.330	0.193	0.746	0.646	0.387	3
Estonia	0.312	0.137	0.620	0.499	0.319	8	0.318	0.121	0.616	0.443	0.304	18	0.314	0.131	0.618	0.476	0.313	16
Finland	0.267	0.226	0.734	0.612	0.368	4	0.254	0.206	0.732	0.585	0.351	9	0.260	0.217	0.733	0.599	0.360	5
France	0.210	0.207	0.688	0.597	0.334	7	0.214	0.226	0.703	0.580	0.340	14	0.211	0.215	0.695	0.589	0.337	7
Germany	0.251	0.135	0.677	0.557	0.314	9	0.314	0.154	0.693	0.565	0.346	11	0.280	0.144	0.685	0.560	0.329	9
Greece	0.119	0.157	0.596	0.448	0.246	27	0.220	0.122	0.592	0.483	0.276	26	0.167	0.141	0.594	0.464	0.260	26
Hungary	0.162	0.152	0.621	0.471	0.266	21	0.209	0.168	0.612	0.483	0.290	24	0.183	0.158	0.617	0.475	0.276	23
Ireland	0.192	0.255	0.675	0.599	0.344	5	0.307	0.232	0.676	0.596	0.376	5	0.249	0.245	0.676	0.598	0.360	6
Italy	0.145	0.237	0.633	0.525	0.302	13	0.272	0.236	0.639	0.546	0.351	10	0.206	0.237	0.637	0.534	0.326	13
Latvia	0.269	0.159	0.543	0.484	0.301	14	0.269	0.098	0.553	0.479	0.279	25	0.268	0.136	0.545	0.480	0.292	19
Lithuania	0.228	0.145	0.614	0.460	0.284	19	0.269	0.155	0.607	0.437	0.296	21	0.244	0.148	0.612	0.448	0.288	20
Luxembourg	0.148	0.180	0.690	0.615	0.306	11	0.215	0.261	0.691	0.647	0.365	7	0.181	0.218	0.692	0.630	0.335	8
Malta	0.084	0.174	0.639	0.495	0.253	25	0.237	0.183	0.629	0.498	0.310	17	0.161	0.179	0.634	0.496	0.282	21
Netherlands	0.227	0.202	0.719	0.607	0.344	6	0.293	0.234	0.729	0.628	0.383	4	0.260	0.217	0.725	0.616	0.363	4
Poland	0.129	0.132	0.598	0.488	0.249	26	0.237	0.105	0.595	0.468	0.273	27	0.178	0.121	0.596	0.478	0.260	27
Portugal	0.180	0.140	0.615	0.541	0.282	20	0.249	0.142	0.614	0.567	0.312	16	0.213	0.141	0.615	0.552	0.296	17
Romania	0.122	0.135	0.563	0.403	0.226	28	0.231	0.121	0.571	0.433	0.267	28	0.174	0.129	0.569	0.415	0.246	28

(continued)

Table 13.10 (continued)

	Women						Men						Total					
Country	Emp	Soc	Liv	Cap	AAI	Rank	Emp	Soc	Liv	Cap	AAI	Rank	Emp	Soc	Liv	Cap	AAI	Rank
Slovakia	0.162	0.140	0.605	0.475	0.261	23	0.255	0.130	0.605	0.479	0.291	22	0.202	0.136	0.604	0.475	0.274	25
Slovenia	0.095	0.162	0.689	0.509	0.261	24	0.175	0.171	0.676	0.506	0.290	23	0.135	0.166	0.683	0.507	0.275	24
Spain	0.160	0.188	0.644	0.559	0.298	16	0.252	0.173	0.653	0.574	0.329	15	0.205	0.181	0.648	0.565	0.313	15
Sweden	0.357	0.210	0.745	0.688	0.410	1	0.392	0.222	0.741	0.693	0.428	1	0.374	0.216	0.745	0.690	0.419	1
United Kingdom	0.337	0.197	0.689	0.605	0.377	2	0.405	0.186	0.699	0.615	0.400	2	0.370	0.192	0.693	0.609	0.388	2

Table 13.11 The Active Ageing Index by age group and sex

	Women						Men						Total					
Country	55–59	60–64	65–69	70–74	75+	Total	55–59	60–64	65–69	70–74	75+	Total	55–59	60–64	65–69	70–74	75+	Total
Austria	0.509	0.308	0.276	0.226	0.196	0.306	0.572	0.393	0.286	0.245	0.225	0.352	0.540	0.351	0.281	0.235	0.206	0.328
Belgium	0.473	0.365	0.276	0.247	0.195	0.312	0.505	0.401	0.300	0.282	0.205	0.341	0.489	0.382	0.288	0.263	0.199	0.326
Bulgaria	0.476	0.327	0.227	0.178	0.165	0.284	0.455	0.394	0.250	0.200	0.176	0.303	0.466	0.357	0.238	0.187	0.170	0.292
Cyprus	0.449	0.369	0.236	0.211	0.155	0.293	0.563	0.484	0.310	0.246	0.186	0.366	0.504	0.426	0.273	0.228	0.169	0.329
Czech Republic	0.478	0.322	0.248	0.234	0.173	0.300	0.526	0.419	0.276	0.233	0.246	0.346	0.501	0.366	0.260	0.233	0.193	0.321
Germany	0.491	0.412	0.245	0.212	0.189	0.314	0.525	0.443	0.280	0.236	0.208	0.346	0.506	0.425	0.262	0.223	0.197	0.329
Denmark	0.530	0.466	0.343	0.299	0.231	0.376	0.570	0.465	0.376	0.308	0.253	0.399	0.550	0.466	0.360	0.302	0.240	0.387
Estonia	0.486	0.407	0.282	0.237	0.167	0.319	0.443	0.374	0.249	0.241	0.166	0.304	0.467	0.391	0.275	0.239	0.167	0.313
Greece	0.350	0.301	0.218	0.172	0.157	0.246	0.409	0.336	0.247	0.186	0.171	0.276	0.375	0.318	0.231	0.178	0.164	0.260
Spain	0.409	0.358	0.286	0.224	0.194	0.298	0.495	0.393	0.279	0.241	0.200	0.329	0.451	0.373	0.284	0.232	0.197	0.313
Finland	0.559	0.475	0.296	0.273	0.204	0.368	0.539	0.414	0.292	0.241	0.216	0.351	0.550	0.445	0.295	0.260	0.208	0.360
France	0.529	0.372	0.297	0.253	0.208	0.334	0.543	0.356	0.305	0.246	0.217	0.340	0.536	0.364	0.301	0.250	0.211	0.337
Croatia	0.378	0.298	0.247	0.199	0.162	0.262	0.424	0.374	0.242	0.209	0.195	0.299	0.401	0.332	0.243	0.204	0.173	0.279
Hungary	0.441	0.303	0.220	0.183	0.153	0.266	0.460	0.318	0.251	0.225	0.163	0.290	0.450	0.307	0.231	0.195	0.157	0.276
Ireland	0.479	0.409	0.307	0.266	0.213	0.344	0.490	0.452	0.335	0.326	0.245	0.376	0.484	0.431	0.322	0.294	0.226	0.360
Italy	0.431	0.343	0.278	0.238	0.200	0.302	0.555	0.380	0.309	0.269	0.202	0.351	0.491	0.361	0.292	0.253	0.201	0.326
Lithuania	0.489	0.328	0.242	0.182	0.145	0.284	0.454	0.368	0.255	0.193	0.164	0.296	0.474	0.345	0.246	0.187	0.151	0.288
Luxembourg	0.441	0.356	0.283	0.232	0.201	0.306	0.552	0.386	0.322	0.291	0.234	0.365	0.499	0.372	0.301	0.258	0.215	0.335
Latvia	0.485	0.354	0.251	0.229	0.156	0.301	0.395	0.344	0.263	0.199	0.163	0.279	0.444	0.350	0.257	0.219	0.158	0.292
Malta	0.357	0.246	0.232	0.210	0.180	0.253	0.527	0.321	0.246	0.243	0.180	0.310	0.445	0.282	0.238	0.224	0.180	0.282
Netherlands	0.545	0.404	0.284	0.274	0.203	0.344	0.566	0.484	0.317	0.263	0.240	0.383	0.555	0.444	0.300	0.269	0.217	0.363
Poland	0.380	0.258	0.209	0.186	0.160	0.249	0.430	0.312	0.225	0.180	0.170	0.273	0.404	0.283	0.216	0.184	0.163	0.260
Portugal	0.409	0.336	0.251	0.197	0.182	0.282	0.478	0.355	0.264	0.223	0.196	0.312	0.442	0.345	0.257	0.209	0.188	0.296
Romania	0.347	0.231	0.194	0.166	0.136	0.226	0.437	0.306	0.216	0.176	0.154	0.267	0.390	0.267	0.204	0.171	0.144	0.246
Sweden	0.589	0.548	0.369	0.310	0.233	0.410	0.605	0.556	0.369	0.313	0.262	0.428	0.597	0.552	0.369	0.312	0.244	0.419
Slovenia	0.397	0.258	0.252	0.200	0.171	0.261	0.458	0.306	0.260	0.207	0.186	0.290	0.428	0.281	0.255	0.204	0.176	0.275
Slovakia	0.450	0.288	0.197	0.168	0.143	0.261	0.475	0.352	0.219	0.192	0.145	0.291	0.460	0.314	0.206	0.178	0.144	0.274
United Kingdom	0.525	0.430	0.364	0.294	0.252	0.377	0.532	0.460	0.400	0.300	0.269	0.400	0.529	0.445	0.381	0.298	0.259	0.388

Table 13.12 Macro variables in EU-28 countries (2012)

Country	Log of GDP per capita (in pps)	Gini index	Social protection expenditures in old age (% of GDP)	Log of pension per capita (in pps)	Statutory average retirement age	Pension coverage (% of eligible population)	Age dependency ratio (%)	Share of females in the 55+ population
Austria	10.41	27.60	13.00	8.51	62.50	85.60	47.83	55.44
Belgium	10.33	26.50	9.60	8.22	65.00	83.90	52.32	54.56
Bulgaria	9.40	33.60	7.50	6.98	61.50	97.45	47.54	56.59
Croatia	9.66	30.90	5.80	7.41	62.50	64.65	49.22	56.90
Cyprus	10.06	31.00	10.50	7.69	65.00	78.60	41.47	52.57
Czech Republic	9.94	24.90	9.30	7.66	61.45	100.00	44.65	55.78
Denmark	10.38	28.10	14.40	8.39	65.00	100.00	53.90	53.04
Estonia	9.81	32.50	6.70	7.28	62.00	98.00	49.74	62.28
Finland	10.29	25.90	11.50	8.23	65.00	100.00	52.90	54.77
France	10.23	30.50	12.90	8.37	60.00	100.00	55.54	55.42
Germany	10.36	28.30	9.40	8.31	65.00	100.00	51.20	54.61
Greece	9.88	34.30	15.40	8.14	62.50	77.30	52.33	54.27
Hungary	9.74	26.90	9.90	7.44	62.00	92.65	45.72	58.98
Ireland	10.40	29.90	6.40	7.68	65.00	83.15	50.32	52.34
Italy	10.15	31.90	15.30	8.36	62.50	84.60	53.48	55.40
Latvia	9.71	35.70	7.50	7.22	62.00	100.00	48.92	63.41
Lithuania	9.81	32.00	6.90	7.28	61.25	100.00	48.97	62.39
Luxembourg	11.11	28.00	6.70	8.67	65.00	78.20	45.14	53.51
Malta	10.00	27.10	8.70	7.65	60.50	64.75	45.37	53.47
Netherlands	10.39	25.40	11.30	8.37	65.00	100.00	50.49	53.06
Poland	9.75	30.90	8.70	7.63	62.50	97.45	40.68	57.42
Portugal	9.87	34.50	12.00	7.93	65.00	100.00	51.43	56.30

(*continued*)

Table 13.12 (continued)

Country	Log of GDP per capita (in pps)	Gini index	Social protection expenditures in old age (% of GDP)	Log of pension per capita (in pps)	Statutory average retirement age	Pension coverage (% of eligible population)	Age dependency ratio (%)	Share of females in the 55+ population
Romania	9.52	33.20	7.60	7.08	61.25	94.00	45.96	56.47
Slovakia	9.87	25.30	7.00	7.44	62.00	100.00	39.25	57.52
Slovenia	9.97	23.70	10.10	7.79	62.00	92.95	45.13	55.44
Spain	10.10	35.00	9.20	7.93	65.00	72.00	48.10	54.84
Sweden	10.38	24.80	12.40	8.23	65.00	100.00	55.10	52.91
United Kingdom	10.19	31.30	12.70	8.06	62.50	99.60	52.34	53.58
Total	10.06	29.63	9.94	7.86	63.07	90.89	48.75	55.83

Sources: Eurostat and International Labour Organization for statutory retirement age and pension coverage

Notes

1. It will become more evident in the empirical section, but it can be said that these differences are taken into account when linear regressions are performed for a sample where the unit of analysis is the cohort as defined above.
2. See http://www1.unece.org/stat/platform/display/AAI/V.+Methodology
3. This is the very last available revision of EU-SILC-2012 (01 Aug 2014) at the time of production of this study.
4. The average values in each domain (employment, participation, independent and capacity) of the official and simulated AAI are (0.279; 0.181; 0.706; 0.544) and (0.231; 0.176; 0.652; 0.542), respectively.
5. This result is perhaps driven by the indicator 3.2 of Table 13.1, which measures the proportion of individuals living in single or couple households. It is much more common to observe older individuals living in these types of households. In fact, it is also more likely that those who struggle with independent living are more likely to move into residential care. In any case, it must be noted that people living in residential care homes are not included in the sample of the EQLS.
6. The log of pension per capita is dropped from model 6 because this presents the largest contribution to the overall multicollinearity measured with the Variable Inflation Factor (VIF). The VIF of that variable is 15.74.

References

Fenger, M. (2007). Welfare regimes in Central and Eastern Europe: Incorporating post-communist countries in a welfare regime typology. *Contemporary Issues and Ideas in Social Sciences, 3*(2), 1–30.

Kammer, A., Niehues, J., & Peichl, A. (2012). Welfare regimes and welfare state outcomes in Europe. *Journal of European Social Policy, 22*(5), 455–471.

Lancee, B., & Van de Werfhorst, H. G. (2012). Income inequality and participation. A comparison of 24 European countries. *Social Science Research, 41*(5), 1166–1178.

Marx, I., Nolan, B., & Olivera, J. (2015). The welfare state and anti-poverty policy in rich countries. In A. B. Atkinson & F. Bourguignon (Eds.), *Handbook of income distribution*. Amsterdam: Elsevier.

Sapir, A. (2006). Globalization and the reform of European social models. *Journal of Common Market Studies, 44*(2), 369–390.

Esping-Andersen, G. (1990). *The three worlds of welfare capitalism*. Princeton, NJ: Princeton University Press.
World Health Organization. (2002). *Active ageing—A policy framework, a contribution of the World Health Organization to the Second United Nations World Assembly on Ageing*. Madrid, Spain, April 2002.
Zaidi, A., Gasior, K., Hofmarcher, M. M., Lelkes, O., Marin, B., Rodrigues, R., Schmidt, A., Vanhuysse, P., & Zolyomi, E. (2013). *Active Ageing Index 2012. Concept, methodology, and final results*. Research Memorandum/Methodology Report. European Centre Vienna, March 2013. Retrieved from www.euro.centre.org/data/aai/1253897823_70974.pdf

Javier Olivera is a researcher at the Luxembourg Institute of Socio-Economic Research (LISER) and professor of economics at Pontificia Universidad Catolica del Peru (PUCP). His research interests include public economics, inequality, pensions, ageing, family transfers and economic demography. Previously, he worked at the University of Luxembourg and University College Dublin, and also for different institutions and ministries of the government of Peru. He has also been involved in advisory work for pension and social protection reform in Latin American countries.

14

Active Ageing Typologies: A Latent Class Analysis of the Older Europeans

Bruno Arpino and Valeria Bordone

14.1 Introduction

All countries around the world are facing rapid demographic changes, with population ageing being central. Therefore, it becomes crucial to study the conditions that guarantee that people age "well", that is, that they do not simply live longer but also better, adding "life to years" in the words of the World Health Organization (WHO 2002).

Several terms encapsulate the notion of "ageing well", including successful ageing, active ageing, healthy ageing, positive ageing, productive ageing and competent ageing (Foster and Walker 2013). All the 29 terms identified by Depp and Jeste's review (2006), however, share the idea to go beyond physical health and consider the ageing process as complex and multidimensional. Rowe and Kahn (1987, 1997) introduced the concept of "successful ageing" to overcome the long-standing focus on pathological aspects

B. Arpino (✉)
Universitat Pompeu Fabra, Barcelona, Spain

V. Bordone
University of Munich, Munich, Germany

A. Zaidi (eds.), *Building Evidence for Active Ageing Policies*,
https://doi.org/10.1007/978-981-10-6017-5_14

of ageing and the traditional division of people into "diseased" and "normal" that failed to recognise the large heterogeneity within these groups and to distinguish ageing and illness processes (Bowling and Dieppe 2005).

The policy discourse on ageing in Europe is dominated by the framework of "active ageing" (Foster and Walker 2013), defined by the WHO as "the process of optimising opportunities for health, participation and security in order to enhance quality of life as people age" (WHO 2002, p. 12), where "active" also refers to continuing participation in economic, social, cultural, spiritual and civic affairs and it is not limited to being in good health.

In order to operationalise this multidimensional concept and to promote a more active role for older people, in 2012, the European Commission and UNECE launched the project "Active Ageing Index 2012" (AAI) (Zaidi et al. 2013). The AAI considers four dimensions of the contribution older people make to society: (1) employment; (2) participation in society; (3) independent, healthy and secure living; and (4) capacity and enabling environment for active ageing. This chapter focuses on the second dimension that comprises voluntary activities, care provided to children and grandchildren, care provided to older adults and political participation. Given the data availability, we add also participation to educational or training course (fourth domain); participation in a sport, social or other kind of club (third domain); whether the respondent has taken part in a religious organisation (church, synagogue, mosque etc.); and provision of help to a relative, friend or neighbour.

In line with the strands of the European year for active ageing,[1] the AAI recognises the important contribution of older people to the society as carers for others both outside and inside the family. Evidence has also shown that caring for others may have positive effects also for the older carers. For example, Arpino and Bordone (2014) found that looking after grandchildren has positive effects on verbal cognitive functioning. However, the effect of care on the carers may depend on the context of caring, the intensity of the care activity (e.g., Chappell and Reid 2002) and the type of care given (e.g., caring for a disabled adult or caring for grandchildren may have different effects due to their different nature).

Other studies have emphasised the positive effects of participation in social activities on individuals' mental (Engelhardt et al. 2010; Hultsch et al. 1999) and physical health (e.g., Pynnönen et al. 2012 on the risk of mortality associated with social activity).

Leaving aside the consequences of engagement in different activities, in this chapter, we analyse whether there exists interdependence among them (e.g., care and non-care activities), a topic that remains not well understood.

Despite some research has shown that people in later life tend to reallocate their time from participation in non-kin activities to kin activities (Lubben and Gironda 2003), several studies found support for the *cumulation* hypothesis, that is, people engaged in one type of activity are also more likely to be engaged in others. Hank and Stuck (2008), in fact, found a positive association between engaging in volunteer work, providing informal help and caring for older adults, even when controlling for a series of individual characteristics such as age, health, partnership and employment status. They interpreted this correlation as the result of a general (unobservable) motivation for being active.

Similarly, Kohli et al. (2009) found that the relationship between various dimensions of what they refer to as social connectedness (i.e., social relationships of various kinds, within kin and non-kin networks) is cumulative rather than competitive. An exception was the relationship between informal social relations (i.e., having received/given practical help from/to friends, neighbours, colleagues) and family relations (a broad measure that included having at least one cohabiting child and/or having received or given practical help primarily from/to a family member from outside the household including grandchild care). Yet, being interested in social connectedness per se, they did not distinguish whether the individual was the provider or the recipient of care. Arpino and Bordone (2017), instead, focused on regular grandchild care as an important type of help given by grandparents and studied whether or not grandchild care interferes with participation in social activities. Although they did not find support for a cumulation hypothesis, their evidence only partially hints to competition between grandparenting and participation in social (non-care related) activities. In particular, regular provision of childcare had a significant negative effect on volunteering for grandmothers.

Recently, Bulanda and Jendrek (2014) studied the association between grandchild care and volunteering. They found that (non-residential) grandchild care is related to a higher likelihood of volunteering, while the opposite holds for raising grandchildren.

Other studies have found support for the *competition hypothesis*, namely, that, engaging in one type of activity has a negative effect on participation in other activities. For example, Caro and Bass (1995) and Choi et al. (2007) found that family caregiving obligations reduced the likelihood of engaging in volunteer work. In the same way, it could be argued that other types of care activities may reduce willingness, energy, and time availability and limit opportunities to carry out those activities that do not involve kin (Koslowski 2009; Minkler 1999).

Our contribution to this literature is twofold. First, we investigate the interrelationship between engagement in various activities using a different approach from the ones used before. Using Latent Class Analysis (LCA), we are able to identify clusters of older people with homogeneous patterns of engagement in different activities without imposing any a priori classification. We also consider several activities within the active ageing framework, including important non-kin activities (voluntary work, enrolment in educational courses, participation in religious and political organisations) and provision of care to grandchildren, relatives and older adults. We not only study whether older people engage in each of these activities, but we also take into account the frequency of such engagement.

Second, we study the characteristics of the different groups of older people both in terms of their patterns of engagement and relative to some basic socio-demographic characteristics.

14.2 Data

Our analyses are based on the Survey of Health, Ageing and Retirement in Europe (SHARE). SHARE is a multidisciplinary longitudinal survey, representative of the non-institutionalised population aged 50 and over (Börsch-Supan et al. 2005, 2008).

We use data from the first (2004) and second (2006) waves. Wave 3, the so-called SHARELIFE, only contains a retrospective survey. We excluded wave 4 because of the different way of collecting information on participation in social activities. For each respondent, we consider the first interview. Fifteen countries participated in at least one of the first two waves of SHARE (Austria, Belgium, Czech Republic, Denmark,

France, Germany, Greece, Ireland, Israel, Italy, the Netherlands, Poland, Spain, Sweden and Switzerland).

We restrict our sample to respondents aged 50–85 who did not declare to be permanently disabled because disability makes very unlikely a regular participation in some of the activities we consider. This exclusion implies that our analyses cannot be extended to draw conclusions on active ageing of this group of people. We exclude cases with missing values on any of the employed variables. After application of the aforementioned selection criteria, our sample includes 50,511 persons.

We consider engagement in different activities and the intensity of such engagement. The SHARE questionnaire asks: “Have you done any of these activities in the last four weeks?” Respondents can tick several activities from a list including voluntary or charity work; educational or training course; a sport, social or other kind of club; taken part in a religious organisation (church, synagogue, mosque, etc.); a political or community-related organisation; caring for a disabled adult; or providing help to a relative, friend or neighbour. For each activity in which respondents are engaged, the frequency of participation is reported (“almost daily; almost every week; almost every month; less often”).

We also consider engagement in grandparental childcare. In this case, SHARE first asks “During the last twelve months, have you regularly or occasionally looked after your grandchild without the presence of the parents?” If the answer is “yes”, a second question asks about the frequency of care to children of each respondent’s child (“almost daily; almost every week; almost every month; less often”).

Based on this information, we create eight categorical variables coded as follows: 1 = engaged in the activity daily; 2 = engaged weekly; 3 = engaged monthly or less often; and 4 = not engaged.

It is well known that participation in social actives and care are gender specific (Arpino and Bordone 2017; Bordone et al. 2017; Hank and Buber 2009). Therefore, we include gender as a control variable (man is the reference). Additionally, participation in society may be substantially influenced by whether a person is working or not (e.g., Hank and Buber 2009). Retirement can, on the one hand, influence the time available to older people and, on the other, it can also be associated with a change in preferences for allocation of time from non-family to family activities (see, e.g., van Bavel and de Winter 2013). Therefore, we control for activity status (retired—reference—; working; other). We also control for age by including a set dummy variables for five-year intervals (50–55—reference—; 56–60; 61–65; 66–70; 71–75; 76–80; 81–85).

14.3 Method

We use Latent Class Analysis (LCA; see, e.g., Hagenaars and McCutcheon 2002) to identify groups of older people that correspond to different active ageing typologies, that is, characterised by different behaviours with respect to participation in the activities we described above ("manifest variables" in the LCA terminology).

Formally, let i represent a respondent ($i = 1, \ldots, N$), j one of the manifest variables ($j = 1, \ldots, J$) and k one of the possible value of the manifest variable ($k = 1, \ldots, K_j$). Let Y_{ijk} indicate the observed values of the J manifest variables such that $Y_{ijk} = 1$ if respondent i gives the kth response to the jth variable, and $Y_{ijk} = 0$ otherwise. The latent class model can be written as

$$P\left(Y_i \mid \pi,p\right) = \sum_{r=1}^{R} p_r \Pi_{j=1}^{J} \Pi_{k=1}^{K_j} \left(\pi_{ijk}\right)^{Y_{ijk}},$$

where p_r indicates the mixing proportion for cluster r ($r = 1, \ldots, R$) or, equivalently, the prior probability that an individual belongs to that cluster, and π_{ijk} denotes the class-conditional probability that an individual in cluster r reports the kth value on the jth manifest variable. Estimates of p_r and π_{ijk} are obtained by maximum likelihood using an iterative algorithm (the Expectation-Maximisation). After estimates of the model parameters have been obtained, for each individual, the posterior probability of belonging to each cluster r can be calculated. These posterior probabilities can be used to classify individuals into one of the R clusters. Typically, an individual is assigned to the cluster to which she/he has the highest probability of belonging.

LCA assumes that within latent clusters responses to all the manifest variables are statistically independent ("local independence"). In a sense, LCA aims at "explaining" the dependence among the manifest variables observed in the data with the existence of a categorical latent variable (the values of which correspond to the latent clusters).

LCA is a model-based approach that offers a variety of model selection tools to choose the number of clusters (Haughton et al. 2009). Following standard practice, we considered different solutions by gradually increasing the number of clusters and chose the solution that gave the best model fit (as measured by the Bayesian Information Criterion). To increase the robustness of the final solution, we replicated the algorithm 200 times and retained the estimates of the model with the best fit.

Not only can LCA be used as a model-based tool for cluster analysis but it also offers the possibility to include covariates in the model to predict cluster membership. Estimates of the LCA model will be used to analyse the profiling of the clusters, that is, to analyse how individual characteristics are associated with the different active ageing typologies.

We implemented LCA using the package polCA (Linzer and Lewis 2011) in the free software environment R.

14.4 Results

14.4.1 Descriptive Results

Table 14.1 shows descriptive statistics on all the manifest variables by gender. In the first two columns, we report the percentage of respondents who declared to be engaged in each activity (irrespective of the frequency). In the last two columns, we report the percentage of respondents engaged in each activity at least weekly. There are important differences by the type of activity considered and by the gender of the respondent. As it is well known, caregiving activities are more common for women than for men (e.g., Hank and Buber 2009). Also participation in educational courses and engagement in religious organisations tend to be more common among women than men (Arpino and Bordone 2017). Table 14.1 also shows that help to family and friends is in general more common among men, but once we consider it on at least a weekly basis, women tend to be more engaged. Engagement in other types of activities (i.e., sport or social clubs, political organisations and volunteering) is more common among men.

Table 14.1 Descriptive statistics on manifest variables by gender

	% Activity was mentioned		% At least weekly	
Manifest variables	Men	Women	Men	Women
Voluntary or charity	13.6	12.1	8.7	7.7
Education	6.7	7.9	2.9	4.2
Sport or social club	22.3	18.3	16.8	14.4
Religious organisations	9.9	13.4	6.6	9.2
Political organisations	6.1	3.1	2.7	1.2
Care to disabled	5.6	8.3	4.4	6.7
Help to family or friends	21.2	20.4	11.8	13.4
Grandparenting	30.9	38.6	15.4	21.5

When considering the percentage of older people engaged in each activity on an at least weekly basis (last two columns of Table 14.1), we notice that in general the percentages are not dramatically lower than those ignoring the frequency of engagement as presented before. This means that when older people engage in a given activity, they tend to do so often. Two exceptions are education and political participation. Also help to family and friends and grandparenting show much lower percentages of engagement when engagement is considered at least weekly. Grandparenting and participation in sport or social clubs are the activities where women and men, respectively, are more often engaged with at least weekly frequency.

In Table 14.2, we report descriptive statistics for the subsamples of retired and working older people.[2] Patterns of participation in society for older workers and retired persons are substantially different. Retirees show substantially lower prevalence of engagement for most of the considered activities. In particular, enrolment in educational courses is less than one third as compared to working respondents, and participation in political organisations is lowered by 40%. On the contrary, grandparenting is more common among retirees, probably also because of an age effect as retirees are, on average, older than working people and therefore it is more likely that they have grandchildren.

14.4.2 LCA Analysis

In order to simultaneously analyse possible engagement in one or more activities and its frequency, we use LCA on all manifest categorical variables described above.

Table 14.2 Descriptive statistics on manifest variables by activity status

	% Activity was mentioned		% At least weekly	
Manifest variables	Working	Retired	Working	Retired
Voluntary or charity	15.0	12.0	8.7	8.2
Education	13.3	4.2	5.1	2.8
Sport or social club	25.3	18.9	19.4	14.6
Religious organisations	10.0	11.5	6.3	7.9
Political organisations	6.4	3.9	2.7	1.6
Care to disabled	8.5	5.8	6.4	4.8
Help to family or friends	27.6	17.3	16.1	10.7
Grandparenting	31.0	36.7	14.4	19.5

Figure 14.1 shows the results of the LCA model. As explained in the "Method" section, we tried different solutions by changing the number of classes specified and we chose the solution that guaranteed the best fit. This solution is characterised by three classes. Note that the order of the classes is arbitrary. Figure 14.1 represents the three classes we identify. On the vertical axis, "Pr(outcome)" corresponds to the probability that a given manifest variable indicated on the horizontal axis assumes one of the four possible values ("Outcomes", indicated on the third dimension): (1) when daily engagement is reported, (2) when weekly, (3) when less often and (4) when never. The numerical values displayed in Fig. 14.1 are reported in Table 14.4 in the Appendix.

Class 1—the inactive: Most of the older Europeans belong to the first latent class (53% of the population is estimated to be part of this group). This class is characterised by high probabilities (between 88% and 98%) of not being engaged in the considered activities (i.e., outcome = 4). These probabilities of non-engagement are always higher than those in the other classes (with the exception of the variable "political organisations" for which the probability of non-engagement is very similar in classes 1 and 3). Notably, this cluster is not necessarily composed of people who are not engaged in any of the considered activities, but it includes people for whom the probability of being engaged in at least one activity, especially with a high intensity, is very low. Therefore, the label "inactive" serves only to simplify the presentation of results.

Class 2—the active: about 24% of the population belongs to a class with relatively high probabilities of engagement, especially on a moderate level of intensity, for several activities. It is interesting to note that there

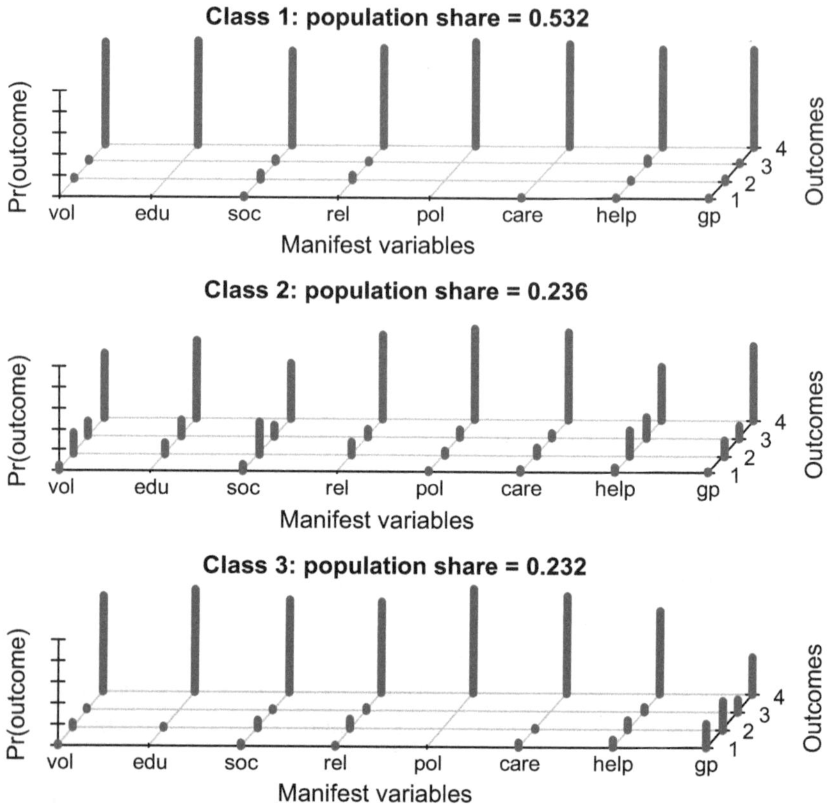

Fig. 14.1 Profile of latent classes with respect to manifest variables. *Note*: *vol* voluntary or charity work, *edu* educational or training course, *soc.* sport, social or other kind of club, *rel* religious organisation, *pol* political or community-related organisation, *care* caring for a disabled adult, *help* providing help to a relative, friend or neighbour, *gp* grandparental childcare. Outcomes: 1 = daily; 2 = weekly 3 = less often; 4 = never. Pr(outcomes) indicates Pr(Y = y), y = 1, 2, 3, 4

is no evidence of "specialisation" in this class, in the sense that both care activities (as help to family and friends) and leisure activities (such as participating in a sport or social club) show high probabilities of engagement (older people in this class have a probability of about 50% to be engaged in the mentioned activities). People in this class are more likely to be able to combine different activities (and of different type).

Class 3—the caregivers: A third latent class (23% of the population) is characterised by a concentration on care activities. In fact, in this class, only the probability of being engaged in care activities is high. For example, in this class, older people have a probability of providing help to friends and family of 21%. The likelihood of being engaged in grandparenting is particularly high: 74%. Importantly, the probabilities of being engaged in care activities on a daily basis are higher than in the other groups (the exception is care to disable for which daily engagement is very low in general and it is similar in the second and third classes) and it is particularly high for grandparenting: 23%.

Table 14.3 reports the effects of the covariates on the probability of belonging to the latent classes. The reference latent class is the first one.

Consistent with the descriptive evidence discussed above, women show a considerably higher probability than men to belong to the caregivers group (doing care activities, especially on a daily basis). Also, retired are more likely to belong to this group than their counterparts in the labour market. Older people with "other" activity statuses (i.e., unemployed and housekeepers) are even more likely to belong to the caregivers group than retired persons.

Somehow contrary to descriptive statistics, women are more likely to be in the active class than men, even though it seems that the effect of gender is smaller than for the caregivers. Therefore, we can think that age and working status mainly explain the gap between men and women observed in Table 14.2. Moreover, we have to keep in mind that the information provided by the LCA also accounts for possible interdependence among activities, while descriptive statistics in Table 14.2 were based on univariate distributions.

Workers are more likely to belong to the active class than retired, while unemployed and housekeepers are less likely to belong to this class.

Importantly, we notice an inverse U-shaped gradient of age for the probability of belonging to the active versus the inactive class: such probability increases for older adults aged between 56 and 70 as compared to older people in the 50–55 age group but then it decreases. A similar pattern is observed also when comparing the caregivers versus the inactive groups but in this case the positive gradient of age is observed up to the 71–76 age group.

Table 14.3 Effect of covariates on the probability of belonging to latent classes

Covariates	Coefficient	Std. error	*P*-value
Active versus inactive			
Female	0.23	0.04	0.00
Working	0.70	0.05	0.00
Other activity status	−0.45	0.06	0.00
Age 56–60	0.25	0.05	0.00
Age 61–65	0.42	0.06	0.00
Age 66–70	0.22	0.06	0.00
Age 71–75	−0.29	0.07	0.00
Age 76–80	−1.02	0.09	0.00
Age 81–85	−1.85	0.14	0.00
Intercept	−1.14	0.06	0.00
Caregivers versus inactive			
Female	0.75	0.05	0.00
Working	−0.73	0.07	0.00
Other activity status	0.15	0.06	0.01
Age 56–60	1.08	0.08	0.00
Age 61–65	1.35	0.09	0.00
Age 66–70	1.11	0.09	0.00
Age 71–75	0.26	0.09	0.00
Age 76–80	−0.96	0.13	0.00
Age 81–85	−14.99	0.00	0.00
Intercept	−1.69	0.10	0.00

14.5 Concluding Remarks

In this chapter, we studied participation in society in middle and later life, an important pillar of the active ageing framework. We focused on understanding to what extent older adults participate in different types of activities. By using LCA we identified three groups, characterised by different patterns of participation in society.

It has been shown that LCA may be useful to summarise relevant patterns of interdependence in the engagement in different activities and its intensity. Our analysis shows evidence in favour of both the competition and cumulation hypotheses. The "active" class, characterised by relatively high probabilities of engagement in different types of activities, favours the cumulation hypothesis: older people in this class may find it easier to combine different types of activities (e.g., grandchild care and participating in a social club). The "caregivers" class, on the contrary, characterised by relatively high probabilities of intensive caregiving favours the competition hypothesis: older people in this group may find it difficult to have sufficient time and energies to engage also in non-care activities.

Therefore, our analysis shows that cumulation and competition are not necessarily mutually exclusive hypotheses, and it highlights that different types of persons may find it more or less likely to combine different types of activities.

Two important results for policymaking in the area of promoting active ageing can be highlighted. First, there is a group of people with very low probabilities of engagement (the "inactive" class). The oldest persons in our sample (aged 76+) are more likely to belong to this group. Also unemployed and housekeepers are at very high risk of non-engagement.

Second, the caregivers—another relevant group—are characterised by intensive engagement in care activities and low probability of engagement in other types of activities (e.g., social groups, volunteering). Women are more likely to belong to this group.

This study thus identified two target populations on which policies to promote active ageing need to focus: the oldest old—very likely to be excluded from active participation in both kin and non-kin activities—and women, who are more likely to be "trapped" in intensive care activities.

While providing care can be beneficial for the carer as being needed by others increases the feeling of control over one's life (Krause 2008), it has to be considered that the effect of caring depends on the context and intensity of caregiving (Krause 1987).

These results emphasise the importance of policy proposals designed to address the three pillars of active ageing: health, participation and security. In particular, this study suggests that attention has to be given to those that are targeted specifically to older people and women: encouraging people to participate fully in family and community life as they grow older; and reducing inequities in participation by women (WHO 2002).

While the first aspect could be reinforced by creating a society for all ages and through promoting a positive image of ageing, the gender-specific target should aim to recognise the important contribution that older women make to families and communities through caregiving and participation in the informal economy. Equal chances should also be guaranteed to women and men to participate also in non-family-related activities.

We plan to further extend the current analyses by testing whether the patterns of participation in society that we identified based on the first two waves of SHARE are associated with health outcomes measured at future points in time. We also plan to refine the analyses by including additional control variables and by analysing heterogeneities across countries.

Appendix

Table 14.4 Conditional item responses probabilities for each manifest variable and latent class

	Item responses probabilities			
Latent classes	Pr(Y=1)	Pr(Y=2)	Pr(Y=3)	Pr(Y=4)
		Voluntary or charity		
Class 1	0.63	1.45	1.87	96.05
Class 2	5.84	18.79	12.88	62.49
Class 3	1.29	4.16	2.45	92.10
		Education		
Class 1	0.32	0.87	0.49	98.32
Class 2	1.13	9.61	14.55	74.72
Class 3	0.20	1.70	0.25	97.84
		Sport or social club		
Class 1	1.90	6.63	3.00	88.47
Class 2	6.58	30.25	10.51	52.66
Class 3	2.18	7.65	2.16	88.02
		Religious organisations		
Class 1	0.74	4.47	2.68	92.11
Class 2	1.45	11.04	7.18	80.33
Class 3	1.26	8.68	2.98	87.09
		Political organisations		
Class 1	0.20	0.48	0.96	98.36
Class 2	1.53	4.37	8.04	86.07
Class 3	0.05	0.37	0.87	98.71
		Care to disabled		
Class 1	2.09	0.73	0.33	96.85
Class 2	4.49	7.00	4.51	84.00
Class 3	4.28	1.97	0.78	92.97
		Help to family or friends		
Class 1	1.19	2.36	4.51	91.94
Class 2	5.91	23.38	19.53	51.18
Class 3	7.35	9.48	4.47	78.70
		Grandparenting		
Class 1	2.09	2.93	2.74	92.23
Class 2	2.11	15.26	11.69	70.94
Class 3	23.41	28.16	12.59	35.84

Notes

1. http://ec.europa.eu/archives/ey2012/ey2012main9ef0.html
2. Here we exclude unemployed and housekeepers because of their relatively small sample sizes. They are however included in the LCA analysis.

References

Arpino, B., & Bordone, V. (2014). Does grandparenting pay off? The effect of child care on grandparents' cognitive functioning. *Journal of Marriage and Family, 76*(2), 337–351.

Arpino, B., & Bordone, V. (2017). Regular provision of grandchild care and participation in social activities. *Review of Economics of the Household, 15*(1), 135–174.

Bordone, V., Arpino, B., & Aassve, A. (2017). Patterns of grandparental child-care across Europe: The role of the policy context and working mothers' need. *Ageing & Society, 37*(4), 845–873.

Börsch-Supan, A., Brugiavini, A., Jürges, H., Kapteyn, K., Mackenbach, J., Siegrist, J., & Weber, G. (2008). *Health, ageing and retirement in Europe (2004–2007). Starting the longitudinal dimension.* Mannheim: Mannheim Research Institute for the Economics of Aging (MEA).

Börsch-Supan, A., Brugiavini, A., Juerges, H., Mackenbach, J., Siegrist, J., & Weber, G. (2005). *Health, ageing and retirement in Europe. First results from the Survey of Health, Ageing and Retirement in Europe.* Mannheim: Mannheim Research Institute for the Economics of Aging (MEA).

Bowling, A., & Dieppe, P. (2005). What is successful ageing and who should define it? *BMJ, 331*(7531), 1548–1551.

Bulanda, J. R., & Jendrek, M. P. (2014). Grandparenting roles and volunteer activity. *Journals of Gerontology, Series B: Psychological Sciences and Social Sciences.* First published online April 10, 2014.

Caro, F., & Bass, S. (1995). Dimensions of productive engagement. In S. A. Bass (Ed.), *Active & Aging.* Baltimore: Johns Hopkins.

Chappell, N. L., & Reid, R. C. (2002). Burden and well-being among caregivers: Examining the distinction. *Gerontologist, 42*(6), 772–780.

Choi, N. G., Burr, J. A., Mutchler, J. E., & Caro, F. G. (2007). Formal and informal volunteer activity and spousal caregiving among older adults. *Research on Aging, 29*(2), 99–124.

Depp, C. A., & Jeste, D. V. (2006). Definitions and predictors of successful aging: A comprehensive review of larger quantitative studies. *American Journal of Geriatric Psychiatry, 14*(1), 6–20.

Engelhardt, H., Buber, I., Skirbekk, V., & Prskawetz, A. (2010). Social involvement, behavioural risks and cognitive functioning among older people. *Ageing & Society, 30*(5), 779–809.

Foster, L., & Walker, A. (2013). Gender and active ageing in Europe. *European Journal of Ageing, 10*(1), 3–10.

Hagenaars, J. A., & McCutcheon, A. L. (Eds.). (2002). *Applied latent class analysis*. Cambridge: Cambridge University Press.

Hank, K., & Buber, I. (2009). Grandparents caring for their grandchildren: Findings from the 2004 survey of health, ageing, and retirement in Europe. *Journal of Family Issues, 30*(1), 53–73.

Hank, K., & Stuck, S. (2008). Volunteer work, informal help and care among the 50+ in Europe: Further evidence for linked productive activities at older ages. *Social Science Research, 37*(4), 1280–1291.

Haughton, D., Legrand, P., & Woolfrod, S. (2009). Review of three latent class cluster analysis packages: Latent GOLD, poLCA, and MCLUST. *The American Statistician, 63*(1), 81–91.

Hultsch, D. F., Hertzog, C., Small, B. J., & Dixon, R. A. (1999). Use it or lose it: Engaged lifestyle as a buffer of cognitive decline in aging? *Psychology and Aging, 14*(2), 245–263.

Kohli, M., Hank, K., & Künemund, H. (2009). The social connectedness of older Europeans: Patterns, dynamics and contexts. *Journal of European Social Policy, 19*(4), 327–340.

Koslowski, A. S. (2009). Grandparents and the care of their grandchildren. In D. Kneale, E. Coast, & J. Stillwell (Eds.), *Fertility, living arrangements, care and mobility: Understanding population trends and processes* (Vol. 1). Dordrecht: Springer.

Krause, N. (1987). Life stress, social support, and self-esteem in an elderly population. *Psychology and Aging, 2*, 349–356.

Krause, N. (2008). *Aging in the church: How social relationships affect health*. West Conshohocken, PA: Templeton Foundation Press.

Linzer, D. A., & Lewis, J. B. (2011). poLCA: An R package for polytomous variable latent class analysis. *Journal of Statistical Software, 42*(10), 1–29.

Lubben, J., & Gironda, M. (2003). Centrality of social ties to the health and well-being of older adults. In B. Berkman & L. Harootyan (Eds.), *Social work and health care in an aging society*. New York: Springer.

Minkler, M. (1999). Intergenerational households headed by grandparents: Contexts, realities, and implications for policy. *Journal of Aging Studies, 13*(2), 199–218.

Pynnönen, K., Törmäkangas, T., Heikkinen, R.-L., Rantanen, T., & Lyyra, T.-M. (2012). Does social activity decrease risk for institutionalization and mortality in older people? *The Journals of Gerontology Series B: Psychological Sciences and Social Sciences, 67*(6), 765–774.

Rowe, J. W., & Kahn, R. L. (1987). Human aging: Usual and successful. *Science, 237*, 143–149.

Rowe, J. W., & Kahn, R. L. (1997). Successful aging. *The Gerontologist, 37*(4), 433–440.

Van Bavel, J., & De Winter, T. (2013). Becoming a grandparent and early retirement in Europe. *European Sociological Review, 29*(6), 1295–1308.

World Health Organization (WHO). (2002). *Active ageing: A policy framework*. Geneva: World Health Organization.

Zaidi, A., Gasior, K., Hofmarcher, M. M., Lelkes, O., Marin, B., Rodrigues, R., Schmidt, A., Vanhuysse, P., & Zolyomi, E. (2013). *Active Ageing Index 2012 concept, methodology and final results*. Vienna: European Centre. Retrieved from http://www1.unece.org/stat/platform/download/attachments/76287849/Methodology-Paper%20Final.pdf?version=1&modificationDate=1393836329990&api=v2

Bruno Arpino is an associate professor at the Universitat Pompeu Fabra (UPF) and co-director of the Research and Expertise Centre on Survey Methodology (RECSM, UPF). His main research interests are in the areas of social demography, causal inference and applied statistics. He is particularly interested in studying intergenerational relationships, ageing and health, fertility and immigrants' assimilation. He currently coordinates a 3-year collaborative project titled "Care, retirement and wellbeing of older people across different welfare regimes" (CREW).

Valeria Bordone is lecturer with the Department of Sociology at the University of Munich (LMU). Her main research interests are in the area of social demography, with a focus on intergenerational relationships, ageing and the subjective dimension of ageing. Previously, she worked at the Centre for Research on Ageing at University of Southampton, UK.

15

How Relevant is Active Ageing? Evidence from Portugal

Sara Valente de Almeida and Pedro Pita Barros

15.1 Introduction

Ageing is one of the biggest social and economic challenges of the twenty-first century, particularly for Europe, the continent with the highest proportion of population aged 65 or over. There is now an emerging discussion regarding active ageing policies and how governments should act to assure the sustainable development of society in the future. During the next years, many new challenges will arise in the European countries. Fostering healthy and active ageing is thus indispensable to keep up with the technological and societal evolution, ensuring prosperity for the elder generations in ageing societies. During this process, governments will

We benefited from the collaboration of Comfort Keepers, who provided the data for the current research project. All opinions are our own.

S.V. de Almeida (✉) • P.P. Barros
Universidade Nova de Lisboa, Lisbon, Portugal

A. Zaidi (eds.), *Building Evidence for Active Ageing Policies*,
https://doi.org/10.1007/978-981-10-6017-5_15

have the important role to assure equity in the access to services for the whole population and, when necessary, even adapt the design of public policy.

Several institutions have explored different ways of identifying the features of active ageing to understand how public services can create the necessary conditions for the population to remain active and integrated in society after reaching an advanced age. The broadness of the active ageing concept itself creates limits to the settlement of a universal definition. Following the World Health Organization (2002), "active aging is the process of optimizing opportunities for health, participation and security in order to enhance quality of life as people age". In turn, the European Commission adapts this vision and considers that active ageing occurs when people continue to participate in the formal labour market and engage in other unpaid productive activities (such as care provision to family members and volunteering), while living healthy, independent and secure as they age (Zaidi et al. (2013)). From this last definition emerged a country-level Active Ageing Index,[1] firstly developed in 2012 by the European Commission and the United Nations Economic Commission for Europe (UNECE) to support the design of sustainable responses to the challenges of an aged population.

Meanwhile, also the member states have tried to answer the ageing challenge at national level, and Portugal is not an exception. Recently, the Health Cluster Portugal developed the Ambient Assisted Living for All (AAL4ALL) project. Besides contributing to the development of Ambient Assisted Living (AAL) practices, this project gathers useful information for the major agents in the provision of healthcare services.[2]

As such, after acknowledging the developments made on the subject of active ageing, an immediate question arises: what is the relation between the active ageing and health?[3] Searching for an answer, the objective of this study is also to provide some insight on how the Portuguese population is ageing, what could be the most urgent needs of this social group and how could public policy provide the necessary support.

Using the micro-level data collected from the AAL4ALL survey for final users, our study adapts the approach of the European Commission and the UNECE (Zaidi and Stanton 2015) for the construction of an Active Ageing Index (AAI) at individual level. Besides providing support to public policy design, we also present an innovative analysis of the AAI at the individual level, exploring its manifold applications and its relationship with the self-assessed health (SAH) indicator.

Results show a positive relationship between the individual Active Ageing Index and the level of self-assessed health reported by the participant along with other health-related variables. In addition, there is a need to reduce barriers to participation in social activities and difficulties in the access to healthcare services, which have a negative impact on active ageing.

This study is organized as follows: the next section presents a review of the literature on active ageing indices, national health surveys and self-assessed health estimations; next, the second section describes the methodology adopted; the third section describes the development and analysis of two versions of the individual Active Ageing; and finally, the last section presents a discussion on the main conclusions and policy implications. Annex 1 contains the variables descriptions, Annex 2 the graphic analysis, Annex 3 provides the variables descriptive statistics and Annex 4 some secondary regressions.

15.2 Literature Review

The constant increase of the aged population represents a challenge at several levels, likely to have impact in the structure of healthcare provision, social security systems and labour markets. Moreover, given the difficulties in measuring such effects, identifying the main needs and prioritizing interventions become a complex exercise.

At this moment, Portugal faces demographic changes related to a decreasing birth rate and increasing longevity, structural problems in rural areas and high emigration rates among the young qualified workers. In such a challenging context, the aforementioned Health Cluster Portugal developed the Ambient Assisted Living for All (AAL4ALL) project. This consisted in the construction of an ecosystem of products and services Ambient Assisted Living (AAL) associated to a business model. Namely, it involves a recent set of methods, concepts, (electronic) systems, devices and services created to promote active and healthy ageing, particularly amidst people with cognitive difficulties or in need for additional support in their everyday lives. In order to identify the characteristics and requirements of the current users, dedicated surveys were conducted on final users, informal and formal caregivers.

The informative potential of health surveys as the ones in the AAL4ALL project is manifold. Pfarr, Schmid and Schneider (2012) show that health behaviour, healthcare services utilization, mental and physical health conditions and country characteristics have significant impacts on the reported health of the population aged 50 and above. The authors use the database from the Survey of Health, Ageing and Retirement in Europe (SHARE) and analyse the level of self-assessed health (SAH) as a fair approximation to the current health status of the population. Likewise, using data from the 1958 British birth cohort, Manor, Matthew and Power (1999) show that self-assessed health can be used as a strong predictor of morbidity. More recently, Atherton and Power (2007) show that health selection effects may explain to some extent increases in mortality, while Mensah and Hobcraft (2008) highlight the relevance of cognitive and behavioural development in childhood. Both these studies also use the 1958 British birth cohort and the self-assessed health indicator as a health status measure. Despite being a popular approach, using information on how the participants evaluate their own health as an actual health indicator has been subject to several criticism due to its subjectivity and exposure to peer effects. In our study, we follow our references, bearing in mind the implications behind using such a subjective measure.

Regarding active ageing measurement, Zaidi et al. (2013, 2015) present a composite measure covering four specific domains of activities with potential to enable the right environment for active ageing. The Active Ageing Index ("AAI") was developed with funding from the European Commission's Directorate General for Employment, Social Affairs and Inclusion (DG EMPL) and the Population Unit of the United Nations Economic Commission for Europe (UNECE). This analytical tool was launched in 2012 for 27 countries of the European Union (EU) and provides insight on the overall level of activity, social integration and participation in the labour market of the elderly population. The aggregated indicator is the arithmetic weighted average of the specific indices from four different domains: (1) employment; (2) participation in society; (3)

independent, healthy and secure living; and (4) capacity and enabling environment for active ageing.

Besides cross-country comparison purposes, this methodology has equally useful applications at national level. In a study developed in Bangladesh, Tareque et al. (2013) discuss the relationship between life expectancy and active ageing using the AAI. The authors developed an AAI based on the recommendations by the Active Aging Taskforce of the Western Australian Government and find a positive correlation between active ageing and healthy life expectancy. In turn, Paúl et al. (2012) studied the World Health Organization's model of active ageing. According to findings, by keeping active, older people seem to overcome difficulties more easily and have a more active social life and healthier behaviour. More recently, in Zasimova and Sheluntcova (2014), the authors use an international Study on global AGEing and adult health (SAGE) to construct an Active Ageing Index covering three different domains—older people's health, participation in social activities and secure living. The results provided a general framework of the Russian elderly level of activity, with 41.5% of the sample being considered inactive.

In our study, we use the main dimensions of the AAI developed with the help of funding and support from the European Commission and the UNECE for the construction of an Active Ageing Index, applied to a sample of the Portuguese population over 50 years old taken from the AAL4ALL survey. For this purpose, we developed an individual approach similar to Zasimova and Sheluntcova (2014) in order to achieve an index applicable on longitudinal databases for the elderly population.

15.3 Data Source and Methodology

The AAL4ALL final user's survey is composed of eight main sections, which provide information at several levels, from the participants' household composition, to their main needs or their attitude towards technology. The survey was simultaneously conducted in 18 different regions of Portugal between July and September 2011.The result is a final database composed by 1174 Portuguese citizens over the age of 50. Subjects were randomly selected from residential facilities, attending day-care centres or living in their own home.

Such a broad range of information enabled us in finding suitable indicators to the framework in Zaidi et al. (2013) and proposing the construction of an objective measure for active ageing at individual level. According to the baseline formulation, only some specific questions were selected to develop the AAI (for the full list of selected questions see Table 15.6 in Annex 1). The individual AAI was firstly constructed according to the following expression:

$$AAI^{a} = 0.35\,D1^{a} + 0.35\,D2 + 0.1\,D3 + 0.2\,D4 \qquad (15.1)$$

Table 15.1 explains the contents in our Eq. (15.1). To have the same range of values for each of the inputs and ensure a fair comparison between individuals, all non-dummy variables were subject to normalization.[4] Before being normalized, when necessary, the variables were also categorized from 1 to 5, so that 1 was the minimum and 5 the maximum activity level. Note that, as in Zaidi et al. (2013), the first domain corresponds to employment status, here represented by a binary individual variable with the value 1 if the person is employed and 0 otherwise. In other words, the participants that are not employed at the moment of the interview scored 0 for this domain.

However, the adaptation of the index from national to individual level is not straightforward. While at national level the employment rate may relate to the level of active ageing in the society, the same is not necessarily true for individuals. If, on one side, being employed before the retirement age may not be necessarily a sign of active ageing, on the other, the individual index should also acknowledge those who retire and maintain a good activity level. As such, we propose that instead of having null value for the employment domain, the retired seniors are ranked proportionally to their social activity.[5] To achieve this new version of the index, our first equation was modified such that

$$AAI^{b} = 0.35\,D1^{b} + 0.35\,D2 + 0.1\,D3 + 0.2\,D4 \qquad (15.2)$$

Table 15.1 Individual active ageing indicator—Eq. (15.1)

Domain	Domain weight (%)	Indicators	Indicator weight (%)
D1[a]—Domain 1: Employment	35	Currently working	100
D2—Domain 2: Participation in society	35	Participation in recreational activities	35
		Participation in social activities	35
		Not lonely	30
D3—Domain 3: Independent, healthy and secure living	10	Income	30
		Autonomy in DLA	40
		Independent	30
D4—Domain 4: Capacity and enabling environment for active ageing	20	Cognitive difficulties	35
		IT use to socialize	20
		Current health status	45

Source: Adapted from Zaidi et al. (2013)

Note: In addition, the investment must focus the quality of the opportunities provided so that there is a true incentive for people to continue engaged in society after entering retirement; Superscript "a" means D^a refers to the first version of the Active Ageing Index and thus, first version of domain 1.

With this second framework, we chose to increase the original weight for social participation (Table 15.2). The new version implies that the AAI of seniors in retirement suffers a penalty relative to those in employment that is proportional to their engagement in social and recreational activities. We believe that the AAI^b provides fairer evaluation of individual activity levels, thus being better suited to the purpose of our study.

Using linear regressions, we then estimate both individual indices as a function of six main groups of variables: socio-demographic (SD_i) and healthcare characteristics (H_i), personal features associated with well-being, activity and quality of life (DLA_i), social activity (Soc_i), access to services ($Serv_i$) and using AAL devices (AAL_i).[6] Apart from age and weight, all variables were converted into binary dummy variables, including self-assessed health. The breaking points to apply this procedure were established according to the positive and negative connotation of each answer. For example, self-assessed health variable is equal to 1 if the answer is at least "good" and 0 otherwise.

Table 15.2 Individual active ageing indicator—Eq. (15.2)

Domain	Domain weight (%)	Indicators	Indicator weight (%)
D1[b]—Domain 1: Employment and active retirement	35	Currently working	50
		Retired × Social Activity*	50
D2—Domain 2: Participation in society	35	Participation in recreational activities	35
		Participation in social activities	35
		Not lonely	30
D3—Domain 3: Independent, healthy and secure living	10	Income	30
		Autonomy in DLA	40
		Independent	30
D4—Domain 4: Capacity and enabling environment for active ageing	20	Cognitive difficulties	35
		IT use to socialize	20
		Current health status	45

Source: Adapted from Zaidi et al. (2013)

Note: Superscript "b" means D^b refers to the second version of the Active Ageing Index and thus, second version of domain 1.

*$Social\ activity = Part.in\ recreational\ activities \times Part.in\ social\ activities$

The result is a relationship expressed as follows:

$$AAI_i^{a,b} = \beta X_i + \varepsilon_i \tag{15.3}$$

where X_i includes the binary variables described in Table 15.3, β_k are the regression coefficients and ε_i the error term.[7]

The indicators were chosen in agreement with the purpose of the study and focusing those that might be subject to public policy intervention.

15.4 The Individual Active Ageing Index

The individual active ageing indicator includes a group of ten indicators chosen to represent the four domains of the original index—employment; participation in society; independent, healthy and secure living; and capacity and enabling environment for active ageing. The selection of the indicators themselves took into account both the context with these domains and the number of responses given, in order to maximize the dimension of the sample and the reliability of the analysis.

Table 15.3 Variables used for the estimation of AAI

Coefficient	Variables	Description
β	Constant	
$\beta_{i,i=1,\ldots,6}$	SD_i	Sex, age, age^2, weight, education and marital status
$\beta_{i,i=7,\ldots,11}$	H_i	Self-assessed health, unemployed for health reasons, not needing monitoring or specialized support, needing home support
$\beta_{i,i=12,13,14}$	DLA_i	Independence in performing daily life activities, difficulties using home devices, insecurity at home
$\beta_{i,i=15,16}$	Soc_i	Frequently attending social spaces, health limitations to social life
$\beta_{i,i=17}$	$Serv_i$	Difficulty in the access to healthcare services
$\beta_{i,i=18}$	AAL_i	AAL user

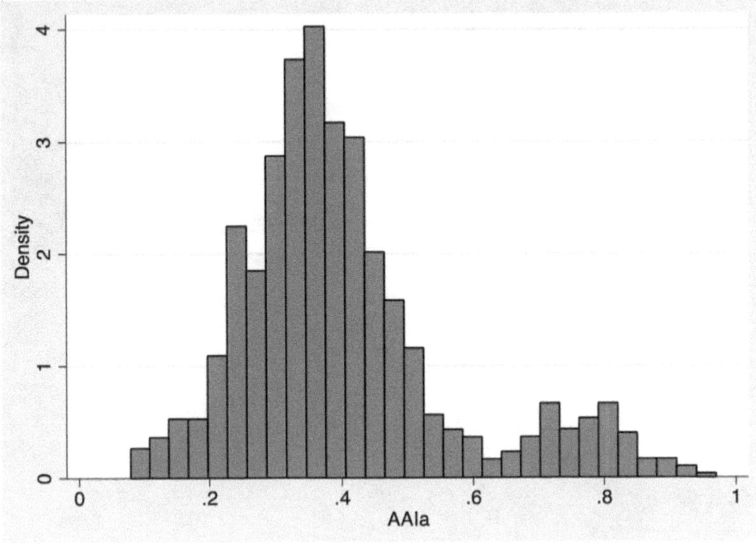

Fig. 15.1 Histogram Active Ageing Index a

The result is a weighted measure of the level of activity for 1021 Portuguese seniors, which presents a quite normal distribution across the sample (Fig. 15.1). However, dividing the index into age groups (Fig. 15.2) unveils a bimodal distribution for the participants between 50 and 64 years old. This evidence suggests the existence of two distinct tendencies with overlapping distributions.

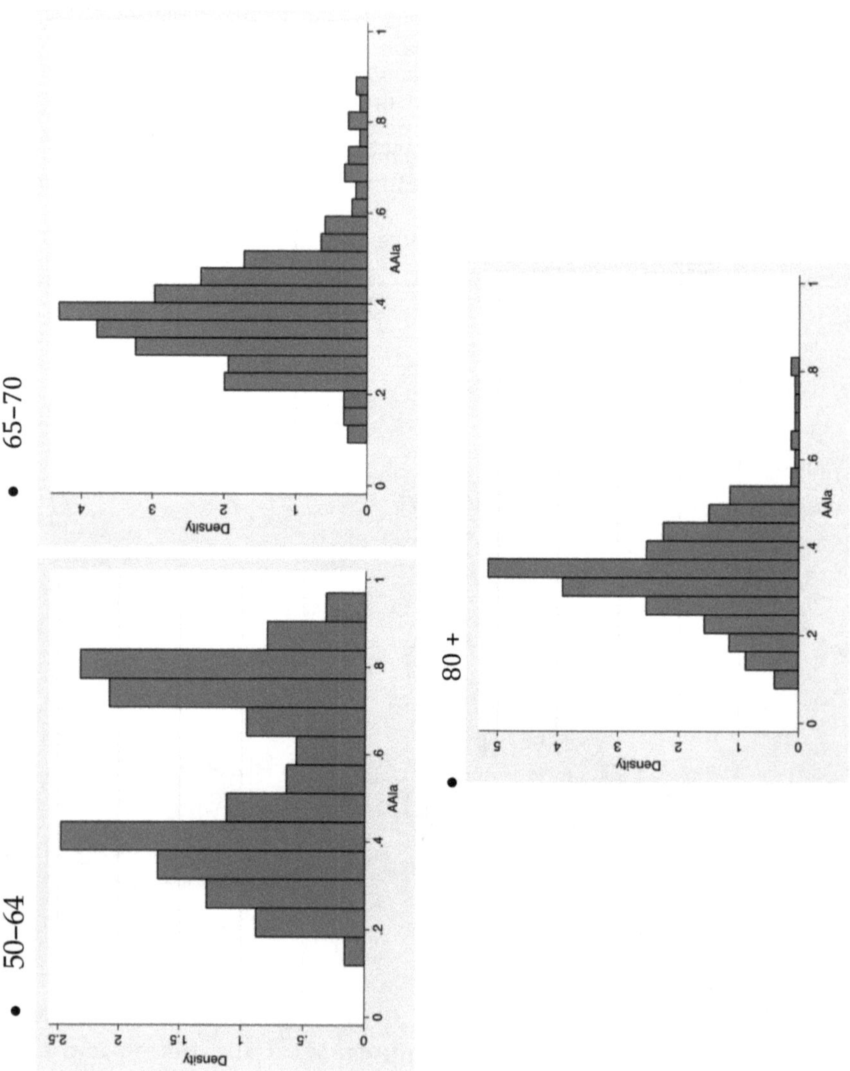

Fig. 15.2 Histogram Active Ageing Index a, by age groups

As presented in the methodology section, in Eq. (15.1), the employment status has a high weight in the construction of the index. As a consequence, the impact of being employed is producing two different modes across the data: one higher for those who are employed and one lower for those who are not (Fig. 15.3). The relevance given to the employment status with this framework happened to be high enough to create bimodal distributions.

After adapting the first domain of the individual AAI according to Eq. (15.2), the bimodal distribution vanishes. The result is a series of normal distributions across all age groups that become progressively skewed to the left as the subjects get older.[8] The general trend of AAI^b is similar to AAI^a but with activity levels more equally distributed across the sample.

Looking further at the distributions, with both approaches, the level of activity decreases with age, and women present lower activity levels than men, as shown in Table 15.4.

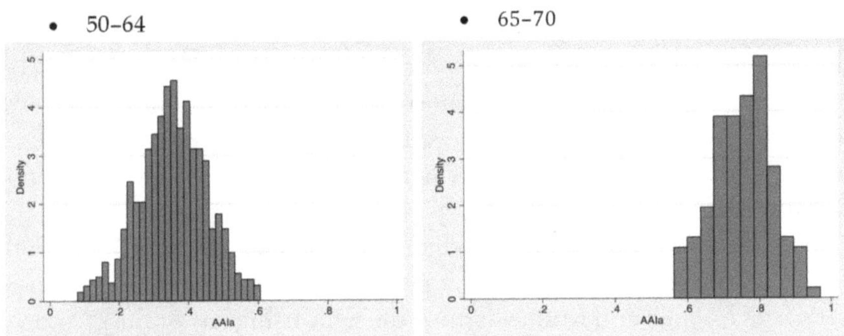

Fig.15.3 Histogram Active Ageing Index a, by employment status

Table 15.4 Average Active Ageing Index, by sex and age group

Age Group	Sex	AAI[a]	AAI[b]	Zaidi et al. (2013)
55–64	W	0.53	0.48	
	M	0.60	0.52	
65–79	W	0.38	0.40	
	M	0.41	0.43	
80+	W	0.33	0.37	
	M	0.38	0.41	
Total	W	0.39	0.40	0.31
	M	0.44	0.44	0.36

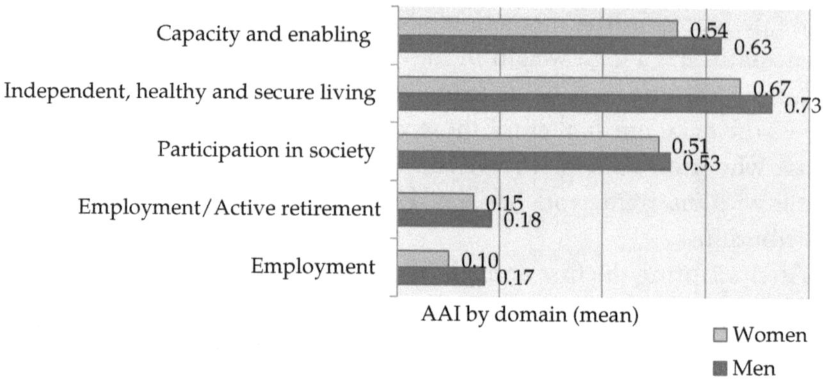

Fig. 15.4 Average activity level by domains of the AAI and sex

Figure 15.4 shows that this evidence persists generally in all domains of the two individual AAIs constructed.

As it happens, women only have a higher average score than men in the domain of social participation for the age group between 50 and 64 years old.[9] The fact that women constitute 70% of our sample may have some influence in these results. Nevertheless, this raises relevant concerns regarding the situation of older women in Portugal, especially taking into account their higher life expectancy.

Our evidence follows the findings in Zaidi et al. (2013) at both European and national levels. The percentage point difference between the scores for men and women is *fairly* the same using any of the presented measures—AAI^a, AAI^b and *AAI* (Zaidi et al. (2013)). Moreover, with the aggregate indicator, the authors also found that in Portugal the domain for participation in society is the most equally distributed across genders, against the employment domain with 11.2 percentage points of difference (Zaidi et al. (2013)). In our case, the highest inequalities are also present in the first domain but only for the first formulation of the individual AAI. With AAI^b, it is the fourth domain with the highest gender gap, which covers the level of cognitive difficulties, the use of IT to socialize and the current health status.[10]

15.4.1 The Determinants of Active Ageing

Using *AAI*[a,b] as a function of selected socio-demographic characteristics (not used in their construction), the estimation of an Ordinary Least Squares regression allowed us to understand the potential predictors of a good level of activity at an advanced age.

The main results are presented on Table 15.5.[11] Both individual AAIs decrease with age until a certain point, but they become relatively stable for the later stages of life. Self-assessed health is significant and positively related to both AAIs, that is, people with higher health status also have higher active ageing indices. Although we do not have enough information to define the causality relationship between both indicators, the individual AAI turns out as a useful informative tool to follow up the

Table 15.5 Factors associated with the *AAI*[a] and *AAI*[b] (OLS estimation)

Variables	(1) AAI[a]		(2) AAI[b]	
Woman	−0.0245**	(0.011)	−0.0276***	(0.010)
Age	−0.0488***	(0.006)	−0.0163***	(0.006)
(Age^2)/100	0.0299***	(0.004)	0.00939**	(0.004)
Weight	−0.000425	(0.000)	−0.000461	(0.000)
Married	0.00875	(0.010)	−0.0115	(0.010)
Self-assessed health	0.0464***	(0.010)	0.0425***	(0.010)
Low education	−0.0320***	(0.011)	−0.0284***	(0.011)
Freq. to social spaces	0.0217**	(0.010)	0.0495***	(0.009)
Unemployed for health reasons	−0.173*	(0.094)	−0.190**	(0.087)
Dif. IADL	−0.0332***	(0.012)	−0.0362***	(0.012)
Health limit. to social life	−0.0382***	(0.012)	−0.0386***	(0.011)
Home support	−0.0396*	(0.024)	−0.0502**	(0.022)
Dif. using HC services	−0.0156	(0.014)	−0.0286**	(0.013)
Dif. using devices	−0.00999	(0.016)	−0.000801	(0.015)
Monitorization n.a.	−0.0112	(0.012)	−0.00810	(0.011)
Specialized support n.a.	0.0584***	(0.012)	0.0409***	(0.012)
Insecurity at home	−0.0125	(0.011)	−0.0118	(0.010)
AAL user	0.0246**	(0.010)	0.0136	(0.010)
Constant	2.403***	(0.220)	1.165***	(0.205)
Observations	891		891	
R-squared	0.383		0.260	

Standard errors in parentheses

*, ** and *** represent the variables which are significant at 10%, 5% and 1% significance levels, respectively

individuals' health condition (here measured by self-assessed health). In addition, other health-related variables, such as being unemployed for health reasons or perceiving difficulty performing instrumental activities of daily living (ADL), are also relevant predictors of both AAIs.

Social participation seems to have positive consequences in the quality of living after 50 years old. Namely, frequent attendances to social spaces and having health limitations to engage in social activities have significant impacts on the Active Ageing Index value.

Regarding the use of AAL services, although users were expected to have lower levels of activity, our results show that it might increase the activity levels. However, besides this effect disappearing with the AAI^b, it could otherwise be a consequence of AAL users having higher household income and education.

Finally, the main difference associated with the use of AAI^b instead of AAI^a is the negative effect of the difficult access to healthcare services. AAI^b provides a better assessment of the willingness to remain active, independently of being employed or not. As such, this evidence is important because it raises thoughts on the role of public and healthcare support, particularly during the transition period into retirement.

Looking at each age group separately, for individuals above the age of 65, being a woman is not such a strong determinant of the AAI.[12] In turn, the use of AAL devices is significant after the age of 65, but not for the older participants. As these services are mainly designed for monitoring and prevention purposes, their impact on activity levels is most expected before reaching the more advanced stage of life.

For individuals living with less than €489, difficulty in the access to services has a significant negative effect on the AAI, and the use of AAL services is not relevant.[13] These results evidence a need to improve equity in the provision of healthcare services, which do not seem to be equally accessible for all individuals across the country.

15.5 Conclusions

This project proposes the construction of an individual-level Active Ageing Index for Portugal, using 1021 inquiries from over 50 years old across 18 regions in the country. The analysis covered the domains of socio-demographic characteristics, healthcare services, activities of daily living (ADL), social participation and access to services (including Ambient Assisted Living) in the pursuit for possible policy instruments to efficiently promote active ageing.

After performing the necessary changes, we applied the AAI methodology presented in Zaidi et al. (2013) to assess the activity level of the population in one EU member state—"Portugal". This exercise contributes to the study of ageing in Europe and shows an alternative use of the framework developed by the AAI project of the European Commission and the UN Economic Commission for Europe.

Our results show that active ageing is related to higher levels of self-assessed health, as well as other health-related variables. This being true, it is likely that the promotion of active ageing in the country is followed by a simultaneous improvement in the general health condition of the population. Active ageing could then benefit healthcare users and providers by reducing the budgetary burden of healthcare services.

Moreover, ageing in Portugal appears to be characterized by unequal levels of activity between genders and income levels. According to our analysis, improving social interaction and inclusion, as well as ensuring an equal access to services to promote lifelong activity can contribute to decrease these disparities in terms of active and healthy ageing.

Our results support previous evidence on the importance of updating public outlays on education, family assistance and healthcare for future generations. In addition, public investment must focus on the quality of the opportunities, in order to provide a true incentive for the elderly to continue engaged in society after entering retirement.

For the time being, there is a need to increase the integration of the elderly in society, avoiding feelings of isolation and uselessness in order to assure the long-term well-being of the population and adapt to the new demographic trends.

The construction of an individual active ageing allows for more in-depth analysis of the population's ageing process that can be subject to future research. There is room to further investigate disparities at regional level, as well as to deepen the study of the determinants of the Active Ageing Index, the self-assessed health indicator and potential users of AAL devices.

Annex 1

Table 15.6 Variables description—Active Ageing Index[a,b]

Indicators		Variable	Variable description
1st domain: Employment			
Currently working		empl	Dummy variable with value 1 if the subject is employed and 0 otherwise
Retired × Social participation		retired	Dummy variable with value 1 if the subject is retired and 0 otherwise
2nd domain: Participation in society			
Social Participation	Participation in recreational activities	f1_N	Subject's frequency of participation in recreational activities or hobbies—"Never" = 1 to "Always" = 5
	Participation in social activities	f3_N	Subject's frequency of participation in social activities—"Never" = 1 to "Always" = 5
Not lonely		nlonely_N	Subject's perception of not being lonely—"Never" = 1 to "Always" = 5
3rd domain: Independent healthy and secure living			
Income		income	The household monthly income divided into the following intervals: 1. Until €245 2. From €246 to €489 3. From €490 to €734 4. From €735 to €1223 5. From €1224 to €2445 6. Above €2445
		income_N	Categorized variable with income thresholds mid-values: • €122.5 if income = 1 • €367.5 if income = 2 • €612 if income = 3 • €979 if income = 4 • €1834.5 if income = 5 • €2445 if income = 6

(*continued*)

Table 15.6 (continued)

Indicators	Variable	Variable description
Autonomy in BADL	autonomy_N	The autonomy performing basic activities of daily living (BADL) divided into the following intervals: 1. If at least needs help in all BADL 2. If needs help in at least one BADL 3. Independent in all BADL
Independent	ind	Dummy variable with value 1 if the subject lives alone or with the spouse[1]
4th domain: Capacity and enabling environment for active ageing		
Cognitive difficulties	ncogdif_N	The level of cognitive difficulties divided into the following intervals: 1. If acknowledges to have extreme difficulty in all cognitive status 2. If acknowledges to have extreme difficulty in at least one cognitive status 3. If acknowledges to have extreme difficulty in at least one cognitive status 4. If acknowledges to have moderate difficulty in at least one cognitive status 5. If acknowledges to have slight difficulty in at least one cognitive status 6. If acknowledges no to have difficulty in all cognitive status
IT use to socialize	f7_N	How often the subject believes he could use technology to talk, book events, play games with family and friends without leaving the house—"Never" = 1 to "Always" = 5

(*continued*)

Table 15.6 (continued)

Indicators	Variable	Variable description
Current health status	currenth_N	Categorization of the subject health status according to: 1. Healthy 2. In recovery 3. Acute disease (e.g., stroke, rheumatism, pneumonia, etc.) 4. Chronic disease (e.g., diabetes, cancer, obesity, respiratory disease, etc.).

"_*N*" indicates the variable was subject to normalization according to: $f(x_i)_N = \frac{f(x_i) - f(x_i)_{\min}}{f(x_i)_{\max} - f(x_i)_{\min}}$

Annex 2

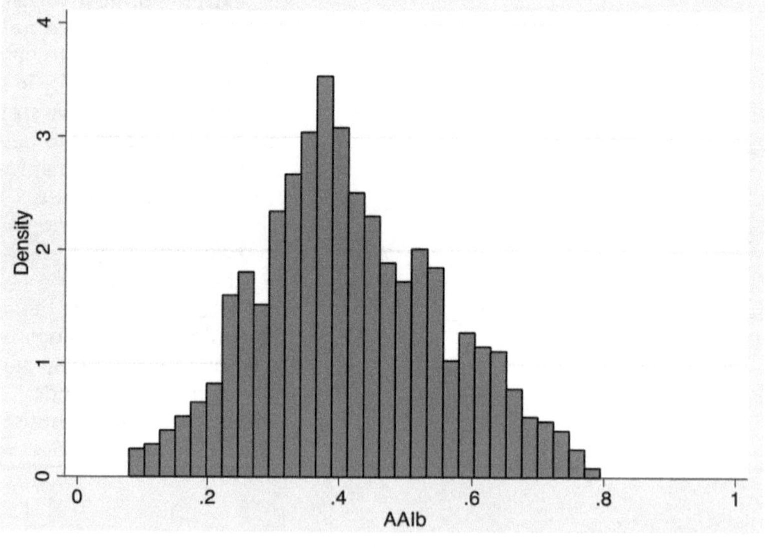

Fig. 15.5 Histogram Active Ageing Index b (AAI^b)

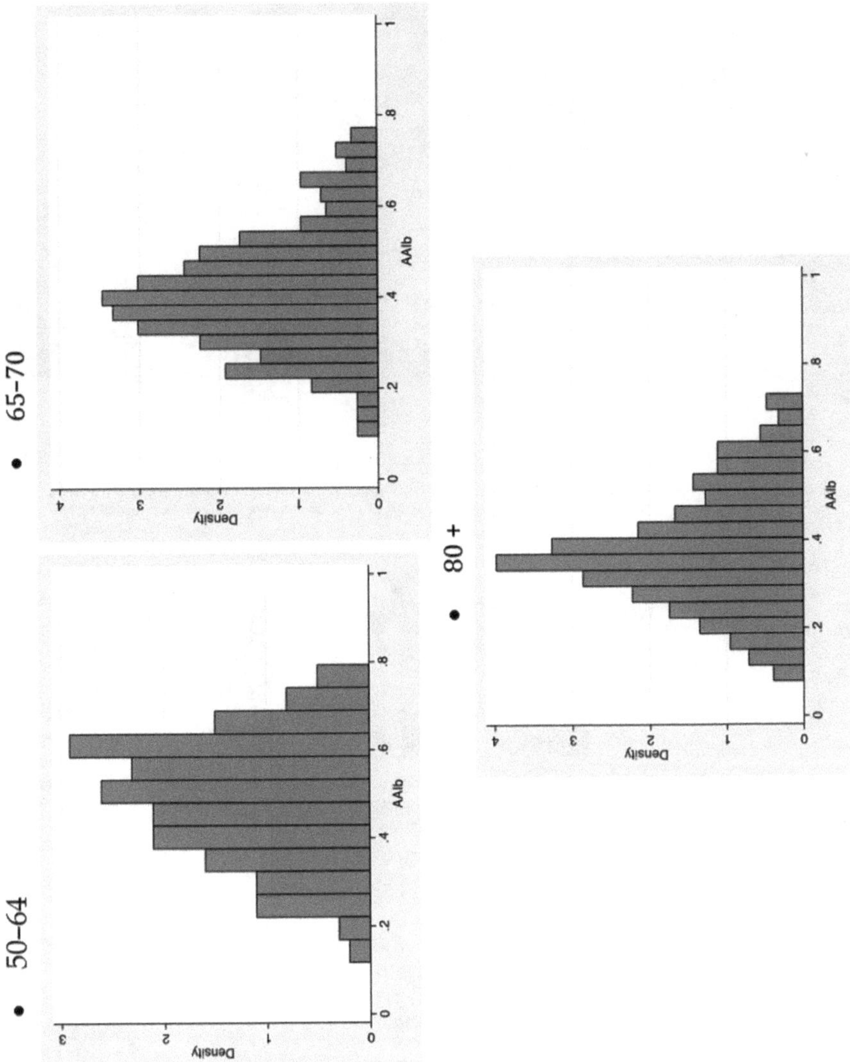

Fig. 15.6 Histogram Active Ageing Index by age group (*AAI*[b])

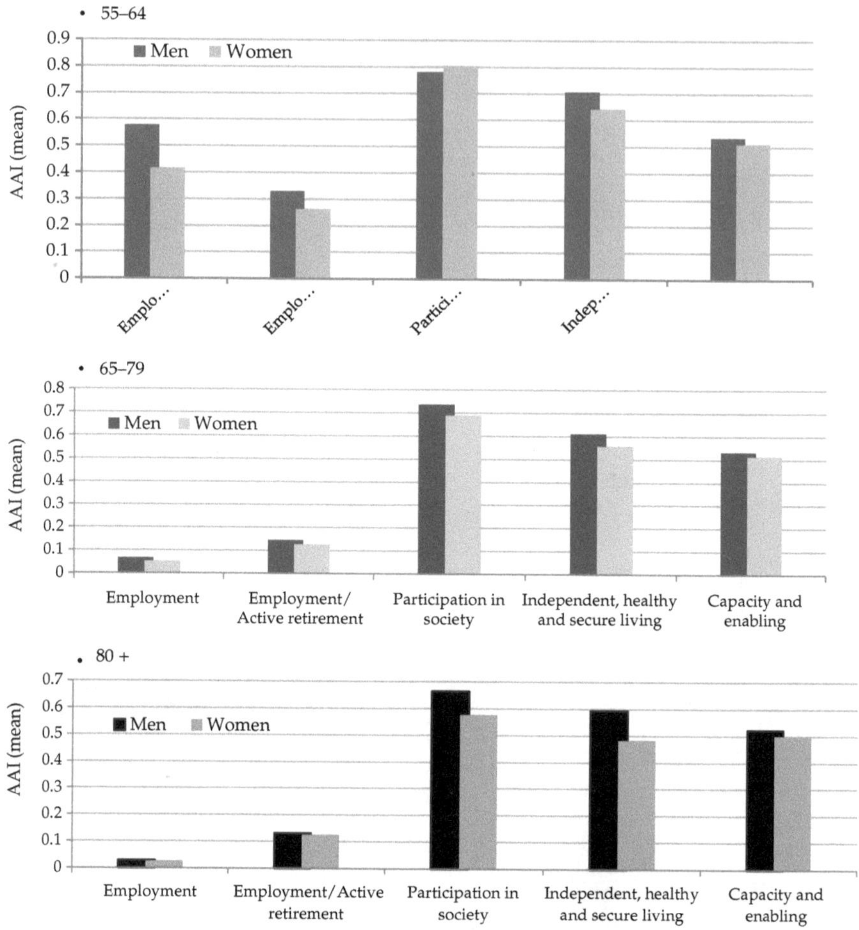

Fig. 15.7 Active ageing domains, by age groups and sex

Annex 3

Table 15.7 Descriptive statistics (before normalization)

Variable	Mean	Std. dev.	Min	Max
Age (years)	74.487	10.333	50	98
Weight (kgs)	69.563	11.660	45	116
Height (cm)	161.823	8.113	150	197
income	3.388	1.474	1	6
empl	0.122	0.327	0	1
hunemp	0.002	0.0412	0	1
ruralarea	2.344	0.759	1	3
nlonely	3.873	1.130	1	5
currenth	2.483	1.348	1	4
autonomy	1.927	0.261	1	2
ind	0.631	0.483	0	1
AAI[a]	0.404	0.162	0.078	0.969
AAI[b]	0.416	0.140	0.078	0.794
female	0.680	0.466	0	1
SAH	0.334	0.472	0	1
iad1	14.629	6.716	9	33
married	0.437	0.496	0	1
socactiv	0.562	0.496	0	1
chronic	0.380	0.486	0	1
acute	0.159	0.366	0	1
socpart	0.578	0.494	0	1
dif_servv	0.175	0.380	0	1
dif_bad1	0.073	0.261	0	1
dif_iad1	0.288	0.453	0	1
dit_mob	0.220	0.414	0	1
soc_activ	0.562	0.496	0	1
soc_part	0.578	0.494	0	1
soc_freq	0.328	0.469	0	1
limithealth	0.198	0.399	0	1
nospec	0.164	0.370	0	1
nomon	0.759	0.428	0	1
nsecurity	0.334	0.472	0	1
hsupport	0.366	0.188	0	1
useAAL	0.663	0.473	0	1
dif_AAL	0.395	0.489	0	1

Annex 4

Table 15.8 Factors associated with the *AAI*[a] and *AAI*[b] (OLS estimation)

Age group 55–64		
Variables	(1) *AAI*[a]	(2) *AAI*[b]
Female	−0.0865***	−0.0572***
	(0.0303)	(0.0209)
Age	0.0863	0.0102
	(0.125)	(0.0864)
$(Age^2)/100$	−0.0874	−0.0143
	(0.108)	(0.0742)
Weight	−0.00182	−0.00128
	(0.00126)	(0.000864)
Married	0.0439	0.0186
	(0.0362)	(0.0250)
Self-assessed health	0.0734**	0.0500**
	(0.0310)	(0.0213)
Low education	−0.00305	−0.00911
	(0.0322)	(0.0222)
Freq. to social spaces	−0.0254	0.000955
	(0.0348)	(0.0240)
Unemployed for health reasons	−0.0838	−0.167
	(0.192)	(0.132)
Dif. IADL	−0.0459	−0.0273
	(0.0513)	(0.0354)
Health limit. to social life	−0.111**	−0.0708**
	(0.0490)	(0.0337)
Home support	0.107	−0.000277
	(0.202)	(0.139)
Dif. using HC services	−0.00900	−0.0281
	(0.0527)	(0.0363)
Dif. using devices	−0.0254	−0.0121
	(0.0675)	(0.0464)
Monitorization n.a.	−0.0789*	−0.0540*
	(0.0405)	(0.0279)
Specialized support n.a.	0.0961***	0.0420*
	(0.0343)	(0.0236)
Insecurity at home	−0.0233	−0.0295
	(0.0365)	(0.0251)
AAL user	0.0518	0.00533
	(0.0337)	(0.0232)

(continued)

Table 15.8 (continued)

Age group 55–64		
Variables	(1) *AAI*[a]	(2) *AAI*[b]
Constant	−1.338	0.532
	(3.641)	(2.507)
Observations	167	167
R-squared	0.384	0.310

Standard errors in parentheses

*, ** and *** represent the variables which are significant at 10%, 5% and 1% significance levels, respectively

Table 15.9 Factors associated with the *AAI*[a] and *AAI*[b] (OLS estimation)

Age group 65–79		
Variables	(1) *AAI*[a]	(2) *AAI*[b]
Female	−0.00810	−0.0185
	(0.0138)	(0.0140)
Age	−0.202***	−0.110**
	(0.0500)	(0.0508)
(Age^2)/100	0.138***	0.0755**
	(0.0346)	(0.0352)
Weight	0.000538	5.81e−05
	(0.000558)	(0.000566)
Married	0.00414	−0.0200
	(0.0129)	(0.0131)
Self-assessed health	0.0475***	0.0497***
	(0.0139)	(0.0141)
Low education	−0.0409***	−0.0456***
	(0.0147)	(0.0150)
Freq. to social spaces	0.0194	0.0459***
	(0.0125)	(0.0127)
Unemployed for health reasons	−0.156	−0.157
	(0.121)	(0.123)
Dif. IADL	−0.0450**	−0.0546***
	(0.0188)	(0.0191)
Health limit. to social life	−0.0390**	−0.0333**
	(0.0160)	(0.0162)
Home support	−0.0260	−0.0272
	(0.0345)	(0.0351)
Dif. using HC services	−0.0222	−0.0178
	(0.0209)	(0.0212)

(*continued*)

Table 15.9 (continued)

Age group 65–79		
Variables	(1) *AAI*[a]	(2) *AAI*[b]
Dif. using devices	0.0122	0.0110
	(0.0255)	(0.0259)
Monitorization n.a.	0.0134	0.00770
	(0.0157)	(0.0159)
Specialized support n.a.	0.0297*	0.0262
	(0.0165)	(0.0168)
Insecurity at home	0.00709	0.0111
	(0.0144)	(0.0146)
AAL user	0.0273**	0.0246*
	(0.0128)	(0.0130)
Constant	7.730***	4.429**
	(1.796)	(1.825)
Observations	434	434
R-squared	0.234	0.205

Standard errors in parentheses
*, ** and *** represent the variables which are significant at 10%, 5% and 1% significance levels, respectively

Table 15.10 Factors associated with the *AAI*[a] and *AAI*[b] (OLS estimation)

Age group 80+		
Variables	(1) *AAI*[a]	(2) *AAI*[b]
Female	−0.0213	−0.0251
	(0.0176)	(0.0201)
Age	−0.0315	−0.0710
	(0.0652)	(0.0745)
$(Age^2)/100$	0.0164	0.0384
	(0.0378)	(0.0432)
Weight	0.000156	0.000201
	(0.000607)	(0.000693)
Married	−0.00440	−0.0140
	(0.0166)	(0.0190)
Self-assessed health	0.0338**	0.0316*
	(0.0155)	(0.0178)
Low education	−0.0131	−0.00129
	(0.0205)	(0.0235)
Freq. to social spaces	0.0493***	0.0865***
	(0.0142)	(0.0162)
Unemployed for health reasons	–	–

(continued)

Table 15.10 (continued)

Age group 80+		
Variables	(1) *AAI*[a]	(2) *AAI*[b]
Dif. IADL	−0.0211	−0.0236
	(0.0145)	(0.0165)
Health limit, to social life	−0.0198	−0.0375**
	(0.0146)	(0.0167)
Home support	−0.0581**	−0.0587**
	(0.0258)	(0.0295)
Dif. using HC services	−0.0204	−0.0399*
	(0.0178)	(0.0203)
Dif. using devices	0.00950	0.0140
	(0.0178)	(0.0203)
Monitorization n.a.	0.00540	0.00729
	(0.0155)	(0.0177)
Specialized support n.a.	0.0452**	0.0422*
	(0.0198)	(0.0226)
Insecurity at home	−0.0315**	−0.0346**
	(0.0147)	(0.0168)
AAL user	0.0223	0.0178
	(0.0135)	(0.0154)
Constant	1.840	3.626
	(2.814)	(3.215)
Observations	290	290
R-squared	0.186	0.229

Standard errors in parentheses
*, ** and *** represent the variables which are significant at 10%, 5% and 1% significance levels, respectively

Table 15.11 Factors associated with the *AAI*[a] and *AAI*[b] (OLS estimation)

Living with less than 489€ per month		
Variables	(1) *AAI*[a]	(2) *AAI*[b]
Female	0.0163	−0.00655
	(0.0178)	(0.0198)
Age	−0.0215**	−0.00439
	(0.0103)	(0.0115)
(Age^2)/100	0.0125*	0.00149
	(0.00682)	(0.00758)
Weight	−0.000483	−0.000520
	(0.000646)	(0.000718)

(continued)

Table 15.11 (continued)

Living with less than 489€ per month		
Variables	(1) *AAI*[a]	(2) *AAI*[b]
Mauled	0.0299	−0.0163
	(0.0203)	(0.0226)
Self-assessed health	0.0554***	0.0555***
	(0.0183)	(0.0204)
Low education	−0.0801***	−0.0598**
	(0.0220)	(0.0244)
Freq. to social spaces	0.0328**	0.0670***
	(0.0162)	(0.0180)
Unemployed for health reasons	−0.178	−0.234*
	(0.124)	(0.137)
Dif. IADL	−0.0244	−0.0225
	(0.0173)	(0.0192)
Health limit. to social life	−0.0368**	−0.0360**
	(0.0164)	(0.0183)
Home support	−0.0502	−0.0661*
	(0.0334)	(0.0371)
Dif. using HC services	−0.0430**	−0.0509**
	(0.0208)	(0.0231)
Dif. using devices	0.0122	0.0194
	(0.0199)	(0.0221)
Monitorization n.a.	−0.00645	−0.00985
	(0.0178)	(0.0198)
Specialized support n.a.	0.0600***	0.0443*
	(0.0206)	(0.0229)
Insecurity at home	0.00317	0.000923
	(0.0157)	(0.0174)
AAL user	0.0223	0.0127
	(0.0140)	(0.0155)
Constant	1.336***	0.724*
	(0.392)	(0.436)
Observations	312	312
R-squared	0.272	0.210

Standard errors in parentheses

*, ** and *** represent the variables which are significant at 10%, 5% and 1% significance levels, respectively

Notes

1. Zaidi et al. (2013).
2. Simões (2012).
3. Here measured by the Active Ageing Index and self-assessed health indicator, respectively.
4. According to $f(x_i)_N = \dfrac{f(x_i) - \min(f(x_i))}{\max(f(x_i)) - \min(f(x_i))}$
5. A first attempt was made only taking into account subjects with the maximum level of social activity and participation in recreational activities. However, only 105 of the retired subjects fulfil these requirements, and the different approach had no impact on the estimation exercise.
6. None of the explanatory variables the variables used to construct the index are included in the estimation.
7. For further details on the econometric models specifications, see Stock and Watson (2003).
8. See Figs. 15.5 and 15.6 in Annex 2.
9. See Fig. 15.7 in Annex 2.
10. In our survey, current health status corresponds to being either healthy, in recovery, having a chronic or an acute disease.
11. See also Annex 1 for variables description.
12. Tables 15.8–15.10 in Annex 4.
13. Table 15.11 in Annex 4.

References

Atherton, K., & Power, C. (2007). Health inequalities with the National Statistics-Socioeconomic classification: Disease risk factors and health in the 1958 British birth cohort. *The European Journal of Public Health, 17*(5), 486–491.

Matthews, S., Manor, O., & Power, C. (1999). Social inequalities in health: Are there gender differences? *Social Science & Medicine, 48*(1), 49–60.

Mensah, F., & Hobcraft, J. (2008). Childhood deprivation, health and development: Associations with adult health in the 1958 and 1970 British prospective birth cohort studies. *Journal of Epidemiology & Community Health, 62*(7), 599–606.

Paúl, C., Ribeiro, O., & Teixeira, L. (2012). Active ageing: An empirical approach to the WHO model. *Current Gerontology and Geriatrics Research, 2012*, 1–10.

Pfarr, C., Schmid, A., & Schneider, U. (2012). Reporting heterogeneity in self-assessed health among elderly Europeans. *Health Economics Review, 2*(1), 21.

Simões, R. (2012). *Analysis of the survey to potential AAL users (seniors and pre-seniors)*. Universidade do Minho.

Tareque, M., Hoque, N., Islam, T., Kawahara, K., & Sugawa, M. (2013). Relationships between the Active Aging Index and disability-free life expectancy: A case study in the Rajshahi District of Bangladesh. *Canadian Journal on Aging, 32*(4), 417–432.

World Health Organization. (2002). *Active ageing—A policy framework*. Geneva: World Health Organization.

Zaidi, A., & Stanton, D. (2015). 'Active ageing index 2014: analytical report', report produced at the Centre for Research on ageing, University of Southampton, under contract with UNECE (Geneva), co-funded by European Commission, Brussels, available at: http://www.southampton.ac.uk/assets/sharepoint/groupsite/Administration/SitePublisher-document-store/Documents/aai_report.pdf.

Zaidi, A., Gasior, K., Hofmarcher, M. M., Lelkes, O., Marin, B., Rodrigues, R., Schmidt, A., Vanhuysse, P., & Zolyomi, E. (2013). *Active Ageing Index 2012. Concept, methodology, and final results*. Research memorandum/methodology report. European Centre Vienna, March 2013. Retrieved from www.euro.centre.org/data/aai/1253897823_70974.pdf

Zasimova, L., & Sheluntcova, M. (2014). Measuring active aging for government policy planning: A case of Russia. *SSRN Electronic Journal*.

Sara Valente de Almeida is a PhD candidate at NOVA School of Business and Economics. Her work focuses on population health and health policy in developing countries. Previously, she worked at the European Commission and the Centre for European Policy Studies, both in Brussels. She is working on a number of projects ranging from active ageing to traditional medicine practices and hospitalization policy.

Pedro Pita Barros is Professor of Economics at Universidade Nova de Lisboa and Research Fellow at the Centre for Economic Policy Research. His research focuses on issues on health economics and on regulation and competition policy. He is also a member of the Portuguese National Ethics Council for the Life Sciences, the Office of Health Economics, the Instituto de Políticas Públicas—Thomas Jefferson—Correia da Serra and the EC Expert Panel on Effective ways of Investing in Health (2017–2020).

16

Study on Active Ageing at Individual Level Based on Active Ageing Index

Coro Piñeiro Vázquez, Jesús Marcial Méndez Magán, Sara Marsillas Rascado, Antonio Rial Boubeta, Teresa Braña Tobío, and Jesús Varela Mallou

16.1 Introduction

A phenomenon known as ageing population appeared throughout the second half of the twentieth century. It resulted in a progressive increase in average longevity and the population over 80 years old. This demographic phenomenon is changing the age structure of the population, with significant consequences for political, economic, social and health standards. In Spain, older people represent 17.6% of the total population, older than 80 are 5.5% and life expectancy is 81.8 years, representing

C. Piñeiro Vázquez
Galicia, Spain

J.M. Méndez Magán
European Committee of the Regions, Bruxelles, Belgium

S. Marsillas Rascado (✉)
Matia Gerontological Institute, Matia, USA

A. Rial Boubeta • T. Braña Tobío • J. Varela Mallou
University of Santiago de Compostela, A Coruña, Spain

A. Zaidi (eds.), *Building Evidence for Active Ageing Policies*,
https://doi.org/10.1007/978-981-10-6017-5_16

one of the highest figures of the European Union. At national level, Galicia is one of the regions leading this process of ageing, with 23.7% of the population over 65 years old and really aged rural areas. Public institutions should know the reality of specific territory in terms of ageing as well as how to enhance the positive aspects of this situation in order to minimize the impact of this phenomenon.

Active ageing is a construct which belongs to the new paradigm of the study of ageing, appeared in recent decades, and which evolved with different nomenclatures (Zamarrón 2007). It was defined as "the process of optimizing opportunities for health, participation and security in order to improve the quality of life as people age" (World Health Organization [WHO] 2002). WHO also proposes a complex and multidimensional model of active ageing, with two broad categories that bring together different variables: on the one hand, contextual determinants associated with the environment of people; and on the other hand, other personal components related to individual conditions and behaviours. All of them are composed by different indicators. This paradigm can be addressed, and therefore evaluated, from two different approaches: at the population level, taking the population as a whole and focusing on the environmental and contextual elements which will permit people to age actively (Zamarrón 2007); and at the individual level, taking the sum of individuals to reach population and concentrating on the personal aspects (biological, social and psychological conditions), which while being enhanced, they permit people to be active agents of their ageing (Bowling 2008; Fernández-Ballesteros et al. 2006).

In recent years, various European institutions and government launched policies to address the current demographic situation and to promote active ageing. A first concern of member states of the European Union (EU) was to know the extent of the problem. Therefore, the European Commission and the United Nations Economic Commission for Europe managed the development of a measurement tool, the Active Ageing Index (AAI) (Zaidi et al. 2013). This tool, aimed at the policy makers, unifies the criteria for evaluating the effectiveness of the strategies developed across the EU and identifies the areas requiring further impetus to ensure the quality of life of older people as well as to create effective policies. The AAI allows comparing the degree of ageing among countries placing them in a ranking by the overall and domain-specific indexes from a population perspective. In addition it quantifies the potential developed by the old people in terms of participation in the economy and social life, healthy life expectancy and independent living, showing the distance between the current and the ideal situation. This is very useful at political and organizational level to meet roughly the situation of a country and to compare it internationally. However, it is built using various data sources derived from European surveys, which affects the information collected and difficults the analysis of more specific areas (regions, cities, etc.). This segmentation is relevant because, for example, in Spain, social policies were transferred at the community level and subsequently divided into territorial departments, hampering to obtain certain measures. Furthermore, as AAI's purpose is to aid countries to ensure the contextual elements which will permit population to age actively, its indicators are mostly addressed at the macro level and they do not consider certain personal areas. In this sense, it would be of interest to supplement the European AAI with variables of individual and subjective character for a better approach to the real needs of the older people.

We started from the need of Active Ageing Indexes both in global or macro and micro or personal terms. Accordingly, we tried to solve the problem of measuring the Galician level of ageing and then, to complement the AAI population measurement with its individual equivalent. Therefore, the aim is to apply and to adapt the AAI and its indicators to an individual level, using its conceptualization and comparative framework as well as incorporating new measurements to cover other spheres. Besides providing an individual index of active ageing comparable with the AAI, we will have two measurements of our region, contextual and personal ones. Their comparison will indicate whether the new tool is the individual equivalent of AAI. To the extent that both results are similar and reflect the same trends, the new tool would also be considered accurate. Moreover, we want to descent into more specific territories. The new tool will be able to segment the results according to socio-demographic variables of interest. Finally, we assigned active ageing scores to each subject in order to develop a personal profile and even try to reach an individual *diagnosis* of active ageing.

16.2 Method

16.2.1 Design

In pursuit of its purpose, a selective methodology was used, carrying out a survey among a representative sample of the 60-year-old residents in Galicia (804,403 persons). For the sampling selection, two-stage sampling was used: "conglomerates" for the selection of the first-level units (municipalities) and "quotas" according to the habitat (urban/semi-urban vs. rural/semi-rural), "gender" and "age group" (60–74 years vs. 75 or older), for the selection of second-level units (individuals).

16.2.2 Participants

The sample was composed by 404 individuals, with ages between 60 and 94 years ($\overline{X} = 72.55$; SX = 8.08). 43.6% were women and 56.4% were men; 57.2% were between 60 and 74 years old and 42.8% older than 75, and 59.2% were residents of a rural/semi-rural area, whereas 40.8% were from urban/semi-urban area.

16.2.3 Variables and Instrument

First, the AAI was applied to Galicia. We obtained Galician data by secondary sources, which provided territorial disaggregation in Spain, but we encountered certain absence of information in some indicators. To have this information, it was necessary to carry out a fieldwork. The distribution of the secondary and primary data sources (fieldwork) is shown in Table 16.1.

The following step was to adapt the AAI tool and its indicators to an individual format, generating a new instrument in which the weight of the indicators was redistributed partially. Moreover, this individual tool should be able to provide information from any segment of interest.

In it, first and second domains maintained the same indicators and criteria; nevertheless, the third and fourth domains were modified. More specifically, the *access to health and dental care* was substituted by the

Table 16.1 AAI Galician data sources

Domains and indicators	AAI Galician data sources
Domain 1: Employment	
Employment rate 55–59	IGE-INE. Labour Force Survey, IV quarter (2013)—Secondary
Employment rate 60–64	IGE-INE. Labour Force Survey, IV quarter (2013)—Secondary
Employment rate 65–69	Own fieldwork
Employment rate 70–74	
Domain 2: Participation in society	
Voluntary activities	Own fieldwork
Care to children/grandchildren	
Care to older adults	
Political participation	
Domain 3: Independent, healthy and secure living	
Physical exercise	Own fieldwork
Access to health and dental care	INE. Living Conditions Survey (2012)—Secondary
Independent living	Own fieldwork
Relative median income	INE. Families Living Conditions Survey (2012)—Secondary
No poverty risk	INE. Families Living Conditions Survey (2012)—Secondary
No severe material deprivation	Own fieldwork
Physical safety	
Lifelong learning	IGE-INE. Labour Force Survey, IV quarter (2013)—Secondary
Domain 4: Capacity and enabling environment	
Remaining life expectancy at age 55	Eurostat—European Health and Life Expectancy Information System (2012)—Secondary
Share of healthy life expectancy at age 55	INE—Living Conditions Survey 2012—Secondary
Mental well-being	Own fieldwork
Use of ICT	
Social connectedness	
Educational attainment	

frequency of assistance to the doctor and the dentist, measured by two open-response questions. The criterion was the stipulated frequency of attendance to both health services. A scheduled frequency to make necessary revisions and treatments reports the health monitoring, which is important to lead a life as active and healthy as possible within each personal circumstances.

The indicator *independent living* was replaced by *Barthel Index* (Mahoney and Barthel 1965) and the *Lawton Instrumental Activities of Daily Living (IADL) Scale* (Lawton and Brody 1969). Limitations provided by the expert group related to multigenerational households and the overestimation of dependence were considered in the choice of the indicators. Both scales, widely used in the geriatric field, were included because they report the independence and autonomy of individuals in daily activities. The compliance criterion was the independence situation in both scales with scores between 60 and 100 for Barthel Index and 5/8 in Lawton IADL Scale (5 for male and 8 for female).

The indicator *median relative income* was replaced by the *perceived economic situation*, carried out through the item "Do you think that your current economic situation allows you to live without big economic concerns (monthly expenses, daily expenses …)?". The responses were measured according to the degree of agreement, and the compliance criterion includes from "agree" up to "totally agree." The relative position of people over and under 65, in a crisis or situation of widespread unemployment with low and unstable income of young adults, may not result very precise indicator of financial security. The next question was elaborated to adapt the indicator *no poverty risk*: "Which is your monthly income?", and the criterion was established regarding the poverty threshold is 50% of median referred to national disposable income, which in Galicia correspond to 500 euros. Related to the indicator *lifelong learning*, the time period was extended from four weeks to six weeks, to adapt it to Galician context. It was due to many cases of intermittent attending to training were detected and the period of the last six months diminished this temporary abandonment meanwhile took into account the duration of the beneficial effects of staying in constant learning in old age.

The indicator *remaining life expectancy at age 55* was eliminated, and the indicator *share of healthy life expectancy at age 55* was replaced by the indicator no limitation in daily activity, capturing the essence of the absence of limitation of the original indicator. It was carried out by the item "How does your current medical condition limit your daily activities?" and the compliance criterion was "nothing."

The indicator *mental well-being* was replaced by Affect Balance Scale (Godoy-Izquierdo et al. 2008). It is made up of 18 items. They measure the frequency experimentation of different affective states in the last week, including both positive and negative affectivity. We wanted to include both dimensions because they are relatively independent and the presence of positive affect does not imply the absence of the negative one, or vice versa. The criterion was a positive affect balance, equivalent to a score higher than 0, of a −18–18 rank.

The *social connectedness* was replaced by an 11-item scale developed for this study. It measures the satisfaction with different aspects of social and family support. Considering the suggestions of the experts group related to multidimensional homes and social disconnection, this concept was preferred instead of the weekly frequency of contact due to support is really important in older age. It was based on the Duke-UNC-11 questionnaire of Bellón et al. (1996a), which is the Spanish adaptation of Duke-UNC-11 Functional Social Support Questionnaire (Broadhead et al. 1988), as well as the APGAR questionnaire of Bellón et al. (1996b), the Spanish version of *family APGAR* of Smilkstein (1978).

A psychometric study of the reliability (internal consistency) and validity was performed in order to ensure the quality of measurement of the scales included. Psychometric properties of the scales included are shown in table 16.2. Good psychometric properties were obtained in all cases, with a high Cronbach's alpha and adequate explanation of the variance.

Table 16.2 Psychometric properties of the scales included

Indicators	Scale	No. items	Factors—AFE	% of variance explained	Internal consistency
Independent living	Lawton IADL Scale	8	1	78.5	Alpha = 0.94
	Barthel Index	10	1	55.2	Alpha = 0.89
Mental well-being	Affective Balance Scale	16	2 Negative affect Positive affect	39.9	Alpha global = 0.42 Alpha negative = 0.80 Alpha positive = 0.76
Social and familiar support	Social and family support Scale	11	2 Family Friends	58.3	Alpha global = 0.88 Alpha family = 0.87 Alpha friendship = 0.79

Global and specific indexes were calculated by counting the percentage of people who met the criteria for each adapted indicator and multiplying each value by its respective weight of the adapted AAI tool.

Finally, the individual scores for each subject were obtained using a modification of the correction procedure. First a value was assigned to each subject in each variable according to whether they met the criteria defined individually for each indicator (0, unfulfilled, and 1, fulfilled). Subsequently domain-specific and global indexes were calculated for each person. The indicators and domains were weighted by their respective weight in the AAI, obtaining a final score between 0 and 1. Trying an individual *diagnosis*, a threshold was established in each domain and in the overall index from a positive score would be considered. The selected criterion was the average obtained in the domain and overall indexes of all European regions evaluated with the AAI, different by gender, choosing those scores that left above the percentage corresponding to the European average in that specific index and gender.

16.2.4 Procedure

The information needed to adapt the AAI derived from the fieldwork. 404 in-depth individual structured interviews were conducted, lasting approximately 20 minutes and following an ad hoc questionnaire developed specifically for this work. These were performed in the usual environment of the subjects being always guaranteed anonymity and confidentiality of participants. Previously a training session was held for the management and piloting of the questionnaire, in order to unify criteria and procedures for collecting information.

16.3 Results

Table 16.3 shows the scores obtained by Galicia with the original AAI and the individual adaptation of the AAI, in the overall and domain-specific indexes. Similar values as well as the same pattern of results were achieved with both tools. Specifically the value of $AAI_{Overall}$ of Galicia was 31.8 while 31.7 corresponded to the adapted AAI (possible range of 0–100).

Table 16.3 Comparison of overall and domain-specific indexes for Galicia—AAI and adapted AAI

		Indexes				
		Domain 1: Employment	Domain 2: Participation in society	Domain 3: Independent, healthy and secure living	Domain 4: Capacity and enabling environment	Overall
AAI	General population	23.9	12.8	75.7	56.8	**31.8**
	Men	27.2	11.3	78.0	58.7	**33.0**
	Women	21.3	14.0	74.5	55.6	**30.9**
Individual adaptation of AAI	General population	22.3	12.8	72.0	61.2	**31.7**
	Men	26.8	11.3	73.2	66.6	**34.0**
	Women	17.8	14.0	71.2	57.1	**29.7**

Galicia was placed in the 18th position in the ranking of the 28 areas assessed (Table 16.4). In addition, Galicia's score was lower than Spain's score (32.5), which placed it in 17th position.

Considering the domains separately, the value of domain *employment* showed few differences (22.3 adapted AAI and 23.9 AAI). Although in our sample the population group of 55–59 years old as part of the universe of reference was not included, the indicator on the employment rate of 55–59 was incorporated because a false value was obtained in the simulation exercise performed excluding this indicator and redistributing the weights of the remaining indicators. This value hindered the comparability with the original $AAI_{Employment}$. Both indexes are shown in table 16.5.

In the second domain (participation in society), the values were the same in both indexes because the indicators were obtained from fieldwork. Specifically, the value was 12.8 which corresponded to the 27th position in the European rank. It showed a really low value, which is presented in table 16.6.

In the third domain (*independent, healthy and secure living*), the variation was higher: from 75.7 obtained with the original AAI to 72.0 with adapted AAI. The largest difference occurred in *access to health and dental care*, falling from 86.8 to 57.4. The individual independence indicator showed better results than the original one (77.0 vs. 67.6), confirming the suspicion of the overestimation of dependency among older when considering only the type of household. It also changed the adequacy of older

Table 16.4 Comparison of Galicia scores and rankings in overall and domain-specific indexes

Overall AAI			Employment			Participation in society			Independent, healthy and secure living			Capacity and enabling environment		
Ranking	Region	Value	Ranking	Region	Value	Ranking	Region	Value	Ranking	Region	Value	Ranking	Region	Value
1	Sweden	44.0	1	Sweden	41.0	1	Ireland	25.2	1	Denmark	79.0	1	Sweden	69,5
2	Denmark	40.2	2	Cyprus	36.1	2	Italy	24.1	2	Sweden	78.7	2	Denmark	66.7
3	Ireland	39.4	3	Un. Kingdom	35.5	3	Luxemburg	22.6	3	Netherlands	77.7	3	Netherlands	61.6
4	Un. Kingdom	39.3	4	Portugal	35.3	4	Sweden	22.6	4	Finland	76.6	4	Luxemburg	61.6
5	Netherlands	38.9	5	Estonia	34.4	5	France	22.4	5	Germany	75.8	5	Un. Kingdom	61.4
6	Finland	38.8	6	Denmark	34.0	6	Netherlands	22.4		Galicia	75.7		Galicia	61.2
7	Cyprus	36.3	7	Finland	32.0	7	Finland	22.4	6	Un. Kingdom	75.7	6	Ireland	60.8
8	Luxemburg	35.1	8	Romania	31.4	8	Austria	21.4	7	Ireland	75.7	7	Finland	60.7
9	Germany	35.0	9	Netherlands	31.4	9	Belgium	20.4	8	Luxemburg	74.7	8	Belgium	60.3
10	Austria	34.9	10	Germany	31.2	10	Denmark	20.1	9	France	74.6	9	France	57.8
11	Czech Rep.	34.3	11	Ireland	31.0	11	Un. Kingdom	20.0	10	Slovenia	74.4	10	Austria	57.5
12	France	34.2	12	Latvia	28.3	12	Czech Rep.	19.4	11	Czech Rep.	73.8		Galicia	56,8
13	Portugal	34.2	13	Lithuania	27.4	13	Cyprus	18.7	12	Belgium	73.4	11	Germany	56.2
14	Belgium	33.5	14	Czech Rep.	26.4	14	Spain	18.3	13	Austria	73.0	12	Spain	56.1
15	Italy	33.3	15	Austria	24.6	15	Malta	18.2		Galicia	72.0	13	Malta	56.1
16	Estonia	33.1	16	Bulgaria	24.6	16	Slovenia	16.7	14	Hungary	71.9	14	Czech Rep.	54.4
17	Spain	32.5	17	Greece	24.4	17	Hungary	16.1	15	Lithuania	70.6	15	Italy	52.8
	Galicia	31.8		Galicia	23.9	18	Lithuania	15.3	16	Romania	70.1	16	Bulgaria	51.7
	Galicia	31.7	18	Spain	23.3	19	Germany	14.9	17	Malta	70.1	17	Cyprus	51.1
18	Lithuania	31.6		Galicia	22.3	20	Portugal	14.3	18	Estonia	70.0	18	Portugal	50.8
19	Malta	31.0	19	Slovenia	21.6	21	Greece	14.2	19	Italy	69.9	19	Slovenia	48.8
20	Romania	30.9	20	Luxemburg	21.1	22	Latvia	13.9	20	Cyprus	69.1	20	Lithuania	47.9
21	Slovenia	30.6	21	France	21.0	23	Slovakia	13.7	21	Poland	67.5	21	Estonia	47.1
22	Latvia	30.2	22	Italy	20.9	24	Estonia	13.3	22	Spain	67.3	22	Poland	46.7
23	Bulgaria	30.0	23	Slovakia	20.1	25	Romania	12.9	23	Slovakia	67.0	23	Greece	46.7
24	Greece	29.3	24	Poland	19.8	26	Bulgaria	12.9	24	Portugal	66.7	24	Slovakia	45.9
25	Hungary	28.2	25	Belgium	19.8		Galicia	12.8	25	Greece	65.2	25	Hungary	45.9
26	Slovakia	27.7	26	Malta	18.3		Galicia	12.8	26	Bulgaria	65.2	26	Latvia	45.4
27	Poland	27.3	27	Hungary	17.8	27	Poland	12.2	27	Latvia	63.2	27	Romania	42.0

(light grey)	Original AAI
(dark grey)	Individual adaptation of AAI

Table 16.5 Comparison of original $AAI_{Employment}$ and adapted $AAI_{Employment}$

	Population	Employment rate 55–59	Employment rate 60–64	Employment rate 65–69	Employment rate 70–74	Domain 1 Index
$AAI_{Employment}$	General population	51.3	33.5	5.1	5.8	**23.9**
	Men	55.6	36.9	10.0	6.1	**27.2**
	Women	47.2	30.3	2.1	5.6	**21.3**
Adapted $AAI_{Employment}$	General population	51.3	27.1	5.1	5.8	**22.3**
	Men	55.6	35.4	10.0	6.1	**26.8**
	Women	47.2	16.2	2.1	5.6	**17.8**

Table 16.6 Comparison of original $AAI_{Participation}$ and adapted $AAI_{Participation}$

	Population	Voluntary activities	Care to children/ grandchildren	Care to other adults	Political participation	Domain 2 Index
$AAI_{Participation}$	General population	10.1	16.3	11.6	13.6	**12.8**
	Men	6.2	13.6	11.4	14.8	**11.3**
	Women	13.2	18.4	11.8	12.7	**14.0**
Adapted $AAI_{Participation}$	General population	10.1	16.3	11.6	13.6	**12.8**
	Men	6.2	13.6	11.4	14.8	**11.3**
	Women	13.2	18.4	11.8	12.7	**14.0**

Table 16.7 Comparison of original $AAI_{Independent}$ and adapted $AAI_{Independent}$

	$AAI_{Independent}$			Adapted $AAI_{Independent}$		
	General population	Men	Women	General population	Men	Women
Physical exercise	65.6	69.9	62.3	65.6	69.9	62.3
Access to health services	86.8	86.2	87.2	57.4	57.4	57.5
Independent living	67.6	72.7	64.5	77.0	80.1	74.6
Financial security	93.4	98.0	92.0	73.5	77.8	70.2
No poverty risk	93.7	93.4	93.9	93.3	96.0	91.2
No severe material deprivation	97.3	98.9	96.1	97.3	98.9	96.1
Physical safety	96.5	99.4	94.3	96.5	99.4	94.3
Lifelong learning	1.9	2.3	3.2	25.2	14.8	33.3
Domain 3 Index	**75.7**	**78.0**	**74.5**	**72.0**	**73.2**	**71.2**

people's income, falling from 93.4 to 73.5, reporting worse situation the micro than the macro indicator. *No risk of poverty* just showed minimal differences between macro and micro data (93.7 and 93.3), converting it in its individual equivalent. Both indexes can be found in table 16.7.

Regarding the last domain (*capacity and enabling environment*), the values were 56.8 ($AAI_{Capacity}$), which placed Galicia in the 11th position of the ranking, and 61.2 (adapted $AAI_{Capacity}$). In this domain there were individual equivalents for every original indicator except for *remaining life expectancy at age 55*, which was eliminated. It was necessary to redistribute the weights of the other indicators of the domain, contributing 34% of the equivalent indicator of healthy life expectancy at 55 (*no limitation in daily activity*), 25% *mental well-being*, 11% *use of ICT*, 19% *social and family support* and 11% *educational attainment*. The absence of limitation in daily activity was 67.8% and *mental well-being* increased by measuring the full range of affective state (micro indicator) versus single measurement of positive moods (macro indicator), 76.5 and 70.3, respectively. The social connection showed a decline when evaluating the weekly contact with people outside home (89.4) in relation to the satisfaction with the social and family support (64.1). Both indexes corresponding the fourth domain, original AAI and adapted AAI, are presented in table 16.8.

Finally, the assignment of individual scores provided useful and discriminative information in terms of profile of indicators. The calculation of individual scores, domain indexes and overall indexes allowed

Table 16.8 Comparison of original $AAI_{Capacity}$ and adapted $AAI_{Capacity}$

	Population	Remaining life expectancy at 55	Share of healthy life expectancy at 55	Mental well-being	Use of ICT	Social connectedness	Education attainment	Domain 4 Index
$AAI_{Capacity}$	General population	59.2	40.4	70.3	25.5	89.4	37.1	56.8
	Men	53.8	44.4	77.8	32.4	91.5	47.7	58.7
	Women	64.4	37.0	64.5	20.2	87.7	28.9	55.6
	Population	–	No limitation of activity	Mental well-being	Use of ICT	Social and family support	Education attainment	Domain 4 Index
Adapted $AAI_{Capacity}$	General population	–	67.8	76.5	25.5	64.1	37.1	61.2
	Men	–	72.2	81.2	32.4	68.0	47.7	66.6
	Women	–	64.5	72.8	20.2	61.1	28.9	57.1

allocating the people on a continuum and making a diagnosis based on the thresholds proposed by gender. Table 16.9 shows an example of the individual scores of three subjects from the sample: subjects with the lowest (woman 62 years old) and the highest score (woman 72 years old) in the overall index and a subject whose score matched the median (woman 81 years old) (Table 16.9).

Taking into account the criteria proposed, only the third case would reflect active ageing by obtaining a higher overall index than the threshold for women (0.31). The specific domains showed also values over the threshold (employment, 0.35; participation in society, 0.16; independent living, 0.07; and capacity, 0.13). Based on her scores, the personal active ageing programme should promote the health and dental care, use of ICT and lifelong learning.

Table 16.9 Example of individual scores obtained in indicators, domains and overall indexes

	Lowest	Medium	Highest
Employment Index	**0.00**	**0.00**	**0.35**
Employment	0.00	0.00	0.35
Participation in society Index	**0.00**	**0.11**	**0.19**
Voluntary activities	0.00	0.00	0.00
Care to children/grandchildren	0.00	0.00	0.25
Care to other adults	0.00	0.30	0.30
Political participation	0.00	0.00	0.00
Independent, healthy and secure living Index	**0.01**	**0.08**	**0.07**
Physical exercise	0.00	0.10	0.10
Access to health and dental care	0.00	0.20	0.00
Independent living	0.00	0.20	0.20
Financial security	0.00	0.00	0.10
No poverty risk	0.00	0.00	0.10
No severe material deprivation	0.00	0.10	0.10
Physical safety	0.01	0.10	0.10
Lifelong learning	0.00	0.10	0.00
Capacity and enabling environment Index	**0.00**	**0.05**	**0.18**
No limitation of activity	0.00	0.00	0.34
Mental well-being	0.00	0.25	0.25
Use of ICT	0.00	0.00	0.00
Social and family support	0.00	0.00	0.19
Education attainment	0.00	0.00	0.11
Overall Index	**0.01**	**0.24**	**0.79**

16.4 Discussion

Active ageing is a complex and multidimensional construct that can be studied from two perspectives, the population and the individual level. The data presented in this chapter allow more accurately to describe the situation of our older people, providing valuable information covering both viewpoints. Thus, the results provided by AAI tool informs us of the situation of Galicia at the macro level, with a value 2.0 points below the European average (31.8 vs. 33.8). These results, however, may not result entirely comparable, as they are based on slightly different sets of indicators. Thus, considering both results, the potential of Galician older people should be strengthened in *employment* and *participation in society*. By contrast, it highlights the good performance in *independent, healthy and secure living*, and the results achieved in *capacity and enabling environment for active ageing* reflect an encouraging outlook.

Otherwise, the adapted AAI quantifies the Galician level of active ageing in 31.7, similarly to the original AAI. The results indicate the need of promoting the monitoring of physical and dental health, adapting their economic situation, encouraging the use of ICT and education as well as addressing the social and family support and limitations in daily activity.

The conjoint application of both tools allows us knowing the overall score of a particular territory and its position in the ranking, segmenting results and territories according to the objective as well as assigning a score to each person placing her on a continuum. These tools are based on the same concept of active ageing, and we can use both the population and individual approaches, including contextual and personal elements. The new procedure of the adapted instrument permits an individual approach to macro indicators of AAI and extracts useful information for policy makers and practitioners in gerontological field, without missing its philosophy and comparability. The advantage is that being the equivalent of the original AAI, it reflects the defining elements of the independent, healthy and secure living, and the ability to age actively from the individual perspective, meanwhile, permits increasing the level of detail of the analysis, from countries to regions, provinces, environments and so on. Thus, the awareness of the problematic and needs of older people can be improved.

Thereby, the potential use of the AAI was enabling the evaluation of the desired population segment as well as by advancing to personal *diagnosis*, not in clinical terms, but referring the areas with high and low performance. In these terms, new indicators suppose a step forward to an individual evaluation of active ageing. The objective was to provide professionals with a rapid and individualized tool which not only measures the personal level of active ageing but also monitors and quantifies the evolution. It would result in some policy implications such as a further adaptation of programmes to older people's needs and an improved assessment of the effectiveness of the initiatives developed. The present study can be useful to assess many initiatives which are implemented from sociocultural centres, municipalities, county councils, provinces or autonomous regions. Even the smaller organization could assess their own programme or initiative related to active ageing. The use of both tools could therefore carry more policy implications such as a better representation of the reality, a quicker and more detailed detection of problems and areas of improvement as well as a higher capacity of reaction in case of decline in any area. In conclusion, their conjoint use would facilitate the identification and improvement of opportunities to age actively from both perspectives.

However, we are aware of the limitations of this first approach to a personal index of active ageing, such as the relative importance of certain elements in the final score of the subjects. We are currently working on the inclusion of other indicators of active ageing based on the scientific literature that were not taken into account by now, such as cognitive status, diseases, lifestyles—food, sleep, toxic habits—or leisure (Depp and Jeste 2006; Fernández-Ballesteros et al. 2004; Paúl et al. 2012). We are also working on the construct of active ageing, specifically on the redistribution of the weights assigned to domains, through modelling different variables with Structured Equation Modelling and supporting by experts. Our objective is to achieve a tool which will result in a quantitative index and a personal profile to inform of the situation of each person in relation to their ageing process. Ultimately, we are also intending to create an observatory to collect the temporal evolution of indicators, contributing to the continuous improvement of public policies.

References

Bellón, J. A., Delgado, A., Luna, J., & Lardelli, P. (1996a). Validez y fiabilidad del cuestionario de apoyo social funcional Duke-UNC-11. *Atención Primaria, 18*(4), 153–163.

Bellón, J. A., Delgado, A., Luna, J., & Lardelli, P. (1996b). Validez y fiabilidad del cuestionario de función familiar Apgar-familiar. *Atención primaria, 18*(6), 289–295.

Bowling, A. (2008). Enhancing later life: How older people perceive active ageing? *Aging and Mental Health, 12*(3), 293–301.

Broadhead, W. E., Gehlbach, S. H., Degruy, F. V., & Kaplan, B. H. (1988). The Duke-UNC functional social support questionnaire: Measurement of social support in family medicine patients. *Medical Care, 26*(7), 709–723.

Depp, C. A., & Jeste, D. V. (2006). Definitions and predictors of successful aging: A comprehensive review of larger quantitative studies. *American Journal of Geriatric Psychiatry, 14*(1), 6–20.

Fernández-Ballesteros, R., Zamarrón, M. D., Rudinger, G., Schroots, J. J., Hekkinnen, E., Drusini, A., Paul, C., Charzewska, J., & Rosenmayr, L. (2004). Assessing competence: The European Survey on Aging Protocol (ESAP). *Gerontology, 50*(5), 330–347.

Fernández-Ballesteros, R., et al. (2006). Estudio Longitudinal sobre Envejecimiento Activo (ELEA). Madrid, IMSERSO, Estudios I+D+I, n° 38. [Fecha de publicación: 01/08/2007]. Retrieved from http://envejecimiento.csic.es/documentos/documentos/imserso-estudiosidi-38.pdf

Godoy-Izquierdo, D., Martínez, A., & Godoy, J. F. (2008). La «Escala de Balance Afectivo». Propiedades psicométricas de un instrumento para la medida del afecto positivo y negativo en población española. *Clínica y Salud, 19*(2), 157–189.

Lawton, M. P., & Brody, E. M. (1969). Assessment of older people: Self-maintaining and instrumental activities of daily living. *The Gerontologist, 9*(3), 179–186.

Mahoney, F., & Barthel, D. (1965). Functional evaluation: The Barthel Index. *Medical Journal, 14*, 61–65.

Paúl, C., Ribeiro, O., & Teixeira, L. (2012). Active ageing: An empirical approach to the WHO model. *Current Gerontology and Geriatrics Research, 2012*, 1–10.

Smilkstein, G. (1978). The family APGAR: A proposal for a family function test and its use by physicians. *The Journal of Family Practice, 6*(6), 1231–1239.

World Health Organization. (2002). *Active ageing: A policy framework*. WHO, Geneva.

Zaidi, A., Gasior, K., Hofmarcher, M. M., Lelkes, O., Marin, B., Rodrigues, R., Schmidt, A., Vanhuysse, P., & Zolyomi, E. (2013). *Active Ageing Index 2012. Concept, methodology, and final results. Research memorandum/methodology report*. European Centre Vienna, March 2013. Retrieved from www.euro.centre.org/data/aai/1253897823_70974.pdf

Zamarrón, M. D. (2007). Estudio bio-psico-social sobre personas de 90 años y más (90 y +). Proyecto Investigación I+D+I IMSERSO. Retrieved from http://www.imserso.es/InterPresent2/groups/imserso/documents/binario/idi35_06uam.pdf

Coro Piñeiro Vázquez worked as Director General for Older People and People with Disabilities of the Regional Ministry of Social Policy, Xunta de Galicia. Her work is related to different topics, such as childhood, family, people with disabilities, dependency or older people.

Jesús Marcial Méndez Magán is Internal Control Officer at the European Committee of the Regions. His work is related to project management or strategic planning of the regions including older people. Previously, he worked as Deputy Director General of Planning and Innovation at the Galician Government, encouraging the implementation of ICT solutions for social services in Galicia with a special focus on older people.

Sara Marsillas Rascado is Researcher at the Matia Gerontological Institute. Her work is related to older people, with a special focus on active ageing and its measurement. Previously, she worked as postdoctoral researcher at the University of Santiago de Compostela, where she has also been a PhD student. Her current research focus is on Active Ageing and Age-Friendly Environments, in order to diagnose and implement action plans to make the municipalities more accessible for older people.

Antonio Rial Boubeta is Lecturer of Methodology of Behavioral Sciences at the University of Santiago de Compostela. His work spans a range of issues, including Internet use or sport practice. His current research focus is on alcohol consumption, leading a study of the National Drugs Plan.

Teresa Braña Tobío is Lecturer of Methodology of Behavioral Sciences at the University of Santiago de Compostela, specialized in non-experimental research designs. Her research work is related to methodological issues, marketing and consumers, in addition to Internet use, drugs and alcohol consumption and older people.

Jesús Varela Mallou is Professor of Methodology of Behavioral Sciences at the University of Santiago de Compostela. His work spans through a broad range of issues, from marketing and methodology research to Internet use, drugs and alcohol consumption and older people.

Part V

Use of Active Ageing Index in Non-EU Countries

17

Comparative Study of Active Ageing in China and the EU Countries

Qian Xiong and Arkadiusz Wiśniowski

17.1 Introduction

Due to the dramatic decline in fertility rates and rapid increase in life expectancy, Chinese population is estimated to age much faster than in any other developed region in the world in next decades. By 2013, China has the largest size of population aged 60 or more in the world (193 million persons), which amounts to approximately 13.8% of total population (ranking the 67th in the world in terms of percentage). This proportion is predicted to increase to 28.1% by 2040, while the older population in more developed regions will only increase by 8.8% (United Nations 2013). Moreover, in 2013, the life expectancy (LE) at birth in China was 75 years, which was close to the level of developed countries (Population Reference Bureau 2013). Data analysis of 2010 census

Q. Xiong (✉)
Lancaster University, Lancaster, UK

A. Wiśniowski
The University of Manchester, Manchester, UK

A. Zaidi (eds.), *Building Evidence for Active Ageing Policies*,
https://doi.org/10.1007/978-981-10-6017-5_17

indicated that 83.15% of Chinese older people aged over 60 or more reported to have generally good health, with the general health status of males being better than that of females (Du 2013).

In the Second United Nations (UN) World Assembly on Ageing held in April 2002, World Health Organization (WHO) provided a policy framework promoting healthy and active ageing (WHO 2002). This active ageing agenda highlights older population's self-independence, contribution to society and assistance to younger generations (Kalache and Gatti 2003). This framework shed light on the active ageing policy in China, as the policy for ageing in China has been designed for enabling the old people to teach, learn, contribute and enjoy their ageing life (China State Council 2006; Mu 2002).

The study of active ageing population is vital for every family in China. The traditional familial care for elderly members is likely to be challenged by the fertility transition brought by the birth control policy and social and economic development since the late 1970s. Chinese philosopher Mencius (372–289 BC) emphasized that in an ideal society everyone should "honor the old people as their own aged parents, and care others' children as their own children". In 1989, the Chongyang Festival[1] was designated as the official Senior's Day; in 2012, China's National People's Congress (NPC) passed the revised *Law on Protection of Rights and Interests of Seniors* to enforce legitimate care and respect for the older population. All the efforts by the individuals and authorities in China have consistently eulogized the social norm of filial piety, that is, the culture of respecting and caring for the older people. The older family members are expected to receive very good care from younger generations. However, the familial care for the elderly members might be weakened since the family structure in China (particularly in cities) has transited from a multi-generation to a "four-two-one" family model. It means that young couples possibly have to take care of four parents (and grandparents) and their own child, so that they are known as "the sandwich generation". Additionally, more and more young people who migrated to new places for jobs cannot fully assume the responsibility of caring for the older (Zhang and Goza 2006). Especially in rural areas, the household support system seems to be more vulnerable due to the small family size led by the fertility transition, the low levels of income and weak social security system.

In this chapter, we explore the potential of the older population in China by means of a recent methodology created by European experts. In particular, we apply the Active Ageing Index (AAI) to examine the lifestyles of the older population, since one of the goals of the AAI is to measure their potential in society by exploring their participation in employment, social life, independent living and capacity and enabling environment for active ageing (Zaidi et al. 2013). Additionally, we investigate a new large-scale survey on the older population in China, the China Health and Retirement Longitudinal Study (CHARLS). We compute the indicators for four domains and obtain the AAI composite score. To permit international comparisons, we select measures as similar as possible to the European datasets, but with some innovations, depending on the data availability. Eventually, we highlight policy implications based on the evidence from the most recent data.

17.2 Data: China Health and Retirement Longitudinal Study

The China Health and Retirement Longitudinal Study (CHARLS) is a biennial multistage longitudinal survey of the mid-aged and older population. Figure 17.1 presents 150 counties and districts and 450 villages and urban communities from 28 provincial-level divisions where the survey was conducted. The overall response rate was 80.51% (94.15% in rural and 68.63% in urban communities). The CHARLS has a similar questionnaire design to other surveys on ageing population, such as the Health and Retirement Study in the USA, English Longitudinal Study of Ageing and the Survey of Health, Ageing and Retirement in Europe (Zhao et al. 2013).

In 2011–2012, the survey covered 17,708 individuals and 10,257 non-institutional households with members aged 45 or above. Due to the specific focus on older population, for our analysis we selected a

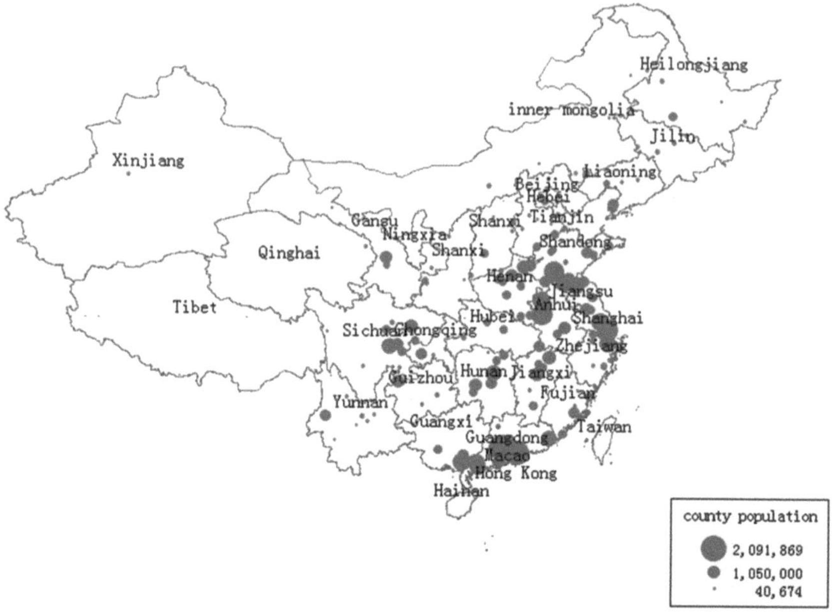

Fig. 17.1 CHARLS—distribution of sampled counties and districts (Zhao et al. 2013, p. 16)

subsample of persons aged 55+ (the same as for the EU) from the CHARLS 2011–2012, that is 11,198 respondents (5583 males and 5615 females) with the average age of 64.9 and the highest age of 101.

We measure the quality of data in two ways. First, Whipple's Index is used to measure the extent to which the reported ages are concentrated at digits ending in 0 or 5. The reported age in CHARLS ranges from 22 to 101 and the Whipple's Index is 95.6. If the index is smaller than 105, it indicates highly accurate data (Swanson et al. 2004). Second, we compare the age and sex structure of all CHARLS samples to 2010 Chinese Census. The distribution is almost identical for population aged 70 and more, but the age group 55–70 is over-represented, whereas 45–55 is generally under-represented (Fig. 17.2). Thus, the overall quality of age reporting in CHARLS data is reasonable, but the younger group of the older population (45–70) are slightly overrepresented comparing with the overall older population in China.

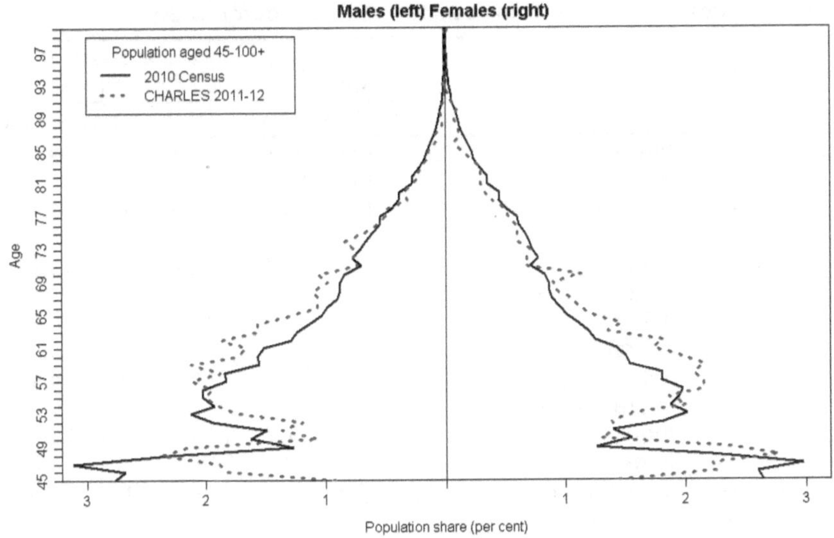

Fig. 17.2 Population pyramids of 2010 census and CHARLS (2011–2012)

17.3 AAI and Four Domains in China

The AAI score for China is much lower than the average 2010 AAI in the EU (Table 17.1), reaching for the total population (55+) 26.7, the average for the EU being 33.8. If considered in the ranking together with the EU countries, China would be last, slightly behind Poland.

The AAI for males and females are 28.9 and 24.4, respectively. The employment of males contributes more than twice than that of females, but females generally perform better in the other domains. This difference, together with the explicit weight of 35% of employment, leads to a large difference in the AAI between two sexes. The marginal gender difference is also present in the EU (Zaidi et al. 2013, pp. 20–21), particularly in Southern Europe.

In terms of the implicit weights of domains to the AAI in China (Fig. 17.3), the fourth domain, capacity and enabling environment for active ageing, contributes 39%, followed by the independent, healthy and secure living (25%), the participation in employment (19%) and the participation in society (17%).

Table 17.1 Results of AAI in China comparing to EU countries

		China			The average of EU countries		
		Both	Male	Female	Both	Male	Female
AAI	Index	26.7	28.9	24.4	33.8	35.9	32
	Rank	28	28	28			
Employment	1.1 Employment rate 55–59	27.3	40.9	14.1	60.7	68.7	52.9
	1.2 Employment rate 60–64	15.1	22.7	7.4	30.4	37.8	23.6
	1.3 Employment rate 65–69	9.4	12.7	6	11.2	14.7	8.2
	1.4 Employment rate 70–74	5.2	6.7	3.4	6.1	8.3	4.3
	Index	14.2	20.7	7.7	27.1	32.4	22.2
	Rank	28	28	27			
Participation in society	2.1 Voluntary activities	1	1.2	0.9	14.9	16	14
	2.2 Care to children, grandchildren	43.6	42.5	44.8	32.4	30.5	33.9
	2.3 Care to older adults	4.8	5.1	4.5	12.8	11.8	13.6
	2.4 Political participation	1.4	1.5	1.3	12.1	15.5	9.2
	Index	12.9	12.7	13	18.1	18.3	17.9
	Rank	27	22	28			
Independent, healthy and secure living	3.1 Physical exercise	83.6	81.7	85.4	11	10.7	11.3
	3.2 No unmet needs of health and dental care	67.1	70.9	63.5	89.6	91.9	91.1
	3.3 Independent living arrangement	78.8	79.5	78.3	84.3	84.4	84.2
	3.4 Relative median income	72.4	76.4	67.7	84	88	82.1
	3.5 No poverty risk	88.8	88.9	88.7	92.5	94.1	91.3
	3.6 No material deprivation				90.8	92.2	90
	3.7 Physical safety				79	82.9	75.9
	3.8 Lifelong learning	0.2	0.3	0.1	4.7	3.7	5.6
	Index	67.0	68.4	65.7	71.7	72.8	71.1
	Rank	24	24	25			

(*continued*)

Table 17.1 (continued)

		China			The average of EU countries		
		Both	Male	Female	Both	Male	Female
Capacity and enabling environment for active ageing	4.1 RLE achievement of 50 years at age 55	48	46	50	53.4	48.3	57.8
	4.2 Share of healthy life years in the RLE at age 55	86.0	85.6	86.5	53.4	57.4	50.5
	4.3 Mental well-being	17.1	16	18.2	63.9	67.6	60.9
	4.4 Use of ICT	1.4	1.7	1	38.3	42.6	34.3
	4.5 Social connectedness	30.2	27.5	33	51.5	50.8	52.2
	4.6 Educational attainment	9.2	12.7	5.6	58.6	63.5	54.1
	Index	52.5	51.9	53.0	54.2	54.6	54.1
	Rank	16	18	15			

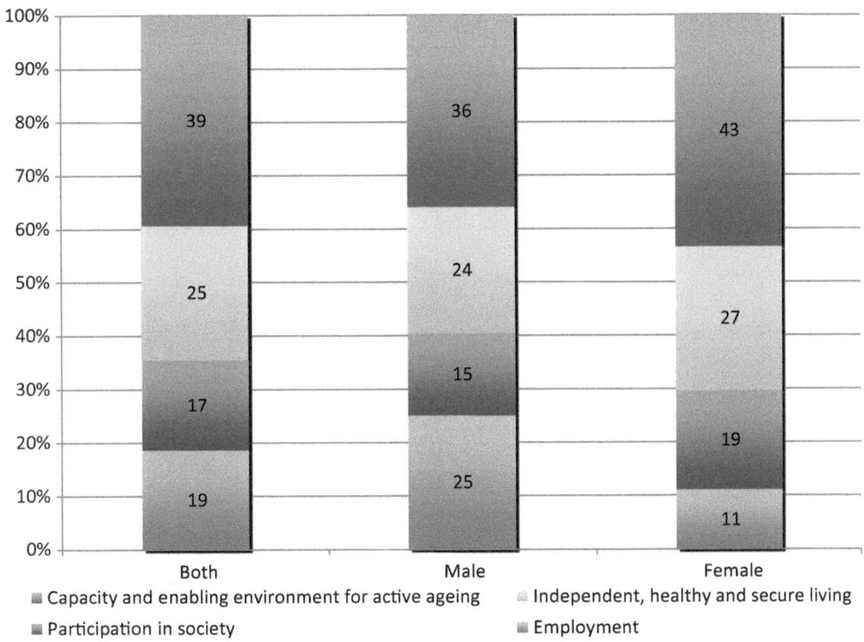

Fig. 17.3 Contribution of domains to the overall Active Ageing Index, China

17.3.1 Employment

The CHARLS data permit us to apply the same definition of employment in China as in Zaidi et al. (2013). The employment rate decreases with age from 27.3% for age group 55–59 to 5.2% for 70–84. Moreover, males always have higher employment rates than females for every age group, but the disparity decreases with age (Table 17.1). The contribution of employment by the youngest group 55–59 is the largest in the first domain (Fig. 17.4).

Comparing to the EU average employment rate, Chinese rate is merely half, but the differences narrow with age. Additionally, large gender disparity regarding the employment rates is similar to the EU results (Table 17.1).

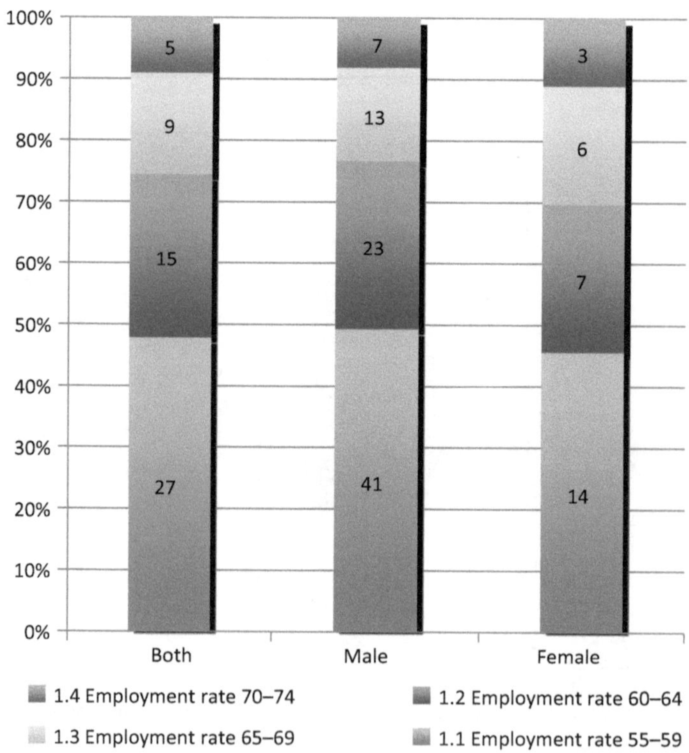

Fig. 17.4 Contribution of indicators to employment, China

17.3.2 Participation in Society

Measures in this domain are the same as in Zaidi et al. (2013), except for the percentage of the older population providing care to their grandchildren and to the other elderly relatives for at least eight weeks last year (Zaidi et al. use "at least once a week in the last year").

The most prominent contribution of the older population to society is through caretaking of grandchildren: 85%, similarly for males and females (Fig. 17.5). The participation in voluntary activities is very low in China, only 1%. Almost half of the older people are care providers to their grandchildren (43.6% for total, 42.5% for males and 44.8% for females), but only 4.8% of them (5.1% males, 4.5% females) provide care to other older adults. The rate of participation in community-related activities is only 1.4% (Table 17.1). Older Europeans, on average, are much more likely to participate in voluntary and political activities, and care for other elderly, while Chinese to provide care to grandchildren (only Cyprus reports a higher percentage).

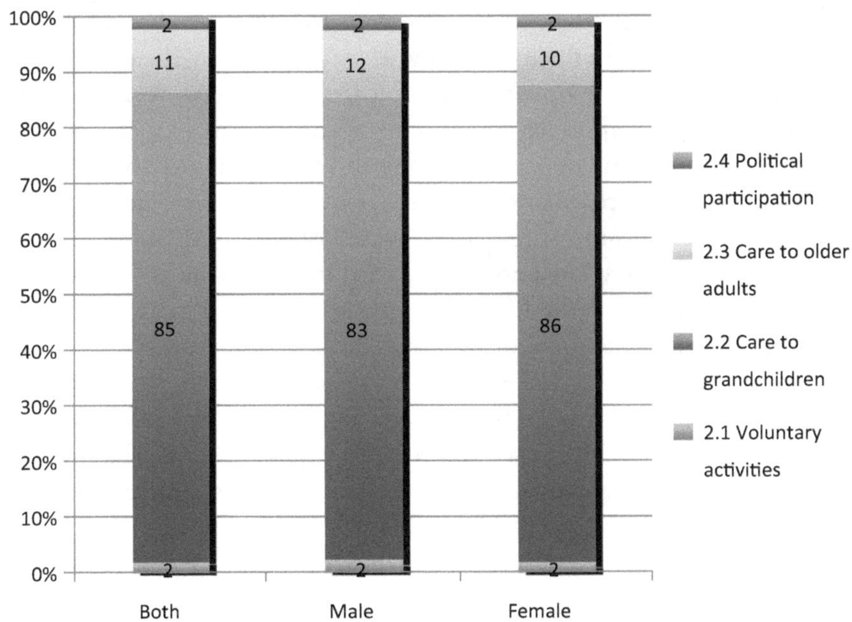

Fig. 17.5 Contribution of indicators to participation in society, China

17.3.3 Independent, Healthy and Secure Living

The first indicator of this domain is the percentage of the older population performing various moderate physical activities (similar to Zaidi et al. 2013). About 83.6% of Chinese older population (81.7% for males and 85.4% for females) reported physical exercise (Table 17.1).

The second indicator in this domain is the percentage of the older population who reported seeking and finding medical treatment while they were ill last month (comparing to 12 months in Zaidi et al. 2013). According to this definition, 67.1% of Chinese older population (70.93% of males and 63.5% of females) reported no unmet medical treatment last month.

The third indicator, living independently (i.e. aged 55+ and without children) was reported by 78.8% (79.5% males and 78.3% females). The fourth indicator relates to the relative median expenditure, which amounts to 72.4% (76.4% males and 67.7% females). These two indicators are measured in the same way as in Zaidi et al. (2013).

The lack of poverty risk in China is measured by the percentage of the older population not receiving the *Wubaohu* or *Tekunhu* subsidy from the government. These are the living allowances provided by the government to support the older people who have lost the ability to work, have no source of income and no legal guardians whatsoever to support them, or whose legal guardians do not have the ability to support them. About 88.8% of the older population is at no risk of poverty by this measurement. However, this measure may be biased due to some population living under the poverty line and in need, but not receiving any subsidies.

The last indicator is the lifelong learning, measured in the CHARLS by attending an educational or training course last month. Here the percentage is very low: only 0.2%.

Since the data on material deprivation and physical safety are not available in the CHARLS, we redistribute the weights of this domain uniformly. After weighting, the independent living arrangement constitutes 28% of this domain, followed by the relative fulfilled needs of healthcare, physical exercise, median income and no risk of poverty (Fig. 17.6).

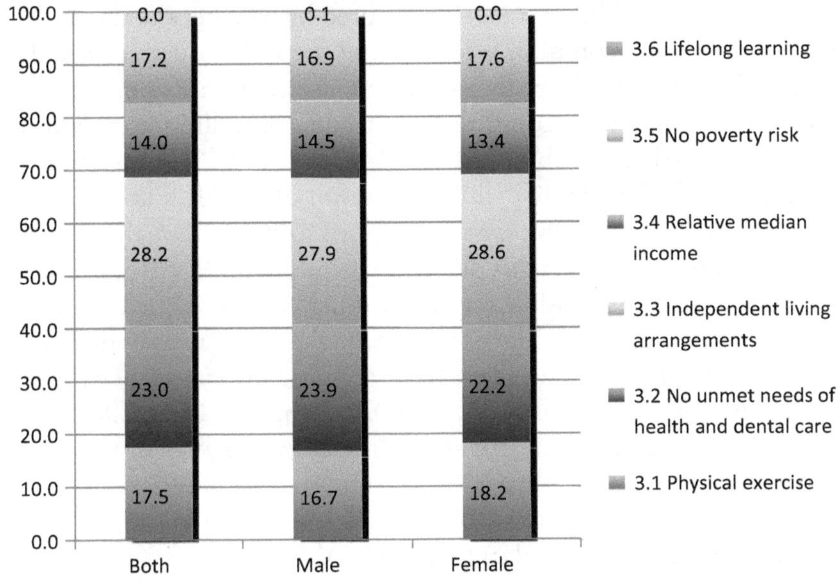

Fig. 17.6 Contribution of indicators to independent, healthy and secure living

Compared to the EU average (Table 17.1), Chinese older people are much more active physically, but they are at larger risk of poverty and have lower income. They are also less likely to live independently (though the value of this indicator is higher than in, e.g. Poland, Slovakia or Spain) and receive almost no training or education (similar to Greece and Hungary).

17.3.4 Capacity and Enabling Environment for Active Ageing

The first indicator of this domain is remaining life expectancy (RLE) at age 55. It is 24 years (23 for males and 25 for females), according to the data from the Global Health Observatory (WHO 2014). In addition, we apply a widely used Sullivan Method (Sullivan 1971) to estimate the healthy life expectancy (HLE). We combine activities of daily living and compute the disability rate as a summary measure of health status in the population. The disability status is measured by whether the respondents have problems to accomplish at least one of the six basic activities of daily living (ADL) without help, that is, eating, dressing, functional mobility, bathing and showering, using the toilet and controlling urination and defecation. The life table by WHO (2014) for China is used as a basis for constructing the healthy life table. Our result of the HLE at ages 60–64 is similar to the estimation by Du and Li (2006). Although the LE in China increased from 40 years old in the 1950s to 75 in 2013, the health of Chinese older population improved much slowly (Du and Andrews 2003; Du and Li 2006). The extension in the LE is mainly due to the improvement in infant mortality; the remaining life expectancy for population aged 60 has increased only slightly (Jiang and Chen 2004).

Based on our calculation, the share of healthy life years in the RLE at age 55 is 86% (85.6% for males, 86.5% for females). The expected number of years free of disability is 20.6 for Chinese people aged 55–59 (19.7 for males and 21.6 for females) (Table 17.2).

Mental well-being is measured by 11 questions on mental health, asking the frequency of negative feelings or behaviours during the last week. Respondents indicate answers on the four-point scale. We use the total of the reversed scale scores (rescaled to 0–100), so that the higher the score, the better mental well-being. The average scores for all, males and females are 70, 72 and 68, respectively.

Social connectedness of Chinese is measured by whether the older population had interactions with friends, visited families or community clubs or

Table 17.2 Healthy life expectancy at age of 55 or over in China

	Healthy life expectancy			Life expectancy		
Age group	Both	Males	Females	Both	Males	Females
55–59	20.6	19.7	21.6	24	23	25
60–64	16.6	15.8	17.4	19	18	21
65–69	12.9	12.3	13.6	16	15	17
70–74	9.7	9.3	10.2	12	11	13
75–79	7.0	6.7	7.3	9	9	10
80–84	4.7	4.5	4.8	7	6	7
85–89	3.2	3.0	3.3	5	4	5
90–94	2.1	1.8	2.2	3	3	3
95–99	0.8	0.0	0.9	2	2	2
100+	0.8	0.0	0.8	2	2	2

Data source: World Health Organization 2014 and self-calculation

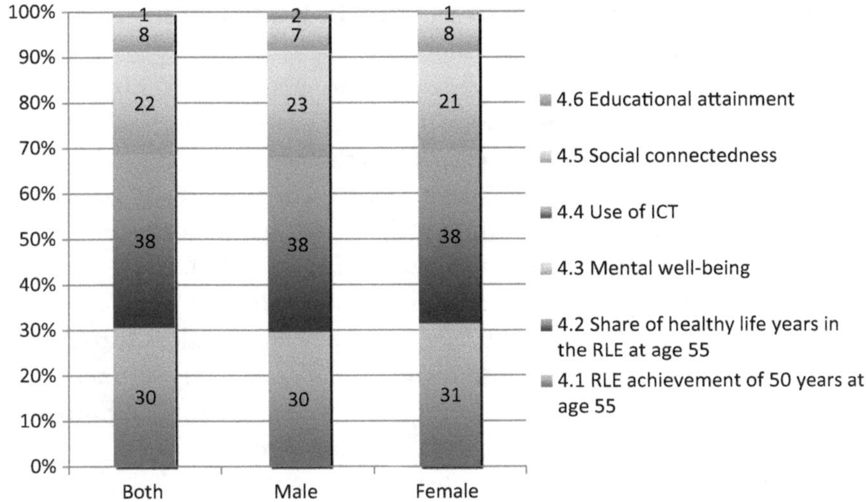

Fig. 17.7 Contribution of indicators to capacity and enabling environment for active ageing

played games (mahjong, chess or cards) in the last month. Only 30% of the older population in China reports social connection with others. In addition, the use of Internet by them is almost non-existent (less than 2%) and the educational attainment is very low (less than 10%).

After applying the weights, the share of HLE (38%) and the RLE achievement (30%) represent the major components of the fourth domain (Fig. 17.7). Compared to the EU average (Table 17.1), Chinese older people have similar life expectancy, a much higher share of HLE (which results from using different measures than in Zaidi et al. 2013), but the use of Internet, social connectedness, higher educational attainment and mental well-being are substantially lower.

17.4 Policy Implications

Our findings have shown the unrealised potential of the older population in China by comparing them with the results for the EU. In this section, we discuss the current policies related to ageing in China, and suggest future directions for policymaking.

17.4.1 Employment of the Older Population in China

The low employment rate of Chinese older population is likely due to the earlier retiring age, lower educational attainment and the possible social exclusion in employment. The statutory retirement age is 55 for females and 60 for males (for some blue-collar female workers: 50 and male: 55). Actually, urban workers who could get secured pensions may decide to retire even earlier than the legal age (Du and Yang 2010). In comparison to China, most of the EU countries have statutory and effective retirement age equal or higher than 60 for both males and females. Introducing flexible retirement ages (e.g. in Finland and Sweden) created incentives for the older people to stay longer on the labour market (The European Commission and The Economic Policy Committee 2012).

The older population in China tends to have the capacity to support their family and community and possibly to participate in the formal workforce. Our results show that Chinese older people are physically active and report high shares of years in good health status after retirement. It is very popular that grandparents (especially females) provide care for grandchildren in China. In addition, the agricultural population still actively participates in the production even when getting old. Data from the 2009 Labour Force Survey show that people are more likely to engage in agricultural than nonagricultural activities. Moreover, they are more likely to be self-employed as they get older (NBSC 2014a). More than 80% of the employed older people (aged 60 or more) are self-employed. Around half of the employed older people are engaged in agricultural activities, comparing to less than 40% for the young.

Despite the agricultural labour force, the reemployment rate of the nonagricultural older people in China is very low and is highly concentrated. They are active in professions such as marketing, consulting, management, healthcare or teaching, which require higher education and skills (Wang and Yang 2005). The employed older people also have lower rate of higher educational attainments (less than 10%) than the young population (who has more than 15%) (NBSC 2010, Table 3.4). The lower educational attainment implies the need of providing continuing education for the older population, but they scarcely receive lifelong training or education and seldom use the Internet.

17.4.2 The Old-Age Security System in China

Until recently, there was very low welfare coverage of older people and an unequal distribution of pension schemes in China (China State Council 2006). Until the restructuring of state owned enterprises (SOEs) in the 1990s, China's pension system covered primarily the workers with SOEs or some large collective enterprises, or employees of government agencies and public institutions. The government reformed the old-age social security system in the 1990s. The new pension system is composed of individual accounts and social pooling, which means that both individuals and employers contribute to the pooling funds together (Du and Yang 2010; Dong and Wang 2014). From 1989 to 2013, the number of urban participants in the basic old-age insurance China increased from 57.10 million to 322.18 million, with the participation rate rising from 19.33% to 44.07% (NBSC 2014). Of those participants, the number of urban retirees increased from 8.93 million to 80.41 million, almost ninefold that of 1989 (NBSC 2014b). In addition, the employment opportunities in SOEs decreased due to restructuring, and more and more workers entered the informal economy (i.e. sector of economy that is not recognised or protected under the legal and regulatory frameworks). About one third of the urban workers together with all rural-to-urban migrant workers are employed by the informal economy. They are likely to face poverty when getting old, because the current law still does not require all workers or employers in informal economy to join the pension system (they can join the pension system voluntarily) (Du and Yang 2010; Dong and Wang 2014).

As for older residents in rural areas, they have been mainly depending on family support (Wang 2006). Only the childless and disabled older people (known as *Wubaohu* or *Five Guarantee persons*) have been provided with food, clothing, medical care, housing and coverage of their burial expenses. In recent years, China has built up the new rural old-age security systems, including an insurance system based on family support, healthcare and personal care services (China State Council 2006). The new rural cooperative medical system started in 2003, and the new old-age social insurance system in rural areas started in 2009. According to the Ministry of Human Resources and Social Security of China (MHRSS), 76.8% of the rural residents have joined the old-age social insurance system, i.e. around 483.70 million by the year 2012 (Ministry of Human Resources and Social Security of China (MHRSS) 2013).

Furthermore, in addition to the social security system, the government has been developing civil services for the older population since 1980s, which may have contributed to their active ageing. The older people enjoy special discounts or receive free public services, such as transportation, museums and cultural and sport facilities. Community centres, including nursing homes and hospitals, have been established to provide supplemental care for the older people being cared for by family members (China State Council 2006).

17.4.3 Challenges and Policy Recommendations

Even though the social security system is improving in China, there are still many challenges for the government. Firstly, according to the official reports (MHRSS 2013), there are still half of the urban residents and one fourth of the rural residents not being covered by the old-age insurance system. Our findings confirm that Chinese older people are less secured than their European counterparts in terms of healthcare and income. Secondly, the government still explores ways to achieve a sustainable system to face the future of decreasing young labour population. Thirdly, discussion about the older population in China concentrates on healthcare and pension system reforms, but neglects the potentials of promoting independent living in older age, and inequality among regions and between the rural and urban residents.

Based on our findings, we conclude that it is important to provide and protect rights of the older population to work. Policies should aim at improving the voluntary participation in labour force, especially for women who retire earlier than men. To improve participation of the older population in the labour force, a more flexible retirement age should be introduced, taking into account health status, educational attainment and human capital. Also, training, education and networking opportunities should be facilitated for the older people who wish to be active and remain part of the workforce. The lifelong learning policy might help older people to maintain ties with the society, thus promoting healthy and active ageing.

Additionally, we find that the young generation continues to play a major role in supporting the older family members in the household. Also, the older family members provide support to younger generations by taking care of grandchildren. It can be suspected that both the older population and young families would face higher risks of poverty, if there were less intergenerational transfer. We suggest that programmes should be launched up at the local community level to support families who provide the familial care for the older population in need at home.

17.5 Conclusion and Future Work

In this chapter, we discussed briefly the quality of the CHARLS data, computed the AAI for China and compared it to the AAI in the EU. The overall score for the older population in China is slightly lower than that in Poland, which has the lowest AAI in the EU. Moreover, the labour force participation rate of the older population is lower in China, but older Chinese have similar high LE as Europeans; they are more physically active and report a much higher share of the HLE. Males are advantaged over females in active ageing, especially in the labour force participation. This complies with the findings for the EU.

Therefore, we suggest that social policies aim at enabling older people, with a special attention paid to females, to maintain quality of life and contribute to the family, economy and society. We suggest that active ageing of the older population can be improved by: (1) providing flexible retirement plans to improve the voluntary participation in the labour market and (2) promoting lifelong learning system and community care system for helping families to guarantee secure and healthy ageing for their elderly at home. Further analysis of the policy implications may require a detailed, international comparison of the retirement systems and their reforms (see, e.g. The European Commission (DG ECFIN) and The Economic Policy Committee (AWG) 2012). At last, we recommend that further research should concentrate on the social and economic determinants of active ageing, including reemployment, participation in social activities and lifelong learning in China, which can inform and support the effective policymaking for active ageing in the future.

Notes

1. Chongyang Festival (also known as Double Ninth Festival, on the ninth of the ninth month in the lunar calendar) is a holiday for the elderly. The festival has existed for around 2000 years. Family members gather to show respect to the elderly. Number "9" also symbolizes longevity in China due to the similar pronunciation ("*jiu*") in Mandarin to the word "eternality".

References

China State Council. (2006). *The development of China's undertakings for the aged.* Beijing: Information Office of the State Council of the People's Republic of China. Retrieved June 4, 2017, from http://www.china.org.cn/english/aged/192020.htm

Dong, K., & Wang, G. (2014). China's pension challenge: Adaptive strategy for success. *Public Administration and Development, 34*(4), 265–280.

Du, P. (2013). An analysis on the health status of the older persons in China (in Chinese). *Population and Economics, 6*, 3–9.

Du, P., & Andrews, G. (2003). Successful aging: Examples of the elderly from Beijing (in Chinese). *Population Research, 27*(3), 4–11.

Du, P., & Li, Q. (2006). The trend of life expectancy of self-care of Chinese elderly (1994–2004) (in Chinese). *Population Research, 30*(5), 9–16.

Du, Y., & Yang, M. (2010). Demographic ageing and employment in China. *Employment Sector employment working paper No. 57*. Geneva: International Labour Office.

Jiang, X., & Chen, Y. (2004). Questions on postponing the current retirement age (in Chinese). *Population Research, 5*, 69–74.

Kalache, A., & Gatti, A. (2003). Active ageing: A policy framework. *Advances in gerontology-Uspekhi gerontologii/Rossiiskaia akademiia nauk, Gerontologicheskoe obshchestvo, 11*, 7–18.

Ministry of Human Resources and Social Security of China (MHRSS). (2013). *The report on social security in 2012.* Retrieved December 12, 2014, from http://www.mohrss.gov.cn/SYrlzyhshbzb/dongtaixinwen/shizheng-yaowen/201306/t20130618_105477.htm

Mu, G. (2002). Discussion on the potential of the elderly: The framework for successfully challenging the aging society in the 21st century (in Chinese). *Population Research, 26*(6), 29–37.

NBSC. (2014a). *China statistical yearbook (2014)*. Retrieved February 8, 2015, from http://www.stats.gov.cn/tjsj/ndsj/2014/indexch.htm

NBSC. (2014b). *National database*. Retrieved December 20, 2014, from http://data.stats.gov.cn/workspace/index?m=hgnd

Population Reference Bureau. (2013). *2013 World population data sheet*. Retrieved April 29, 2014, from http://www.prb.org/pdf13/2013-population-data-sheet_eng.pdf

Sullivan, D. F. (1971). A single index of mortality and morbidity. *HSMHA Health Reports, 86*(4), 347–354.

Swanson, D., Siegel, J. S., & Shryock, H. S. (2004). *The methods and materials of demography*. New York: Elsevier Academic Press.

The European Commission and The Economic Policy Committee. (2012). The 2012 Ageing Report Economic and budgetary projections for the 27 EU Member States (2010–2060). *European Economy, 2*(2), 1–472.

United Nations. (2013). *World population ageing 2013*. Retrieved June 4, 2017, from http://www.un.org/en/development/desa/population/publications/ageing/WorldPopulationAgeingReport2013.shtml

Wang, D. (2006). China's urban and rural old age security system: Challenges and options. *China & World Economy, 14*(1), 102–116.

Wang, S., & Yang, Y. (2005). The strategy of exploring human capital of the elderly (in Chinese). *Population Research, 3*, 63–69.

WHO. (2002). *Active ageing: A policy framework: A contribution of the second United Nations World Assembly on Ageing*. Retrieved October 1, 2014, from http://apps.who.int/iris/bitstream/10665/67215/1/WHO_NMH_NPH_02.8.pdf

WHO. (2014). *Life expectancy data by country: China*. Retrieved October 1, 2014, from http://apps.who.int/gho/data/view.main.680

Zaidi, A., Gasior, K., Hofmarcher, M., Lelkes, O., Marin, B., Rodrigues, R., Schmidt, A., Vanhuysse, P., & Zolyomi, E. (2013). *Active Ageing Index 2012: Concept, methodology and final results*. Retrieved from www.euro.centre.org/data/aai/1253897823_70974.pdf

Zhang, Y., & Goza, F. W. (2006). Who will care for the elderly in China?: A review of the problems caused by China's one-child policy and their potential solutions. *Journal of Aging Studies, 20*(2), 151–164.

Zhao, Y., Strauss, J., Yang, G., Giles, J., Hu, P., Hu, Y., Lei, X., Liu, M., Park, A., Smith, J. P., & Wang, Y. (2013). *China health and retirement longitudinal study, 2011–2012 National baseline user's guide*. Retrieved January 10, 2016, from http://charls.pku.edu.cn/uploads/document/2011-charls-wave1/application/CHARLS_nationalbaseline_users_guide.pdf

Qian Xiong is a Lecturer in Ageing in the Centre for Ageing Research, Division of Health Research, Lancaster University. Her research aims to contribute to the understanding of population ageing and social, economic and health inequalities in later life through the lens of demographic and sociological theories and the life course perspective. Her research seeks evidence from both developing and developed countries to inform policy changes and promote the well-being of older people.

Arkadiusz Wiśniowski is a Lecturer in Social Statistics at the University of Manchester. His research interest is in developing statistical methods for combining various sources of data through Bayesian inference, especially in the area of measuring and forecasting populations and migration.

18

Quality of Life of the Elderly and Applicability of the Active Ageing Index to Latin American Countries

Javiera Fanta

18.1 Introduction

Since the mid-twentieth century, Latin America and the Caribbean (LAC) have experienced an abrupt rise in the share of older people in the total population. Between 1950 and 2000, the average life expectancy at birth increased from 52 to 73 years, and the proportion of senior citizens[1] grew from 5.6% to 8.1% (ECLAC 2015). Currently, there are nearly 70 million people aged 60 and over in the region, representing 11.1% of the total population. Due to the accelerated pace of ageing, this percentage is expected to double by 2040, matching the number of children aged 15 and under (ECLAC 2015).

As is the case in the majority of developing countries, transformations in the demographic structure have unravelled independently of the industrialization and modernization processes. The drops in mortality and birth rates observed in the second half of the twentieth century—and, by extension, the increase in the number of older people—were essentially a

J. Fanta (✉)
Córdoba National University, Córdoba, Argentina

A. Zaidi (eds.), *Building Evidence for Active Ageing Policies*,
https://doi.org/10.1007/978-981-10-6017-5_18

consequence of the capacity that governments had for (a) incorporating available technology at low cost in the areas of health and medicine; (b) improving preventive health care, thus reducing the level of infant mortality; and (c) increasing coverage of family planning services (Guzmán 2002). However, these measures were not accompanied by actions towards improving the living conditions and quality of life of the population, particularly for the elderly. In fact, before the signing of the Madrid International Plan of Action on Ageing (MIPAA) in 2002, poverty levels among older people increased from 37.5% in 1997 to 37.8% in 2002[2] (ECLAC 2013), and there were no regulatory frameworks aimed at addressing the needs of this group, neither as a regional strategy nor as a state policy.

In the light of the priorities defined by the Regional Strategy (RS) for the implementation in the Latin American and Caribbean countries of the MIPAA,[3] the assessment of the various aspects related to ageing, protection and welfare of the elderly have become paramount topics in the regional agenda. Specifically, the RS places the issue of quality of life in older age at the core of the discussion (ECLAC 2003). While in European countries the construct used to monitor changes regarding the goals of MIPAA is Active Ageing, in LAC this concept is still contingent upon achieving other immediate priorities, as today national governments in general have not been able to generate sustained basic standards of living for a large proportion of the elderly (Acevedo 2014; Huenchuan 2014). The latter does not contradict the fact that Active Ageing, as a policy discourse, may be useful to achieve improvements in these areas and that the Active Ageing Index (AAI), as a measurement tool, may also serve this purpose. Indeed, one of the guidelines set forth by the MIPAA-RS refers to the promotion of active ageing, as it 'fosters people's self-esteem and dignity and the full exercise of all their human rights and fundamental freedoms' (ECLAC 2003: 2).

Given that knowledge and analytical work on this matter is rather recent, policy makers and stakeholders face two main limitations: the scarce availability of data on ageing, and the lack of tools necessary to monitor changes in the quality of life of the elderly (QLE). In fact, out of 22 signatory countries, only three have incorporated ongoing evaluations on QLE through continuous surveys.[4]

In an attempt to generate follow-up guidelines for the assessment of senior population, the Latin American and Caribbean Demographic Centre (CELADE 2006) drafted a methodological and operational manual containing baseline indicators on QLE. Today, this document represents a reference point from which to develop statistical information on the matter.

Based on the indicators proposed by CELADE and on the procedure followed in the construction of the AAI, this study contributes to create a composed index for assessing the quality of life of older adults. In addition, we analyse the applicability of the AAI to six countries of Latin America (Argentina, Bolivia, Chile, Ecuador, Guatemala and Peru). The question that guides this second task is whether the AAI can be used as a proxy to QLE. Throughout this article, QLE refers to the situation where people aged 60 and over can access and enjoy resources and services for a dignified old age; the latter includes (a) economic security, (b) healthy living and (c) satisfactory environmental conditions (CELADE 2006). The study aim is to develop a portrait of both potential measurement tools and current status on quality of life in older age among the countries concerned.

18.2 Methods

18.2.1 Data Sources

Population censuses from the 2010 round and national surveys conducted between 2010 and 2015 from 19 countries[5] were examined. Recognized flawed sources were discarded, and a starting minimum set of indicators of both AAI and Quality of Life of the Elderly Index (QLEI) was determined.

Appendix 1 lists the type and name of the sources used to provide information on active ageing and quality of life in older age for selected countries. The most consistent data on income, consumption, educational level and housing conditions was reported by household surveys and population censuses. It was sought to identify censuses whose omission did not exceed 5% of total population,[6] and household surveys with nationwide coverage.[7] Information on health and well-being was estimated from national health surveys and, when not available, data was derived from multipurpose household surveys. Information on mortality was obtained from UN database (2012) and from population projections by CELADE (2015), since mortality records are documented to be a flawed source in many countries of Latin America.[8] Specific indicators on ageing, such as functional living skills, were derived from special surveys addressing older adults.

18.2.2 Quality of Life of the Elderly Index

The methodological approach for the construction of the QLEI was grounded on the judgement of experts,[9] CELADE recommendations and literature on AAI methodology (Zaidi et al. 2013). Selection criteria for the indicators of the QLEI are set forth in Table 18.1. The indices are

Table 18.1 Selection criteria for QLEI indicators[a]

Selection criteria/ requirements of indicators	Description
Aim	QLEI aims to measure changes in critical aspects of the elderly's welfare, security and global well-being, by taking into account different generations of older people, instead of a life-course perspective.
Parsimony	CELADE lists over 80 indicators to assess the QLE. If all these variables were included in a single index, its robustness would be restricted. That is why the composed index comprises a limited number of indicators, which is sufficient to synthetize the different dimensions involved in the definition of QLE.
Data availability and international comparability	Selected indicators can be derived from available sources for every country concerned in the analysis. Although outcomes on one same variable are obtained from different sources, common definitions and time parameters allow cross-country comparisons.
Data quality	Unreliable and invalid information was excluded. For instance, the variable 'percentage of population aged 60+ who has been victim of abuse or mistreatment' was not incorporated, as its definition was ambiguous, did not include timeline criteria and demonstrated to have misreporting problems.
Access to micro-datasets	It was sought to include accessible micro-data from the national statistical offices' websites, in order to enable further disaggregated analysis by gender, age and—when possible—ethnic group.

[a]These requirements are based on the selection criteria used by Zaidi et al. (2013) to build the AAI for the EU countries

defined for the age group above 60, as this is the age limit used by most countries of the region to define older population.

From the overall list proposed by CELADE, a preliminary set of 44 indicators was drafted. After examining available data by country, an abridged version comprising 32 variables was developed. Eleven out of 19 countries failed to meet this baseline, thus being excluded from further analysis. Aware of the fact that the pace of ageing is highly heterogeneous within the region, six out of eight remaining countries were selected, according to the following demographic transition stages: (a) moderate stage, Bolivia and Guatemala; (b) full transition stage, Peru and Ecuador; and (c) advanced stage, Argentina and Chile (ECLAC 2014).

From the latter dashboard, low-quality data were excluded and 20 indicators distributed in 3 domains were used to build the aggregated index (see Table 18.2). Note that the first domain (economic security)

Table 18.2 Name, description and assigned weights of the indicators selected for the QLEI

Domain/Indicator	Description (CELADE 2006)	Weight (%)
I. Economic Security	**Objective: To measure the capacity to acquire and use regular and sufficient income, in order to ensure economic independency and minimal standards of well-being for the elderly.**	**35**
Formal labour	Includes all remunerative work (i.e., both self-employment and wage employment) that is registered, regulated or protected by existing legal or regulatory frameworks. This indicator comprises people aged 60+ with formal employment as a percentage of the labour force aged 60+.	15
Employment rate (60+)	Number of employed persons aged 60+ as a percentage of the labour force aged 60+ (the total number of people employed plus unemployed).	10
Economic inactivity rate in advanced old age (80+)	Proportion of the population aged 80+ that is not economically active. Active working force refers to all people who supply labour for the production of goods and services during a specified period.	10

(continued)

Table 18.2 (continued)

Domain/Indicator	Description (CELADE 2006)	Weight (%)
Provision of retirement benefit in advanced old age	People aged 80+ who receive retirement pension as a percentage of the total population aged 80+.	15
Retirement income	Average monthly payment made to a pension fund member (or dependents) after retirement, where: – 100% means that the monthly average retirement income exceeds at least five times the monthly amount determined by the poverty line (PL) and – 0% means that the monthly average retirement income is equal to the PL. Poverty line refers to an income level that is considered minimally sufficient to meet daily needs.	15
No poverty	Percentage of the population aged 60+ who are above the line of poverty determined by each country, or who register no unsatisfied basic needs.	15
Extreme poverty ratio	Ratio of people aged 60+ living in extreme poverty to the total population of 60+ living in poverty. Extreme poverty line refers to the minimum income necessary to meet the cost of a basic food basket.	10
Poverty gap	Distance between the value of poverty line per capita (Z) and the average monthly income per capita of poor households with older members (μ_Z), expressed in terms of percentage. This indicator shows the deficit of poor households in relation to the poverty line and is obtained as follows: Poverty gap = $100-(Z-\mu_Z)/Z*100$.	10
II. Health and Well-Being	**Objective: To assess physical, mental and social well-being along the ageing process.**	**35**
Share of healthy life years (HLY) in the remaining life expectancy (RLE) at age 55	Average number of years that a person aged 55 can expect to live in 'full health' by taking into account years lived in less than full health due to disease or injury (Zaidi et al. 2013). This indicator is calculated as follows: $(HLY_0-55)/RLE_{55} \times 100$. HLY_0 represents Healthy Life Years at birth, and RLE_{55} refers to Remaining Life Expectancy at age 55.	10

(*continued*)

Table 18.2 (continued)

Domain/Indicator	Description (CELADE 2006)	Weight (%)
Functional performance	Percentage of people aged 60+ with no functional disability. Functional disability is defined by the presence of at least two limitations in the performance of daily activities. Daily activities include: making purchases, doing household chores, using means of transportation, getting dressed, taking a bath, eating and using the toilet.	15
No permanent disability	Disability denotes any long-term limitation in activity resulting from a condition or health problem (including visual, hearing, physical, cognitive and language impairments). This indicator refers to the percentage of population aged 60+ with no permanent disability.	15
Health coverage	People aged 60+ who report having health insurance or other types of health-care plan.	20
Access to health services	Percentage of people aged 60+ who report no unmet need for medical attention or treatment.	20
Level of education	Average years of total schooling of the population aged 60+, where 100%=12 years of schooling (average) and 0%=0 years of schooling. The limit of 12 years was determined by the time (in years) it takes to complete the primary and secondary schools in Latin American countries (Barro-Lee, 2010).	20
III. Enabling Environments	**Objective: To measure environmental conditions required for a dignified and secure ageing in the home community.**	**30**
Housing tenure	Number of homeowners as a percentage of the total population aged 60+.	18
Quality of building material	Proportion of people aged 60+ who live in dwellings that (a) do not have earthen floor **and** (b) whose walls are built from any material other than cane, palms, tree trunks, mud brick, plastic and cardboard.	15
Basic sanitation facilities	People aged 60+ living in dwellings where the toilet bowl is properly connected to the sanitary sewer.	20

(continued)

Table 18.2 (continued)

Domain/Indicator	Description (CELADE 2006)	Weight (%)
No overcrowding	Percentage of people aged 60+ living in non-overcrowded dwellings. Overcrowding is determined by the average number of people per room. Three or more people per room indicate overcrowding.	19
No lodging households	Percentage of people aged 60+ living in housing units occupied by only one household.	8
Formal settlements	Proportion of the population aged 60+ living in dwellings that (a) enjoy formal ownership of property, (b) have access to electricity **and** (c) have access to water supply network.	20

contains three indicators related to access and quality of employment, that is, formal labour, employment rate (60+) and economic inactivity rate in advanced old age (80+). However, certain key topics, such as economic participation and employment rate by age, have not been included as it was not possible to assign the same normative value judgement. Participation in the labour force in advanced old age might be interpreted as a positive indicator in countries where retirement pensions and subsidies for the elderly are guaranteed; but this criterion does not apply to low- and even medium-income countries, where economic participation in older age is related to subsistence needs and precarious jobs, thus limiting the economic security and dignity of the elderly. The former consideration led to include economic inactivity and provision of retirement pension in advanced old age (80+) as indicators that positively affect the quality of life, on the assumption that people aged 80+ who are still economically active work out of necessity, since (a) they do not receive retirement benefits or (b) their retirement income does not provide economic security. Employment rate (without differentiating by age) was also included in this domain in order to measure the economic security of those who, whether it is out of necessity or personal will, are still active in the labour force.

Lack of sources and limited available data led to exclude several indicators from the aggregated index. For example, note that the indicators listed in the third domain (enabling environments) relate only to housing conditions; they do not address other environmental features, such as physical safety and urban environment, as these aspects have been barely studied at a national level.

Indicators are expressed in terms of percentage, with a lower goalpost of 0 and an upper goalpost of 100. Variables whose outcomes were quantified in absolute terms (i.e., retirement income and level of education) were turned into percentage rates from standardized scales. This methodology allowed assigning positive and normative values to every indicator. Due to the lack of theoretical and empirical grounding on the contribution of each indicator (and thus of each domain) to QLE, weights were assigned on the basis of experts' assessment (Table 18.2).

18.2.3 Active Ageing Index

The methodology chosen to examine the applicability of the AAI to selected countries was based on three basic requirements: (a) conceptual and empirical relevance, (b) data availability and (c) access to micro-datasets. The definition and underlying concepts of each indicator were analysed in accordance with the guidelines of the MIPAA-RS, CELADE recommendations and relevant literature on the matter (Huenchuan and Morlachetti 2007; ECLAC/CELADE 2009; Huenchuan 2011, 2014). The crucial question at this point was whether or not the dashboard of indicators included in the AAI reflects, in fact, aspects of healthy, independent and active ageing within the population concerned.

In addition, data sources from each selected country were examined, and those indicators whose outcomes could be derived from available sources and compared across countries were considered for further analysis.

18.3 Results

18.3.1 Quality of Life of the Elderly Index

Results for the overall index (QLEI) are displayed in Fig. 18.1. Argentina comes at the top of the ranking followed by Chile, with 83.2 and 79.8%, respectively. Let us recall the fact that both countries are

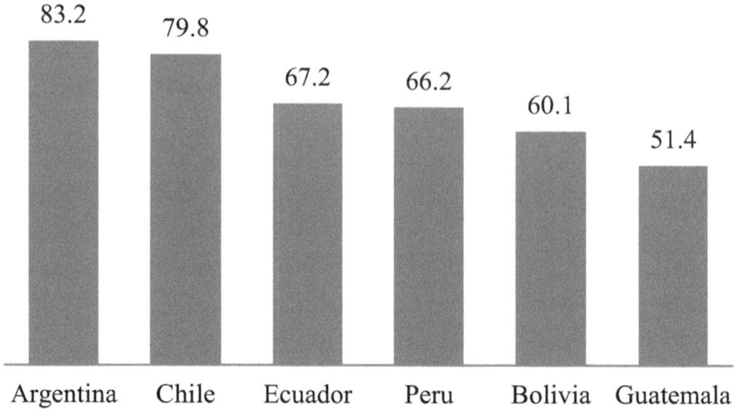

Fig. 18.1 Results for the QLEI by country

crossing an advanced stage of the demographic transition, which is partly related to higher levels of development. By contrast, countries going through moderate stages of this process are the lowest performers (Bolivia = 60.1%; Guatemala = 51.4%), while countries included in the full transition stage are placed in an intermediate position in terms of the aggregated index results (Ecuador = 67.2%; Peru: 66.2%). Outcomes show that even the top performing countries must aim for further improvements, as the overall index comprises baseline standards of QLE.

Results on each domain of the QLEI are presented in Fig. 18.2. It can be observed that countries' performance on individual domains is not necessarily related to their performance in the overall index. With the exception of Argentina, economic security is the field with greater scope for further improvements, especially in the areas of 'formal labour' and 'provision of retirement benefit in advanced old age'[10] (Appendix 2). Formal labour rate is particularly low in Argentina, Bolivia, Ecuador, Guatemala and Peru, with percentages ranging from 11.6% (Bolivia) to 34.2% (Argentina). Only in Chile this indicator scores relatively high (79.5%), but there is still a large part of the employed population of 60+

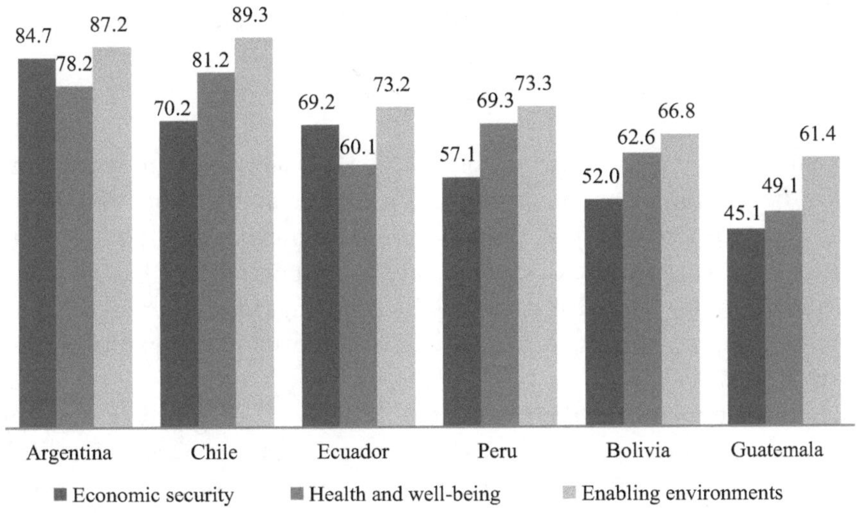

Fig. 18.2 Results of the QLEI by domain

that do not work in formal conditions, indicating that the risk of labour precariousness is latent even in the best performing country.

As is the case with the overall index, the second domain (health and well-being) performs better in Chile and Argentina (Chile = 81.2%; Argentina = 78.2%). The average outcome registered by Bolivia within this field (62.6%) is closer to the results achieved by full-stage transition countries and even exceeds the result obtained by Ecuador (60.1%). Noteworthy is the low percentage showed by 'level of education' within this domain, irrespective of the country, which ranges from 77.2% in Argentina to 28.7% in Guatemala.

All countries perform better in the third domain (enabling environments). Chile and Guatemala, in particular, reflect higher values in this field than in the former domains. 'Basic sanitation facilities' is, in almost all cases, the indicator with greater scope for improvement for countries and where there are substantial cross-country differences (Appendix 2). Specifically, in Peru, Ecuador, Bolivia and Guatemala, at least one third of the older population lacks of basic sanitation facilities (Peru = 60.5%; Ecuador = 62.8%; Bolivia = 60.8%; Guatemala = 45.1%).

18.3.2 Applicability of the AAI to Latin America

18.3.2.1 First Domain: Employment

As life expectancy increases across Europe, pension ages are rising, and thus, being able to remain active in the labour market until retirement reflects a positive aspect. Although life expectancy is rapidly growing in Latin America, too, the discussion on raising pension age has not been included in the regional agenda yet, since putting this matter on the table would bring to the light the problem of mortality inequality within populations.

If this domain and their respective indicators (employment rate by age) were used for assessing active ageing among the countries studied, they should be interpreted with caution. First, consider that, even in medium-income countries of Latin America, labour conditions of the elderly are a critical issue—let us recall the fact that, in five out of six countries surveyed, at least 65% of their active working population aged 60+ works in the informal labour market. Secondly, consulted experts suggest that employment in old age may be driven by the need to alleviate poverty, as a large proportion of older adults have never contributed to a pension scheme (e.g., in Guatemala only 14.7% of the economically inactive population aged 60+ receive pension benefits, according to 2013 data). And third, as can be seen in Table 18.3, there are substantial gender differences in relation to the participation of older adults in the labour

Table 18.3 Employment rate by age group and sex (%)

	Employment rate 55–59		Employment rate 60–64		Employment rate 65–69		Employment rate 70–74	
Country	Men	Women	Men	Women	Men	Women	Men	Women
Argentina	82.2	54.4	72.0	36.8	44.7	18.8	19.9	7.5
Bolivia	89.9	63.4	80.8	57.4	66.0	51.7	59.6	42.1
Chile	82.6	45.7	74.1	32.7	46.1	16.3	28.0	9.0
Ecuador	88.8	52.7	77.4	46.5	70.0	35.3	56.8	24.7
Guatemala	92.1	32.7	81.5	25.2	79.6	15.1	61.2	14.8
Peru	88.2	64.6	82.2	58.0	69.0	45.8	54.3	37.5

Sources: Argentina, EPH (October 2014); Bolivia, EH (2014); Chile, CASEN (2013); Ecuador, ENEMDU (2015); Guatemala, ENEI (2013); Peru, ENAHO (2014)

market, which can be better explained by cultural patterns rather than by individual degree of activity or autonomy.

Given the former considerations, high outcomes in this domain (thus, expressing a better performance of active ageing) might actually be omitting the fact that within the active labour force there is a large proportion of older adults who (a) work in the informal sector, which is associated to precarious labour conditions, and (b) work as a survival strategy, since they do not receive retirement benefits—and if they do, this income does not allow them to meet basic needs. In both cases, neither income security nor the securing of a rewarding employment is guaranteed. Conversely, low outcomes in this field, in average, might be due to gender-based inequalities and not precisely to lower levels of healthy and autonomous living.

18.3.2.2 Second Domain: Participation in Society

The absence of uniform criteria for addressing the variables comprised by this field limits the applicability of the AAI to selected countries. For example, when trying to measure 'unpaid voluntary work through organizations', it was noted that different time parameters are used, impeding cross-country comparisons. Moreover, the definition itself of 'voluntary activities' differs from one country to another: in Ecuador, time-use survey (2012) makes a distinction between 'voluntary activities' and 'community service', applying a specific word from the Quechua language to define the latter—'Minga', activities based on reciprocal relationships between the members of a community—while in Argentina QLE survey (2012) merges both concepts into a single variable.

Questions addressing 'care to children and/or grandchildren' also vary and have different time criteria. Surveys from Chile, Peru and Guatemala do not allow us to identify whether caring was a paid or unpaid activity, and time parameters used to determine the performance of caring activities rank from one week through three months preceding the survey. With regard to 'care to older adults', data on this matter is limited, and could only be derived from the Chilean survey on QLE (2013).

Information about political participation is another critical issue among LAC, especially when concerning older population. In countries like Argentina, Bolivia, Chile, Peru and Guatemala, people currently aged 60 and over experienced military dictatorships and civil wars, where participation in politics was related to serious threats to their own safety. According to expert's opinion, although significant progresses have been made in restoring democracy, fear of declaring political participation still prevails among the elderly, which is why information derived from this indicator is not completely reliable. In fact, from the overall countries analysed only Argentina included a question concerning participation in political parties (QLE survey, 2012).

18.3.2.3 Third Domain: Independent, Healthy and Secure Living

Data allowing cross-country comparisons in this domain are scarce, as survey questionnaires are often unspecific in regard to healthy living of older population. For instance, data on 'physical exercise' provided by the Chilean QLE survey does not differentiate between a six-block walk and doing gymnastics, while the questionnaire from the Guatemalan survey on living conditions (ENCOVI, 2014) combines watching television and exercising into a same question.[11]

The indicator of the AAI 'access to health and dental care' is similar to the one covered by the QLEI 'access to health services', with the caveat that the latter does not include dental care, as this aspect is barely addressed by the countries studied.

With regard to independent living arrangements, cultural considerations must be taken into account when analysing the applicability of this indicator as a measure of active ageing. In some countries of LAC, living in multigenerational households is a desirable feature in terms of network support and recognition of older people as active family members. Also, since retirement pensions do not ensure economic security to a large part of the elderly, living in multigenerational households might have a positive impact on promoting a better QLE. Indeed, even in Argentina, where the average of retirement pension is proportionally much higher than in other LAC countries (it is 5.3 times higher than the amount of poverty line per capita), the share of multigenerational households reaches 48.3%

among older population (ENCAVIAM, 2012). This indicator differentiates from the one covered by the QLEI 'lodging housing', as the latter attempts to capture the possibility of enjoying housing autonomy by identifying more than one household's head in a single dwelling, instead of capturing decisional autonomy regarding one's own life in old age.

Indicators on financial security are conditioned by specific features of each region, which is why applying 'no poverty risk', 'relative medium income' and 'no severe material deprivation' as measures of independent, healthy and secure living to populations of LAC, would lead to flawed outcomes. In LAC, economic security is highly dependent upon achieving basic daily needs, since elementary living conditions—such as being able to meet the cost of a basic food basket, having access to water and electricity, and having adequate sanitary conditions—are still not ensured for a large part of the population. Hence, the subset of indicators comprising the first domain of the QLEI seems to be more appropriate to assess economic security in older age than those contained in the AAI, as the latter respond to a different economic scenario. Under these considerations, 'life-long learning' could also be considered a non-suitable indicator for evaluating active ageing in LAC countries, since the average years of total schooling across the region ranks from 6.2 in Argentina through 1.9 years in Haiti (Barro-Lee 2010).

Data concerning 'physical safety' is available only in certain countries and lacks of unequivocal criteria to enable international comparisons. Throughout the analysis, it was noticed that surveys assessing physical safety address mainly the type of violence known as 'domestic violence'. Further literature review (Adams 2012; Cano et al. 2015; Llanes-Torres et al. 2015) revealed that this matter is an issue of special concern within the discussion on ageing in LAC.

18.3.2.4 Fourth Domain: Capacity and Enabling Environment for Active Ageing

Outcomes on this domain also face limitations regarding data availability and absence of uniform time criteria, particularly with respect to 'social connectedness', 'mental well-being' and 'use of ICT'. Although the latter

Table 18.4 RLE achievement of 50 years at age 55 and HLY in the remaining life expectancy at age 55 (total), by country

Country	RLE achievement of 50 years at age 55	HLY in the remaining life expectancy at age 55
Argentina	42.2	56.9
Bolivia	25.6	31.3
Chile	47.4	63.3
Ecuador	41.2	53.4
Guatemala	35.2	39.8
Peru	38.4	62.5

Sources: UN data, 2012; ECLAC 2015

aspect has been incorporated in most households and QLE surveys, questions on this matter do not always provide information on active ageing. For example, the question included in the Argentinean survey on QLE for assessing computer and internet use was: 'During the last week, did you use internet to relax?', thus excluding the use of information and communication technologies for working purposes.

With regard to 'remaining life expectancy (RLE) achievement of 50 years at age 55'[12] and 'share of healthy life years (HLY) in the RLE at age 55', it was found that both indicators—which intend to measure human assets and health capital among the elderly—can be applied to Latin American countries without the inconvenient of leaving aside cultural and social considerations. In fact, as mentioned above, 'share of HLY in the RLE at age 55' was included in the QLEI within the field of health and well-being, providing standardized information on quality and quantity of life. Results of both indicators are displayed in Table 18.4. It can be noticed that most countries perform accordingly with demographic transition's progression, thus reflecting lower outcomes in Guatemala and Bolivia, which currently experience moderate stages of this process.

18.4 Discussion

Over ten years have passed since MIPAA was agreed. However, until today, accurate and continuous data on ageing are scarce among LAC. Indeed, out of 19 countries studied, only one has implemented special and regular surveys addressed to older population (Chile).

This article attempted to meet the aim of developing a measurement tool for assessing the goals contained within the MIPAA-RS, by constructing an index that might be applied to different countries of the region. QLEI represents an initial approach to the task of developing accessible, synthetic and comparable measurements regarding economic security, well-being and dignified living conditions in old age. Most indicators included in the final dashboard represent critical areas of the quality of life of the elderly, which is why the aggregated index should be interpreted as the relative gap between the current situation of older population and the successful compliance with the requirements for ensuring minimum standards of safety, welfare and dignity for the old aged. Due to the lack of sources, it was not possible to develop an index that could measure 'optimal' instead of 'minimum sufficient' levels of life quality, as it is the case with the AAI.

From the application of the QLEI to six Latin American countries, it was noted that life quality of the elderly has a better performance among countries that are currently placed in advanced stages of the demographic transition. However, the countries mentioned herein are, in general, a long way off from offering generalized conditions of proper life quality for their older citizens. It is likely that the latter statement might also apply to the remaining countries of the region.

Many of the limitations faced in the construction of the QLEI were also encountered when trying to apply the AAI to Latin American countries. Insofar as information regarding living conditions and welfare of the elderly lacks of common standards, unequivocal definitions and available data, the applicability of an accurate and multidimensional index on the matter, such as the AAI, is not possible.

Although today active ageing, as a policy guideline, is circumscribed within a wider concept—as it is life quality of the elderly—its assessment should not be underestimated, since it provides tangible information on how healthy, independently and safe people are ageing. The fact that some indicators of the AAI could not be applied to LAC countries due to social and cultural differences should lead us to seek alternative indicators allowing a closer approach on the matter. In this regard, the experience of the AAI's construction—especially with respect to its methodological procedure—may serve as a basis for further development of measurement tools adjusted to the regional circumstances.

Appendix 1

Data sources by country and type of source

Country	Household surveys	Health surveys	Surveys addressed to people aged 60+ and specific surveys	Population censuses	Population projections
Argentina	Permanent Household Survey (EPH, 3rd term 2013 & 3rd term 2014); National Survey of Household Expenditure (ENGHO, 2012–13)		National Survey on QLE (ENCAVIAM, 2012); Time-Use Survey (2013)	Population census of 2010	Long-term population estimates and projections 1950–2100 (ECLAC 2015); UN Database (2012)
Bolivia	Household Survey (EH, 2013 & 2014)		Survey on Household with Older Adults and People Approaching 60 Years Old (EPAM, 2011)	Population census of 2012	
Chile	National Socioeconomic Survey (CASEN, 2013)		National Survey on QLE (ECVAM, 2013)		
Ecuador	National Survey on Employment, Unemployment and Underemployment (ENEMDU, 2015); National Survey of Household Incomes and Expenditures (ENIGHUR, 2011–12); Survey on Living Conditions (ECV, 2013–14)		Time-Use Survey (2012)	Population census of 2010	

(continued)

(continued)

Country	Household surveys	Health surveys	Surveys addressed to people aged 60+ and specific surveys	Population censuses	Population projections
Guatemala	National Survey on Living Conditions (ENCOVI, 2014 & ENCOVI, 2011); National Employment and Income Survey (ENEI, Oct. 2013)				
Peru	National Household Survey (ENAHO, 2014)	Demographic and Family Health Survey (ENDES, 2014); Survey on Health and Well-being of Older Adults (ESBAM, 2012)			

Appendix 2

Disaggregated results of the QLEI by country

Indicator	Country					
	Argentina	Chile	Ecuador	Peru	Bolivia	Guatemala
Economic Security	**84.7**	**70.2**	**69.2**	**57.1**	**52.0**	**45.1**
Formal labour	34.2	79.5	31.2	17.1	11.6	12.3
Employment rate (60+)	97.0	96.8	98.6	97.7	99.2	99.6
Economic inactivity rate in advanced old age (80+)	97.2	95.6	79.9	76.3	73.7	83.6
Provision of retirement benefit in advanced old age (80+)	96.8	45.2	29.1	30.1	17.6	19.8
Retirement income	100.0	13.7	100.0	41.2	59.3	8.4
No poverty	95.7	91.6	82.4	79.8	64.1	53.9
Extreme poverty ratio	89.9	76.4	78.6	74.2	57.7	66.1
Poverty gap	72.5	88.1	70.5	70.4	60.1	60.3
Health and Well-Being	**78.2**	**81.2**	**60.1**	**69.3**	**62.6**	**49.1**
Share of HLY in the RLE at age 55	56.9	63.3	53.4	62.5	31.3	39.8
Functional performance	78.1	82.1	74.8	89.3	77.0	69.5
No permanent disability	62.7	77.4	87.0	86.7	93.7	91.3
Health coverage	90.6	98.6	42.3	75.9	59.8	14.1
Access to health services	89.2	92.6	72.6	58.3	63.9	62.1
Level of education	77.2	63.7	37.2	49.3	45.4	28.7
Enabling Environments	**87.2**	**89.3**	**73.2**	**73.3**	**66.8**	**61.4**
Housing tenure	86.6	82.4	68.1	77.0	64.9	71.9
Quality of building material	98.5	97.0	86.0	73.8	73.0	63.9
Basic sanitation facilities	76.1	84.3	62.8	60.5	60.8	45.1
No overcrowding	88.7	95.2	78.2	94.9	75.4	49.0
No lodging households	99.1	78.2	98.5	97.8	61.6	98.7
Formal settlements	84.1	93.4	63.8	52.0	63.9	63.4

Notes

1. Throughout this article, *senior citizens*, *elderly* and *older persons* refer to population aged 60 and over, in correspondence with the age limit defined by the Latin American and Caribbean Demographic Centre (CELADE 2006).

2. After 2002, poverty among the elderly has registered a continuous reduction in the region, reaching 25.7% in 2013 (ECLAC 2013).
3. The Regional Strategy was discussed and agreed during the Intergovernmental Conference on Ageing: *Towards a Regional Strategy for the Implementation in Latin America and the Caribbean of the MIPAA*, which took place in Santiago de Chile in November 2003.
4. Chile: National Survey on QLE; Colombia: DHS; Cuba: National Survey on Ageing.
5. The following countries were included in the preliminary analysis: Argentina, Bolivia, Brazil, Chile, Colombia, Cuba, Ecuador, El Salvador, Guatemala, Haiti, Honduras, Mexico, Nicaragua, Panama, Paraguay, Peru, Dom. Republic, Uruguay and Venezuela.
6. Although significant advances have been made in the past three decades to improve the quality of population censuses, in some countries censuses coverage is still deficient. In Guatemala, for example, the census omission reached 14.5% in 1990 and 5.8% in 2002 (Tacla 2006). Due to the lack of funding, the population census of the 2010 round has not been implemented yet.
7. This condition was accomplished in every country with the exception of Argentina, where the Household survey includes urban areas only. In 2015, 92.8% of the Argentinean population aged 60 and over lived in urban areas (ECLAC 2015).
8. For instance, the estimated proportion of unreported deaths between 2000 and 2005 was 52% in Peru and 65% in Bolivia (PAHO/WHO 2007).
9. Professor Leandro González (PhD, Director of the Master's Program in Demography, UNC-Argentina); Prof. Eduardo Chávez Molina (PhD, Instituto Gino Germani).
10. In Bolivia, the proportion of older adults aged 80 and above who receive retirement pension reaches only 17.6%. However, it has to be mentioned that, given the large percentage of the elderly who did not contribute to the pension scheme, the government implemented in 2007 a universal pension for older adults, which according to 2014 data, was given to 96.2% of the population aged 80 and above. Currently, this allowance is equal to US$36 for those who do not receive any income from the social security system.
11. In this case, the question falls within the field of 'time use': 'Yesterday, did you do or were engaged in sports, cultural or leisure activities (such as watching television, going to the cinema, going to the theatre, etc.)?'.
12. This indicator considers RLE at 55 divided by 50 to calculate the proportion of life expectancy achievement in the target of 105 years of life expectancy (Zaidi et al. 2013).

References

Acevedo, W. (2014). El Grupo de Trabajo sobre la Protección de los Derechos Humanos de las Personas Mayores de la Organización de los Estados Americanos. In W. Huenchuan & R. Icela Rodríguez (Eds.), *Autonomía y dignidad en la vejez: Teoría y práctica en políticas de derechos de las personas mayores*. Mexico City: ECLAC/UN.

Adams, Y. (2012). Maltrato en el adulto mayor institucionalizado. *Revista Médica Clínica Las Condes, 23*(1), 84–90.

Barro-Lee. (2010). *Educational attainment dataset* [Online]. Retrieved from http://www.barrolee.com

Cano, S., Garzon, M., Segura, A., & Cardona, D. (2015). Associated factors of the abuse of the elderly people in Antioquia, 2012. *Revista Facultad Nacional Salud Pública, 33*(1), 67–74.

CELADE. (2006). *Manual of indicators on quality of Life in Old Age*. Santiago de Chile: CELADE/ECLAC. Retrieved from http://www.cepal.org/es/publicaciones/manual-sobre-indicadores-de-calidad-de-vida-en-la-vejez

ECLAC. (2003). *Regional strategy for the implementation in Latin America and the Caribbean of the madrid international plan of action on ageing*. Santiago de Chile: CELADE/ECLAC. Available from http://undesadspd.org/LinkClick.aspx?fileticket=RHZ2ZEsjCWY%3D&tabid=330

ECLAC. (2013). *CEPALSTAT databases and statistical publications* [Online]. Retrieved from http://estadisticas.cepal.org/cepalstat/WEB_CEPALSTAT/estadisticasIndicadores.asp?idioma=e

ECLAC. (2014). *Population projections: Demographic observatory 2013*. Santiago de Chile: ECLAC/UN.

ECLAC. (2015). *Population projections and estimations* [Online]. Retrieved from http://www.cepal.org/es/estimaciones-proyecciones-poblacion-largo-plazo-1950-2100

ECLAC/CELADE. (2009). *Envejecimiento, derechos humanos y políticas públicas*. Santiago de Chile: ECLAC/UN.

Guzmán, J. M. (2002). Envejecimiento y desarrollo en América Latina y el Caribe. *Serie Población y Desarrollo*, No 28. Santiago de Chile: ECLAC.

Huenchuan, S. (2011). *La protección de la salud en el marco de la dinámica demográfica y los derechos*. Santiago de Chile: UNFPA/CELADE.

Huenchuan, S. (2014). "¿Qué más puedo esperar a mi edad?" Cuidado, derechos de las personas mayores y obligaciones del Estado. In W. Huenchuan & R. Icela Rodríguez (Eds.), *Autonomía y dignidad en la vejez: Teoría y práctica en políticas de derechos de las personas mayores*. ECLAC/UN: Mexico City.

Huenchuan, S., & Morlachetti, A. (2007). Derechos sociales y envejecimiento: Modalidades y perspectivas de respeto y garantía en América Latina. *Notas de Población, 85*, 145–180.

Llanes-Torres, H., López-Sepúlveda, Y., Vázquez-Aguilar, J., & Hernández-Pérez, R. (2015). Factores psicosociales que inciden en la depresión del adulto mayor. *Revista de Ciencias Médicas de La Habana, 21*(1), 9. [Online] Retrieved from http://revcmhabana.sld.cu/index.php/rcmh/article/view/707/pdf_26

PAHO/WHO. (2007). *Situación de las Estadísticas Vitales, de Morbilidad y de Recursos y Servicios en Salud de los países de las Américas*. Santiago de Chile: HSD/HA.

Tacla, O. (2006). La omisión censal en América Latina, 1950–2000. *Series de Población*, N° 65. Santiago de Chile: ECLAC/UN.

Zaidi, A., Gasior, K., Hofmarcher, M. M., Lelkes, O., Marin, B., Rodrigues, R., Schmidt, A., Vanhuysse, P., & Zolyomi, E. (2013). *Active Ageing Index 2012. Concept, methodology, and final results*. Research memorandum/methodology report. European Centre Vienna, March 2013. Retrieved from www.euro.centre.org/data/aai/1253897823_70974.pdf

Javiera Fanta is a doctoral candidate in Demography at Córdoba National University (Argentina) and research fellow of the National Council for Scientific and Technological Research at Gino Germani Research Institute (Buenos Aires). Previously, she worked at the Third Sector associations with immigrant population in Argentina. Her fields of study include adolescent fertility, women's sexual and reproductive health, ageing and quality of life of older people in countries of Latin America and the Caribbean.

19

Active Ageing Index: A Russian Study

Galina A. Barysheva, Elena A. Frolova,
Veronika A. Malanina, and Ekaterina A. Taran

19.1 Introduction

The demographic changes of past century have caused transformations in population structure, and as a result there is a sustainable tendency to an increasing share of older people, which has reached 23.5% in Russia. Third and fourth age adults are more exposed to certain risks coming with age, such as higher levels of morbidity and mortality, and lower levels of mental health and physical activity. These are the main reasons for investigating how we can change the meaning of old age in modern society, solve the ageism problem, enhance active ageing and improve wellbeing of older adults (Boundly 2013; Foster and Walker 2014; Thompson and Thompson 2001).

In this chapter we will focus mainly on older people wellbeing domains included in the Active Ageing Index. We expect to find differences in subjective wellbeing of the elderly across Federal Districts of Russia.

G.A. Barysheva • E.A. Frolova (✉) • V.A. Malanina • E.A. Taran
Tomsk Polytechnic University, Tomsk, Russia

A. Zaidi (eds.), *Building Evidence for Active Ageing Policies*,
https://doi.org/10.1007/978-981-10-6017-5_19

Defining "national" wellbeing, we must consider both objective and subjective parameters. J. Matheson (National Statistician from the UK) argues: "We must measure what matters—the key elements of national wellbeing. We want to develop measures based on what people tell us matter most" (Office of National Statistics 2014b). In this regard, the UK Office of National Statistics describes national wellbeing through a wellbeing wheel. It consists of ten domains: personal wellbeing, relationship, health, activities, social environment, finance, economy, education and skills, government and natural environment (Office of National Statistics 2014a).

WHO suggests the following definition of active ageing, another concept used in this chapter, "the process of optimizing opportunities for health, participation and security in order to enhance quality of life as people age". The main goal of active ageing is therefore "to extend healthy life expectancy and quality of life for all people, even when they are frail, disabled or in need of care" (WHO 2002).

There are a lot of determinants of active ageing which are similar to national wellbeing measurements: health and social services, behaviour, natural environment, social environment, economy and personal attitudes. In this regard, we can conclude that the active ageing concept and the wellbeing concept are very close to each other, but not identical.

From the point of view of experts from the European Commission, "the essence of emerging modern concept of active ageing combines the core elements of productive ageing with a strong emphasis on quality of life and mental and psychological wellbeing" (Rudawska 2010).

Therefore, active ageing describes conditions for cohesion and intergeneration support to enhance social capital and make strong ties between people (Vincent et al. 2006). These aspects are more important for older people than for younger ones, because many of them feel themselves useless after retirement, when the willingness to live longer and self-esteem decrease.

Walker identified seven principles of active ageing (Walker 2002):

1. Activity should consist of all meaningful pursuits beyond paid work, which contribute to the individual wellbeing.
2. Active ageing must encompass ALL older people, even those who are frail and dependent through creating institutions of formal or family support.

3. Activity is primarily a preventive concept, meaning that people can prevent by themselves some future disabilities during the life course (illness or dependency; improve skills to prevent earlier retirement).
4. Intergenerational solidarity is a crucial part of active ageing (Kalmijn and Saraceno 2008). Adult children provide support to their older parents, thus improving their quality of life. In this way, governments save money for formal support for people who do not have any relatives and cannot receive informal support at all.
5. Active ageing concept means that all people have the same rights but also the responsibility to make their life better during the life course, and to make a better life for people close to them (relatives, friends, neighbours and also strangers, if they need help).
6. A strategy for active ageing should be participative and empowering. People of all ages should take part in developing their own future, making sure that their views are included into active ageing policies.
7. Active ageing has to respect national and cultural diversity. Socio-economic differences, culture, ethnicity, government regulations and nature provide for nationally specific active ageing policies.

A modern effective strategy on active ageing will be based on a partnership of citizens and society. There are a lot of different kinds of activities which can improve quality of life during ageing, but it is necessary to use a life course approach to assess the usefulness of these activities for different age groups (Victor 2005; Phillips et al. 2010; Hutchison 2010).

Active ageing affects not only the age group of 55–60 years and older, it begins with the entry of an individual in the working age. Opportunities in old age are only partially determined by the characteristics of pension systems in national economies and depend a lot on practical skills, habits, educational strategies that individuals create during the life course. The study of the peculiarities of individual choice and social and economic policies that have an impact on that choice in the context of education, profession and the number of children in the family can provide a greater understanding of independence and activity in old age. In the context of a possible increase in the retirement age in Russia, it is important to trace the role of all factors of active ageing at the age of 55+ for women and 60+ for men in the new institutional environment.

19.2 Method

The Active Ageing Index (AAI) was developed to study and measure active ageing across the world. In the European Union 2012 was established as the European Year for Active Ageing and Solidarity between Generations. The methodology on which the AAI is based defines active ageing as "the situation where people continue to participate in the formal labour market as well as engage in other unpaid productive activities such as care provision to family members and volunteering and live healthy, independent and secure lives as they age" (Zaidi et al. 2013).

The methodology of AAI consists of indicators which can be measured on the basis of European Union statistics and it is necessary to develop cross-cultural research to study differences in active ageing across the world (Sidorenko and Zaidi 2013). In this chapter we study the application of AAI domains in Russia and at the subnational level of Russian Federal Districts, to find methodological and real-life challenges in the process of evaluating and enhancing wellbeing of old adults in Russia.

There are some data sources available to study active ageing and to evaluate the application of the Active Ageing Index to Russia. We use AAI methodology and AAI data of completed studies (28 countries), data from Federal State Statistics Service (FSSS) (Russia), from European Social Survey (ESS, wave 6–2012) and the Generations and Gender Survey (GGS, Russia Wave 1, 2010).

In the Table 19.1 data from the various sources is described. For most indicators we used data from ESS and GGS, because of lack of information about active ageing for Russia in other global and national databases.

However, in most cases we have no appropriate data to apply the AAI methodology, because ESS, GGS and FSSS collect data following different methodological approaches. Gender disaggregated data in AAI indicators also is not always available in Russian databases.

19.3 Results

Our research showed that total AAI for Russia accounts for 30.2 points, which corresponds to a 22nd rank among 29 countries (Fig. 19.1).

Table 19.1 Active Ageing Index estimations for Russian Federal Districts

Domain	Indicator	Data from AAI (EU28)	Data from ESS 6–2012, GGS, FSSS Russia and Russian Federal Districts (FO)								
			Russia	Central FO	North-Western FO	Volga FO	South FO	North Caucasian FO	Ural FO	Siberian FO	Far East FO
Employment	Employment 55–59	62.2	59.9	71.1	52.4	57.9	61.9	44.4	55.0	59.0	55.6
	Employment 60–64	61.1	26.1	26.6	26.1	22.5	18.2	33.3	23.5	34.2	12.5
	Employment 65–69	11.6	19.5	27.5	7.1	17.2	16.7	0.0	12.5	22.7	16.7
	Employment 70–74	6.1	4.7	6.8	6.3	1.4	5.4	0.0	4.5	3.6	9.1
Participation in society	Voluntary activities	13.8	4.1	6.3	2.2	5.0	1.2	8.3	1.5	2.6	5.9
	Care to children and grand children	33.9	21.8	28.3	17.8	20.6	16.1	29.2	17.9	15.5	41.2
	Care to older adults	13.7	9.9	10.5	9.6	11.3	8.6	5.4	8.5	9.4	6.6
	Political participation	9.1	3.4	3.1	6.7	3.3	3.7	4.2	6.0	0.7	2.9

(*continued*)

Table 19.1 (continued)

Domain	Indicator	Data from AAI (EU28)	Data from ESS 6–2012, GGS, FSSS Russia and Russian Federal Districts (FO)								
			Russia	Central FO	North-Western FO	Volga FO	South FO	North Caucasian FO	Ural FO	Siberian FO	Far East FO
Independent and secure living	Physical exercise	15.6	39.6	26.5	34.4	37.2	19.8	54.2	46.3	54.2	44.1
	Access to health and dental care	88.2	N.A.	N.A.	N.A.	N.A.	N.A.	N.A.	N.A.	N.A.	N.A.
	Independent living arrangements	84.2	85.3	81.8	100	85.4	88.5	66.7	83.3	92.85	33.3
	Relative median income	86.3	82.6	N.A.	N.A.	N.A.	N.A.	N.A.	N.A.	N.A.	N.A.
	No poverty risk	93	89.9	88.4	90.3	92.2	89.6	90.3	94.2	92.2	91.1
	No severe material deprivation	90	72.6	74.1	79.2	76.7	75.6	69.0	75.6	78.3	71.7
	Physical safety	69.3	48.1	46.6	50	56.1	53.1	58.3	38.8	40.0	44.1
	Lifelong learning	4.5	4.7	6.2	5.5	4.3	7.3	0.0	4.1	3.9	0.0

(*continued*)

Table 19.1 (continued)

Domain	Indicator	Data from AAI (EU28)	Data from ESS 6–2012, GGS, FSSS Russia and Russian Federal Districts (FO)								
			Russia	Central FO	North-Western FO	Volga FO	South FO	North Caucasian FO	Ural FO	Siberian FO	Far East FO
Capacity and enabling environment for active ageing	Remaining life expectancy achievement of 50 years at age 55	53.8	46.1	47.3	46.3	45.4	46.6	48.5	45.3	44.5	43.4
	Share of HLY in the remaining life expectancy at age 55	53.2	N.A.	N.A.	N.A.	N.A.	N.A.	N.A.	N.A.	N.A.	N.A.
	Mental wellbeing	64.6	51.3	57.9	46.7	52.8	38.3	83.3	59.7	43.9	38.2
	Use of ICT	40.8	9.6	12.1	15.6	6.7	1.2	12.5	20.9	4.5	11.8
	Social connectedness	49.0	75.8	50.0	80.0	72.8	74.1	50.0	89.6	79.4	55.9
	Educational attainment of older persons	59.7	91.1	91.7	96.7	86.7	90.1	91.7	86.6	86.5	100.0

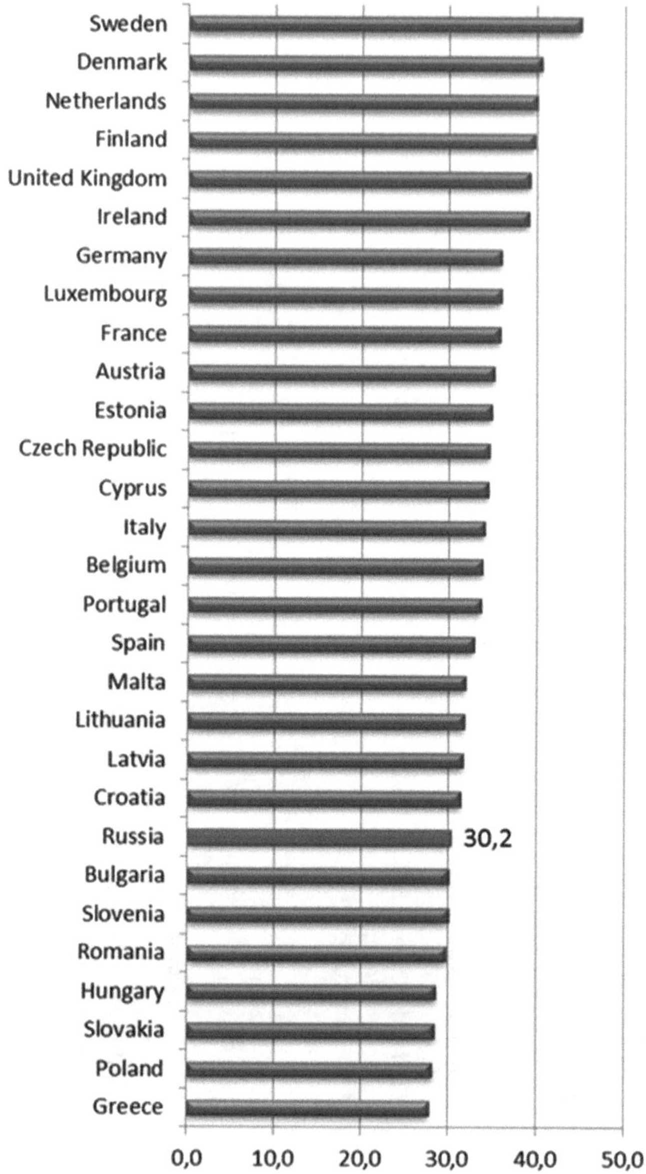

Fig. 19.1 Active Ageing Index in Europe and Russia

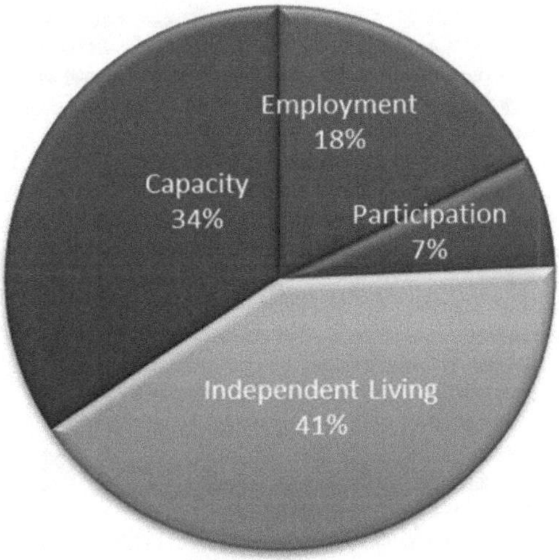

Fig. 19.2 Active Ageing Index in Russia by domains

Considering the contribution of each domain in the total AAI, it is possible to resume that independent and secure living (41%) and capacity and enabling environment for active ageing (34%) are the most contributing domains. Employment accounts for 18%, while participation in society only for 7 points (Fig. 19.2).

Generally, Russian AAI value by all four domains turns to be very close to mean domain values of the index across countries.

19.3.1 Employment of Older People

Federal State Statistics Service keeps statistics for the employment rates of older generations without considering age and gender groups within the aged population. The data available on ESS 2012 shows that Russia is generally within the predictable trend of aged employment. At the starting point (55 years) the employment rate in Russia is a little lower than the European mean, but in the 65–69 interval is 6.6 percentage points higher (Figs. 19.3 and 19.4).

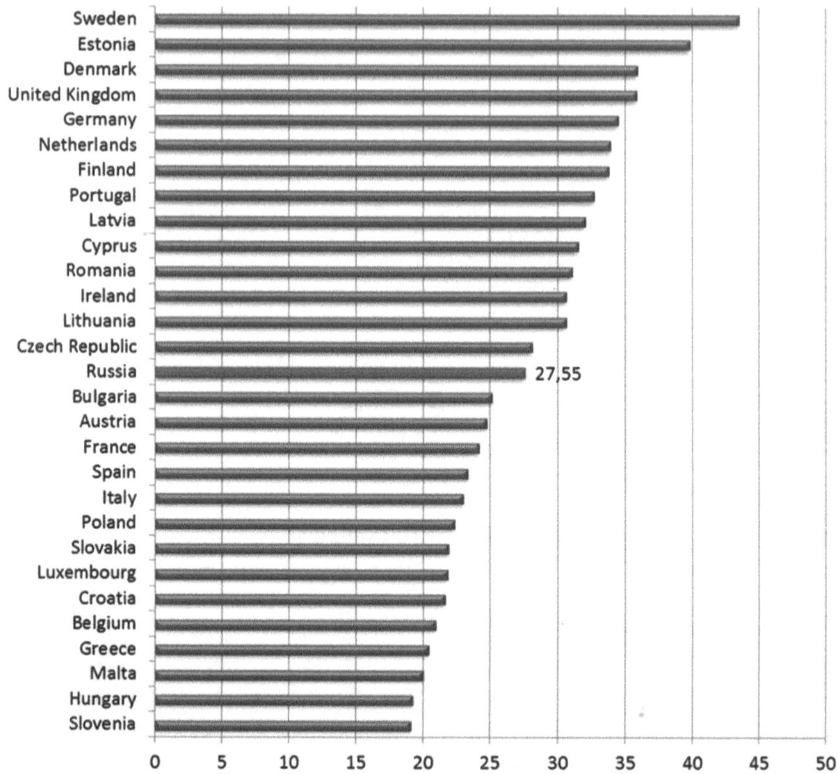

Fig. 19.3 Employment domain of Active Ageing Index in Europe and Russia

According to this analysis, the conclusion we came to is that the distinctive feature of modern Russians is their willingness to take on the risks associated with changes in the career trajectory. This readiness can vary depending upon gender, age, education, professional experience and human health, and may be determined internally, by personal choices.

19.3.2 Participation in Society

Social factors are integral components of wellbeing and are especially important for the elderly. This conclusion was confirmed in the framework of the study. More than 40% of older adults prefer to take part in different social activities.

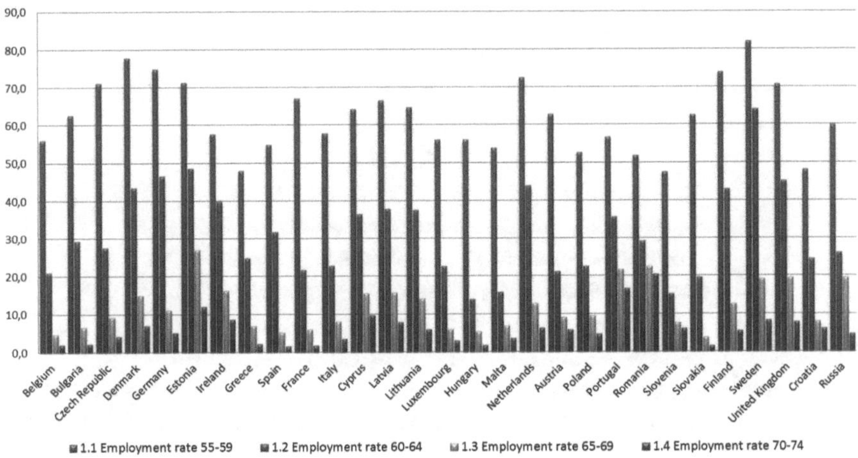

Fig. 19.4 Employment rate in Europe and Russia by age groups (AAI database and ESS 6–2012)

Importance of social factors is confirmed by the willingness and ability to help someone, even when there are significant limitations in performing activities. Most elderly are able and like to care for relatives (children and grandchildren, parents and other relatives). About 3.5% of elders demonstrate a political participation, while only 4% of elders, primarily young old adults, take part in some voluntary activities (Musick et al. 1999; Pilkington et al. 2012). There is a small amount of social communities for elders in Russia, who are less involved in voluntary activities and charity than elders from Ireland, Netherlands or Sweden (Figs. 19.5 and 19.6).

Participation domain received 10.1 points for Russia (max. for EU is 25), which is comparatively low but quite predictable for two reasons:

- low general trust in the society, in political parties and volunteer organizations. As a result, people in Russia prefer helping someone in person, without any intermediaries;
- lack of infrastructure for organized voluntary activities (there are 87,000 non-profit organizations in Russia, including only 6141 volunteer organizations, i.e. only 7% of the total).

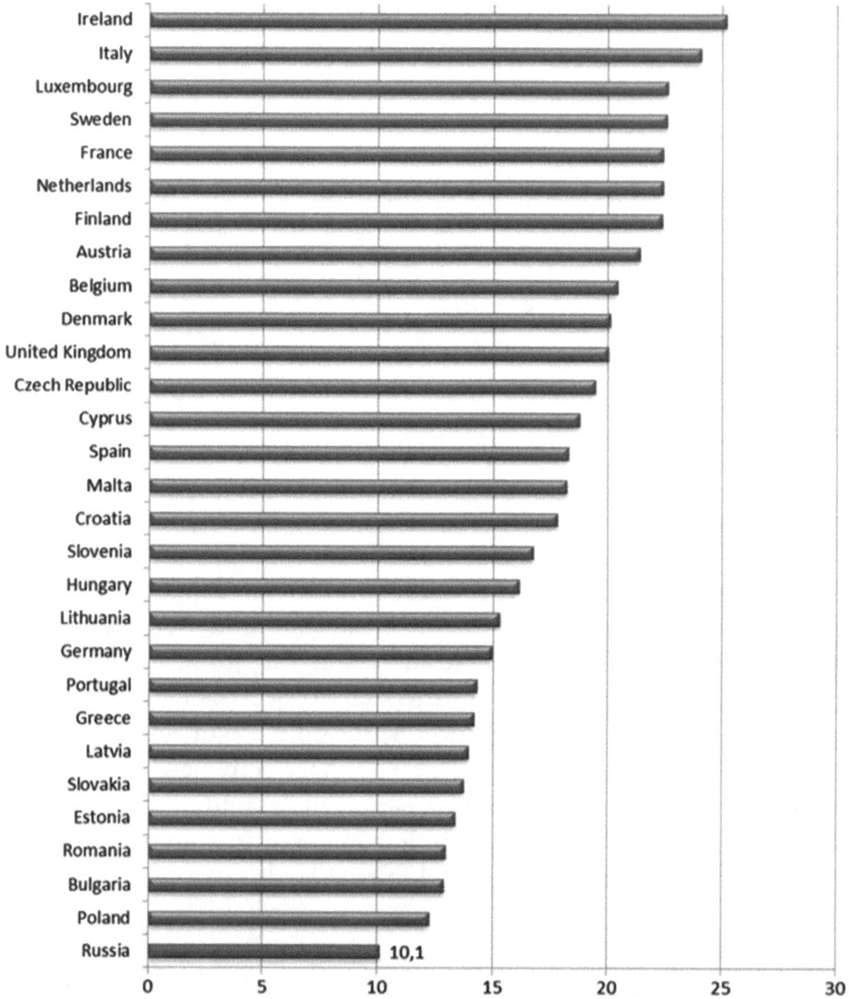

Fig. 19.5 Participation domain of Active Ageing Index in Europe and Russia

On the other hand, the indicator for social activity within families (caring for children and older adults) reports a higher mean than the European one. In Russia grandparents often substitute kindergartens, thus allowing parents to start working after or without taking maternity leave.

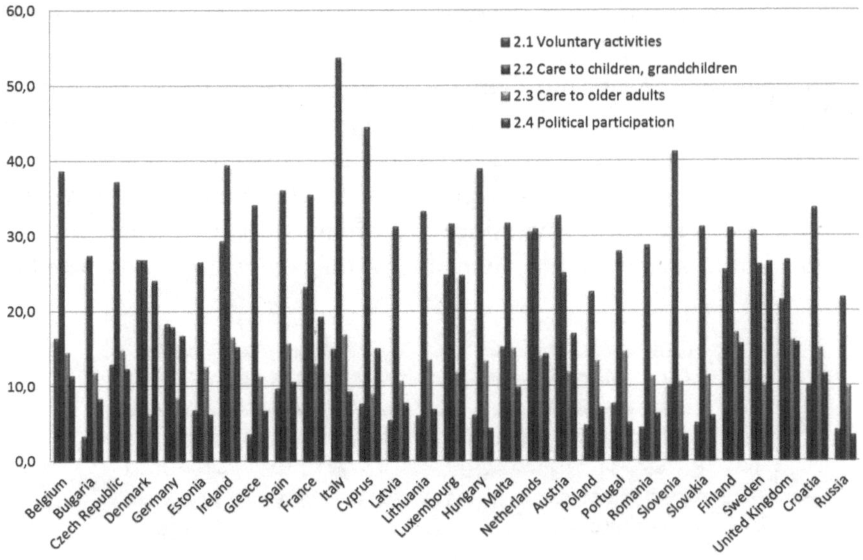

Fig. 19.6 Participation domain in Europe and Russia by sub-indices

The results obtained nationwide are close to those at subnational level. Basing upon the survey conducted in 2014 among elder people in Tomsk, we found that most respondents live independently, but independency leads to higher life satisfaction only when combined with involvement in the decision-making processes that are important for the community. Thus, engaging elder people in community life is a probable way of raising their life satisfaction level and providing active ageing.

19.3.3 Independent and Secure Living of Older People

Unfortunately, we could not obtain any data on access to health and dental care for older people in Russia. Neither official statistics by Federal State Statistics Service nor other Russian surveys can provide such data for Russia now. ESS and GGS do not formulate appropriate questions as well. No appropriate proxy indicators can be used so far, because access to health and dental care is a very specific and exact indicator in the AAI methodology (Fig. 19.7).

Independent and secure living of older people demonstrates a huge difference between European countries and Russia in some indices of this domain. Compared to Bulgaria and Romania, Russia reports lower material deprivation level and lower poverty risk (76.2% and 89.9% respectively). It's quite surprising, since we often consider the majority of population Russian elderly as living below the poverty line. Compared to

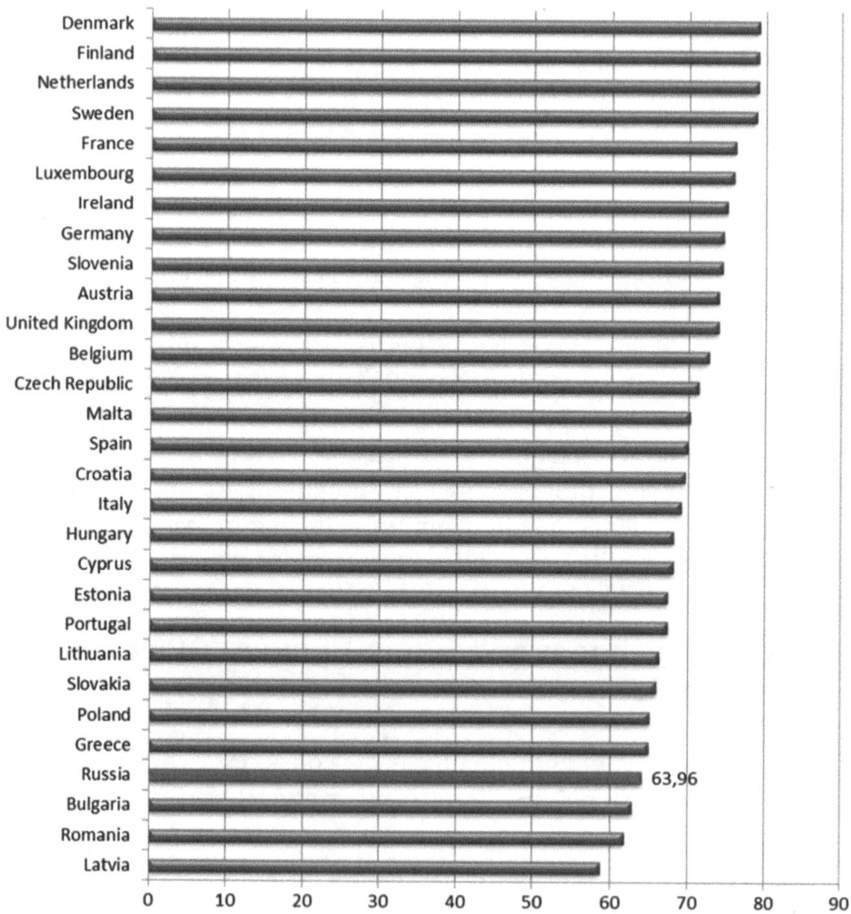

Fig. 19.7 Independent and secure living domain of Active Ageing Index in Europe and Russia

other EU countries, Russian elderly are far more physically active (39.6% in comparison with European mean of 17.2%). In this index Russia falls behind Sweden (42.6%) and Finland (48.9%) only. Lower level of elderly physical safety (48.1%) determines lower level of personal and institutional trust, which is confirmed by the data of the Participation in society domain.

19.3.4 Capacity and Enabling Environment for Active Ageing

Capacity of active ageing for older adults in Russia is one of the most contributing domains. It corresponds to a 16th rank among 29 European countries. The most important element of this domain for Russia is educational attainment, which makes it possible to accumulate and use human capital during many years after retirement. Because of that, in the 65–69 age group the employment rate in Russia is 7.9 percentage points higher than the European mean, although Russian legislation allows early exit from labour market (at 55 for women and at 60 for men) (Fig. 19.8).

However, a lot of difficulties emerge when we measure the share of healthy life years in the RLE at age 55 in Russia. This indicator is very important to study the positive mood of old persons and public expenditures to health care. But global and Russian databases do not include this indicator.

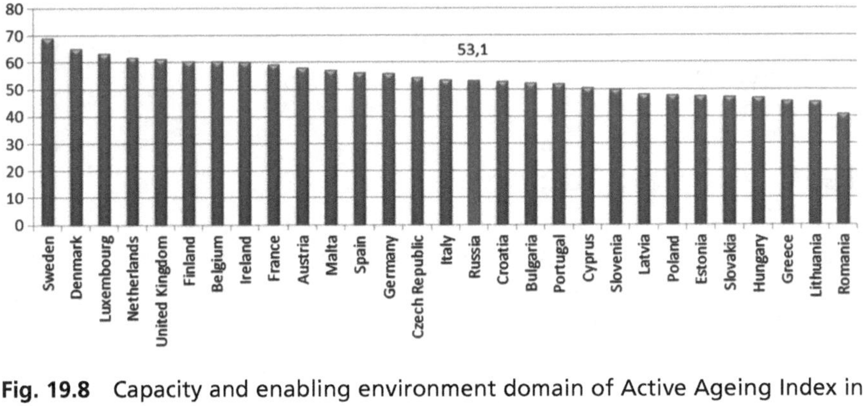

Fig. 19.8 Capacity and enabling environment domain of Active Ageing Index in Europe and Russia

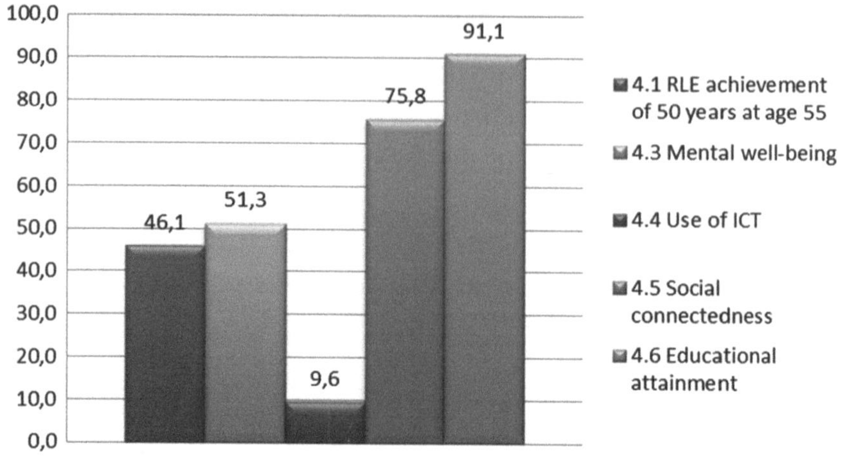

Fig. 19.9 Capacity and enabling environment domain in Russia by sub-indices available

The capacity domain received 53.1 points for Russia, with a considerable contribution of educational attainment indicator (Fig. 19.9), due to:

- age discrimination at workplace in business (where pensioners can only get low-paid positions of operating personnel), forcing the elderly to get formal higher education even if their practical skills are good enough.
- in age-tolerant spheres (primary, secondary and higher education, public sector in general), pensioners still need to be competitive and are motivated to raise their qualification by formal labour contract requirements.

19.4 Russian Regions

When attempting to assess subnational differences across Russian regions,[1] we found a considerable inequality between the eight Federal Districts (Federal Okrugs or FO) in several indices, but a very similar contribution of each domain to the overall subnational AAI (Figs. 19.10, 19.11, 19.12).

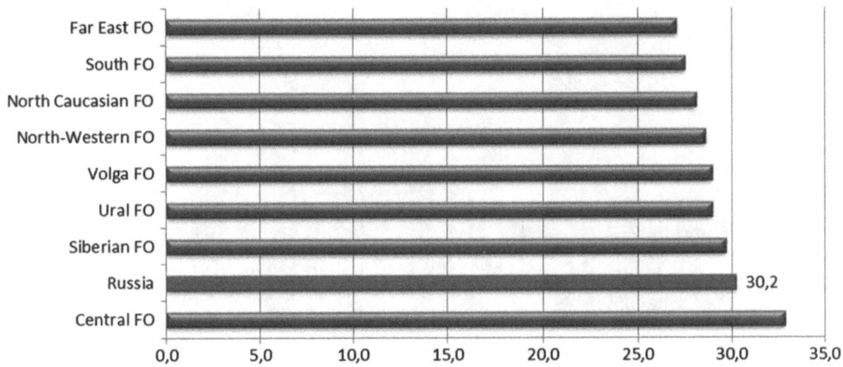

Fig. 19.10 Active Ageing Index for Russian regions (Federal Districts)

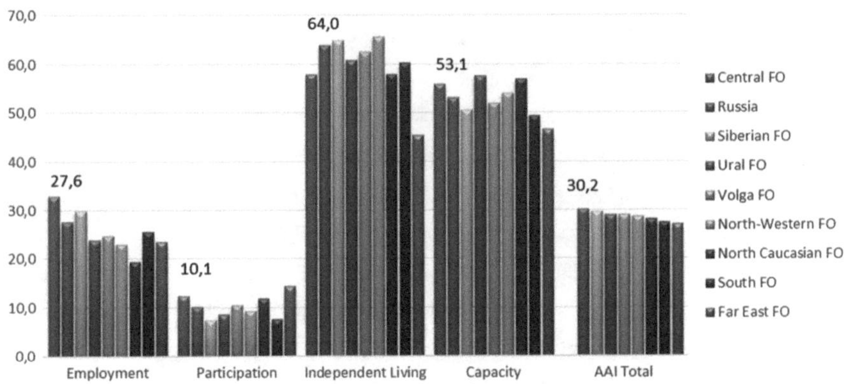

Fig. 19.11 Active Ageing Index domains for Russian regions

In the employment index for age group 55–59 we have an outstanding performance of Central FO (71.1), which provides more opportunities for the elderly as the most investment attracting, populated and well-paid positions region. Still, North Caucasian FO with the lowest employment among all regions (19.4) seems to enjoy the trade-off between lower employment rates and higher mental wellbeing, due probably to a less-stressed life course. All FOs experience serious decline in employment starting with the 60–64 age group, though for North Caucasian FO and Siberian FO it is not as dramatic as for the rest of regions, since this process starts later there (Figs. 19.13 and 19.14).

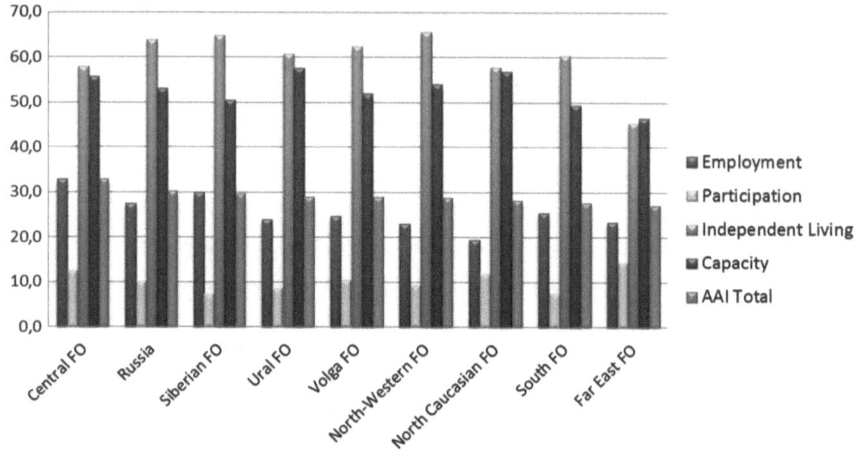

Fig. 19.12 Russian regions' scores by AAI domains

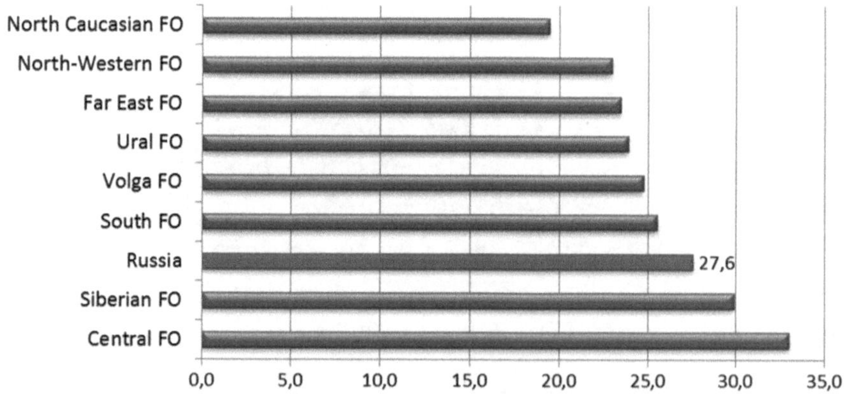

Fig. 19.13 Employment domain for Russian regions

In the participation in society domain, for some regions (North Caucasian and Far East FOs) we see a trade-off between employment and caring for children, probably as a consequence of the need to look after the younger generation due to lack of childcare institutions in these regions. Overall subnational AAI for participation domain varies from the lowest level in Siberian FO (7.5) to the highest in Far East FO (14.5), but with a common result of low voluntary activity and political

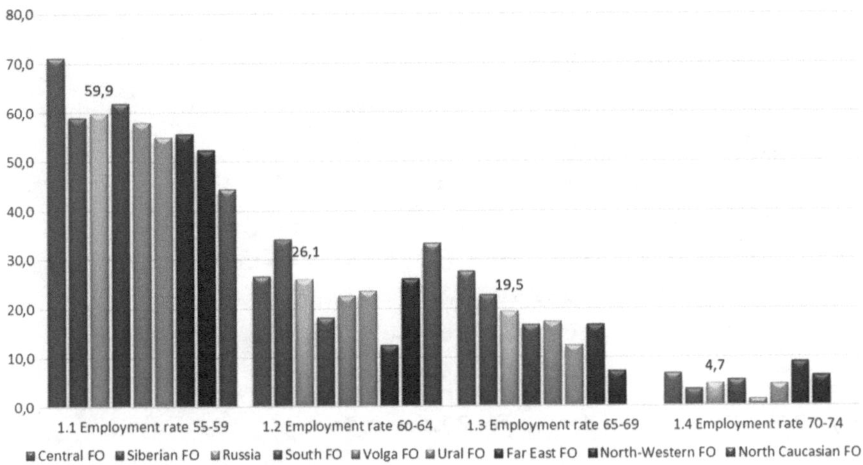

Fig. 19.14 Employment domain for Russian regions by age groups

participation when compared to the EU, and a higher intra-family activity, which still falls short of the EU mean. This questions the widely known stereotype of prevalently multigenerational character of Russian families (Figs. 19.15 and 19.16).

Similar myth-busting findings emerge for the third domain concerning the indicator 3.3 of the independent living domain, as the overall independent living score (85.25) is even higher that the EU mean. Cross-regional differences are considerable from 33.33 in Far East FO (with a hard climate and infrastructural conditions) to 100 in North-Western FO. Yet, in other indicators of the third domain regions are comparatively equal. The Far East FO comes out with high employment and high care for grandchildren, which can be combined in co-habitation living situations (Figs. 19.17 and 19.18).

In the fourth domain of capacity and enabling environment, we observe two outstanding scores for the indices of social connectedness (75.8) and educational attainment of older people (91.1), where all Russian regions are showing much higher values than the EU mean. High Ural and North Caucasian FO values in this regard were quite unexpected, because Ural is supposed to be ecologically and North Caucasian infrastructure disadvantaged regions. Some results need

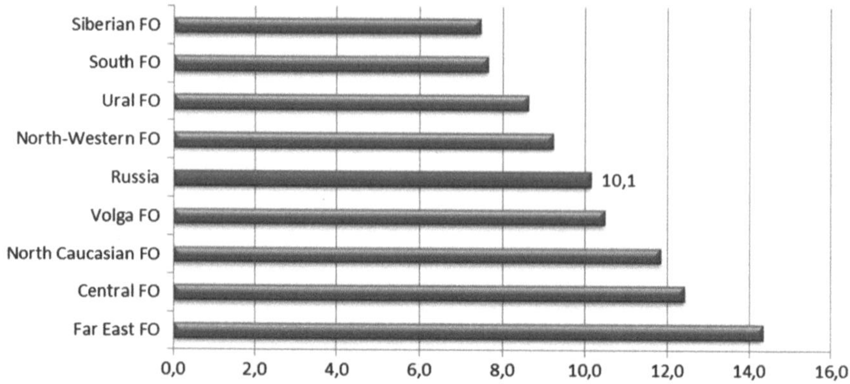

Fig. 19.15 Participation domain for Russian regions

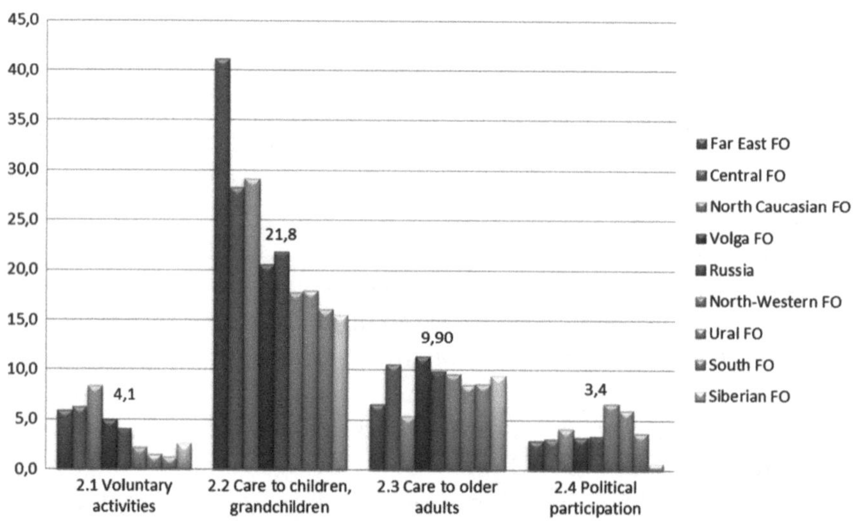

Fig. 19.16 Participation domain for Russian regions by sub-indices

further analysis, like the relatively high overall AAI score for the Siberian FO (second after Central FO), which derives from high employment and independency, partially due to internal migration of younger generations to Central and North-Western FOs. Low physical activity in agricultural South FO may look counterintuitive, but it might be explained by the fact that less effort is needed to successful gardening here when compared

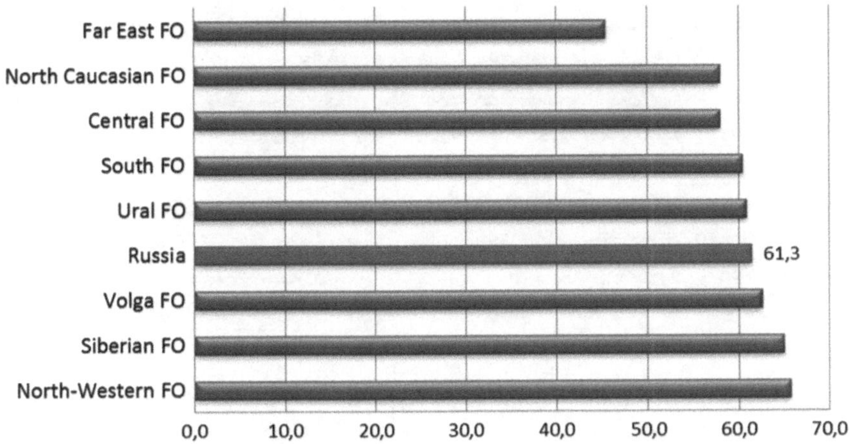

Fig. 19.17 Independent and secure living domain for Russian regions

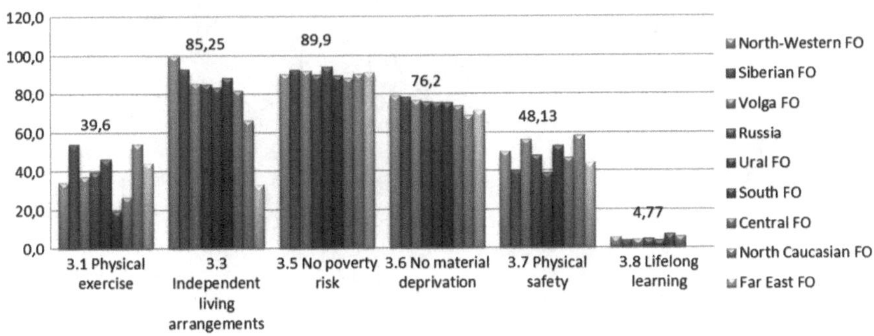

Fig. 19.18 Independent and secure domain for Russian regions by sub-indices

to colder climate regions (Siberia, Ural, Far East). Comparatively low social connectedness in North Caucasian region may derive from family prevalence and lack of elderly contacts outside the family. Highest educational attainment in Far East FO looks counterintuitive, since the region is continuously depopulating. The probable explanation may be that the elderly in this region are educated enough to keep on working, while those who prefer to retire may leave the region to join their children and grandchildren elsewhere (Figs. 19.19 and 19.20).

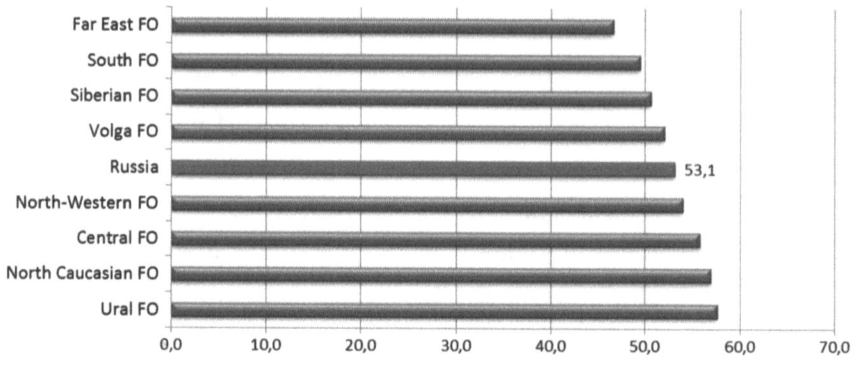

Fig. 19.19 Capacity domain for Russian regions

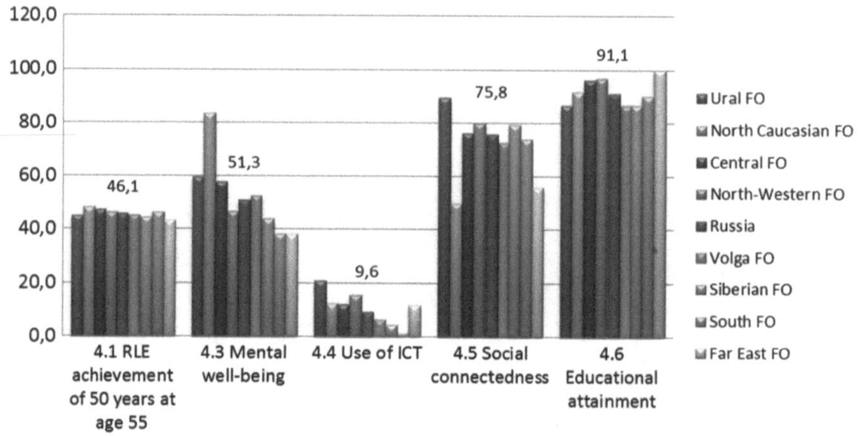

Fig. 19.20 Capacity domain for Russian regions by sub-indices

After having compared Active Ageing Index scores for Russia on national and subnational levels, we can now proceed to the conclusions, including the issue of AAI calculation and application to the Russian context.

19.5 Discussion and Conclusion

When assessing regional features of active ageing process in Russian regions, we found little differences in overall scores. Similar values of subnational AAI (i.e. remaining within a 5 percentage points variation)

confirm the idea of a trade-off between active ageing domains. All regional differences taken into account—in terms of economy structure, variety of nationalities, infrastructural gap between the centre and periphery—gave us no serious regional gaps in overall subnational AAI values. Elderly people finally adapt to the lack of employment opportunities by care to children and grandchildren, which in turn cuts the opportunities for life-long learning and social connectedness outside the family. In all FOs we have almost the same rating of domain's contribution (ranked in ascending order): social participation, employment, capacity and enabling environment, independent living (with the sole exception of Far East FO, where capacity prevails over independent living).

The attempt of estimating AAI for Russia faces methodological problems of availability and quality of statistical information. A variety of indicators needed for the original AAI calculation are not available through the Russian Federal State Statistics Service databases. A range of proxy indicators may be found for less-specific socio-economic variables—like social connectedness (corresponding to ESS "feel close to people in local area") or mental wellbeing (corresponding to ESS "enjoyed life, how often past week")—but no appropriate substitution can be found for access to health and dental care. So, it is necessary to incorporate the AAI methodology into sociological and statistical research, and to some extent we made it possible through our regional Tomsk research.

Aggregating data regardless of age and gender seems to be a very common practice in the official Russian statistics. Though Federal State Statistics Service collects data on the elderly generation, most of it is presented in aggregate form. Other Russian sociological agencies (Russian Public Opinion Research Center (VCIOM) or Public Opinion Fund (POF)) do not provide disaggregated data by age (less often by gender). Thus, we are facing limits to cross-country analysis. Due to statistical methodology differences between the EU and Russia, AAI estimations for the latter cannot be considered full and perfectly comparable. Errors and omissions in estimating some variables are persistent, due to data unavailability or its aggregated form mentioned above. Data collected from various sources may not always be directly comparable.

In our opinion, it is necessary to implement the AAI methodology in statistics and data collection, to ensure comparability of the active ageing process between European countries and Russia. The practical use of the AAI in Russia could be beneficial for the state policy aimed at raising older adults' wellbeing, while the share of aged people in the Russian population is continuously growing and demonstrating the same trend as in the European Union.

Because of rising life expectancy, the average number of years that people spend in retirement increases. During this time, the factors providing a high quality of life are different from the factors that are significant during working age. In most cases, opportunities for employment are reduced, since career developments and long-term labour contracts are no longer available. Priorities are shifting to unpaid activities, such as helping children, grandchildren and other older people, subsistence farming, political and volunteer activity or simply communication with others. Due to reduced capacity of health and physical activity, enabling environment (access to health care, physical accessibility of facilities and transportation for people with disabilities, the criminal situation in the community) becomes more important for active ageing. The beneficiaries of the projects developing the availability and security of the environment are of all ages, which is economically feasible and fit into the expanded concept of "active ageing". Accordingly, there is a considerable potential for older peoples' quality of life improvement through a meaningful impact on the different components influencing the process of active ageing.

Acknowledgements Data analysis was supported by the project "Evaluation and enhancement of social, economic and emotional wellbeing of older adults", under the agreement No.14.Z50.31.0029, Tomsk Polytechnic University. Data collection was supported by the Tomsk Polytechnic University within the framework of Tomsk Polytechnic University Competitiveness Enhancement Program.

Notes

1.
    - Central FO with almost 39 million prevalently urban and Russian population is the leading region in gross regional product.
    - North-Western FO with 14 million is quite similar by the share of urban population and ethnic composition.

- Ural FO plays a role of heavy and mining (including oil and gas) industry centre for Russian economy.
- South FO with 14 million people is far less urbanized, is agriculturally oriented and has more national minorities.
- Volga FO has even greater variety of nationalities among its 30 million population and equal shares of agriculture and industry in its gross regional product.
- Siberian FO occupies one third of the country territory and is known for low density of the population, oil and gas extraction and universities.
- North Caucasian FO differs from all other FOs by its ethnic composition with Russians as a minority, prevalent Muslim traditions affecting family relations and other spheres of living and ageing.
- Far East FO is characterized by continuous depopulation during the last decade and large migration inflow from China.

References

Boundly, K. (2013). Active ageing: From empty rhetoric to effective policy tool. *Ageing and Society, 33*(6), 1077–1098.

Foster, L., & Walker, A. (2014). Active and successful ageing: A European policy perspective. *The Gerontologist* (00 Special Issue), 1–8.

Hutchison, E. (2010). *Dimension of human behavior: The changing life course* [online]. Retrieved January 5, 2015, from http://www.sagepub.com/upm-data/36521_CLC_Chapter1.pdf

International Longevity Centre—UK and New Philanthropy Capital for the Commission on the Voluntary Sector & Ageing. (2014). *Population ageing & the voluntary sector: Key figures & projected trends*. Retrieved January 5, 2015, from http://www.ilcuk.org.uk/index.php/publications/publication_details/population_ageing_the_voluntary_sector_key_figures_projected_trends

Kalmijn, M., & Saraceno, C. (2008). A comparative perspective on intergenerational support. *European Societies, 10*(3), 479–508.

Musick, M. W., Regula Herzog, A., & House, J. S. (1999). Volunteering and mortality among old adults: Findings from national sample. *The Journals of Gerontology: Psychological sciences and Social Sciences, 54B*, 173–180.

Office of National Statistics. (2014a). *Measures of national well-being*. Retrieved January 5, 2015, from http://www.neighbourhood.statistics.gov.uk/HTMLDocs/dvc146/wrapper.html

Office of National Statistics. (2014b). *Measuring national well-being*. Retrieved January 5, 2015, from http://www.ons.gov.uk/ons/guide-method/user-guidance/well-being/index.html

Phillips, J., Ajrouch, K., & Hillcoat-Nalletamby, S. (2010). *Key concepts in social gerontology*. London: Sage.

Pilkington, P. D., Windsor, T. D., & Crisp, D. A. (2012). Volunteering and subjective well-being in midlife and older adults: The role of supportive social networks. *The Journals of Gerontology, Series B: Psychological Sciences and Social Sciences, 67*(2), 249–260.

Rudawska, I. (2010). Active ageing and its impact on labor market. *Economics & Sociology, 3*(1), 9–24.

Sidorenko, A., & Zaidi, A. (2013). Active ageing in CIS countries: Semantics, challenges, and responses. *Current gerontology and geriatrics research*. Vol. 2013. Retrieved January 20, 2015. doi:10.1155/2013/261819.

Thompson, N., & Thompson, S. (2001). Empowering older people: Beyond the care model. *Journal of Social Work, 1*, 61–76.

Victor, C. (2005). *The social context of ageing: A textbook of gerontology*. London: Routledge.

Vincent, J., Phillipson, C., & Dawns, M. (2006). *The futures of old age*. London: Sage.

Walker, A. (2002). A strategy for active ageing. *International Social Security Review, 1*, 121–139.

World Health Organization. (2002). *Active ageing. A policy framework*. Retrieved January 15, 2015, from http://apps.who.int/iris/bitstream/10665/67215/1/WHO_NMH_NPH_02.8.pdf

Zaidi, A., Gasior, K., Hofmarcher, M. M., Lelkes, O., Marin, B., Rodrigues, R., Schmidt, A., Vanhuysse, P., & Zolyomi, E. (2013). *Active Ageing Index 2012. Concept, methodology, and final results*. Research Memorandum/Methodology Report. European Centre Vienna, March 2013. Retrieved January 5, 2015, from www.euro.centre.org/data/aai/1253897823_70974.pdf

Galina A. Barysheva is a Professor, Doctor of Science and Head of Department of Economics at National Research Tomsk Polytechnic University (Tomsk, Russia). She is a member of Expert Council of Governor of Tomsk region and a member of Scientific Council of Tomsk Polytechnic University. Her current research focus is on economics of innovation, quality of life, wellbeing of older people and social policy.

Elena A. Frolova is a Senior Research Fellow at the Lifelong Wellbeing Lab at National Research Tomsk Polytechnic University (Tomsk, Russia). Her work includes issues such as social responsibility, institutional economy, active ageing, quality of life of older people. Her current research is focused on the evaluation and improvement of economic, social and emotional wellbeing of older adults in Russia, where she is leading a study of volunteering, social care and family support for older people.

Veronika A. Malanina has received PhD in economic theory in 2009 and joined the Department of Economics of Tomsk Polytechnic University in 2014. She is a Senior Research Fellow at the International Scientific Educational Laboratory for the Improvement of Wellbeing Technologies of Older Adults. The fields of her scientific interests are represented by, but not limited to, institutional economy, economics, social and emotional wellbeing of older adults and active ageing.

Ekaterina A. Taran joined the Department of Economics of Tomsk Polytechnic University in 2005 as a PhD student and continues to work here at the moment. The field of her scientific interests is economic wellbeing of older adults, including income and savings, participation of older people in labour market. She is a Research Fellow at the International Scientific Educational Laboratory for the Improvement of Wellbeing Technologies of Older Adults.

20

Active Ageing Index (AAI) in India: Is the Approach Used in European Countries Applicable to Developing Countries?

Aravinda Meera Guntupalli
and Suchandrima Chakraborty

20.1 Introduction

The last decade of the twentieth century experienced the emergence of a new paradigm in gerontology which presented a positive perspective towards ageing, evidenced by approaches such as 'Active Ageing' (WHO 2002; Walker 2002; UNECE 2012), 'Healthy Ageing' (WHO 1990), 'Successful Ageing' (Rowe and Kahn 1987, 1997; Baltes and Baltes 1990) and 'Productive Ageing' (Butler and Gleason 1985). For instance, Successful Ageing, a multidimensional concept, focuses on prevention of disease and disability, continuous preservation of cognitive and physical capacity and an active social engagement in later life (Rowe and Kahn 1997). Similarly, Active Ageing is defined by the World Health

A.M. Guntupalli (✉)
The Open University, Milton Keynes, UK

S. Chakraborty
International Institute for Population Sciences, Mumbai, India

A. Zaidi (eds.), *Building Evidence for Active Ageing Policies*,
https://doi.org/10.1007/978-981-10-6017-5_20

Organization (WHO), as the process of enhancing health opportunities and participation in society in order to improve quality of life in old age (WHO 2002).

These approaches describe various experiences of ageing using a multi-dimensional perspective on health status, economic situation, social engagement and capacity to live independently in later life (Beard et al. 2012). When these concepts emerged, they were mostly popular as conceptual tools. However, in recent years, these conceptual tools have been transformed into functional tools to enhance their measurement, standardisation and application for policy purposes. For instance, the active ageing concept has recently been converted into an index by Zaidi et al. (2013). The recently created index is widely accepted, applied and interpreted in Europe (less so in Asia, Latin America and Africa). Hence, the main aim of this chapter is to construct the Active Ageing Index for India, the world's second most populous country, using micro-level data.

As argued by Fernández-Ballesteros et al. (2013), an individual is an agent of his or her ageing process affected by decisions taken during the life course in addition to behavioural characteristics, making the ageing process a non-random phenomenon. In addition to the impact of decisions taken across the life course, the ageing process in developing countries such as India extensively depends on the support provided by families, due to weaker formal state support compared to European countries. Due to limited formal state support, Indian older people, especially women, rely on family support for their financial, social and emotional domains of well-being (Rajan et al. 1999; Rajan 2007). Given this background of weak social policies, a meagre state pension support and a poor public health system, a significant proportion of Indian older people face three disadvantages in the form of avoidable disability, deprivation and dependency, making them more vulnerable in later life as compared to most of the elderly people in developed parts of the world. Such vulnerabilities could result in an irreversible decline in health, impacting both quality and duration of life. Hence, this chapter intends to summarise the relevant issues and capture the quality of later life information for older adults in India, using the Active Ageing Index. One of the objectives of this chapter is to critically evaluate the applicability of this particular index tool in policy making for developing countries such as India.

The following section (Sect. 20.2) describes ageing in India, followed by a discussion of the emergence of the active ageing approach and the Active Ageing Index (AAI hereafter). Section 20.3 highlights the emergence of AAI and critically evaluates the index. Section 20.4 describes our data and methodology, and Sect. 20.5 interprets and discusses our findings, followed by a conclusion.

20.2 Background and Context of Ageing in India

20.2.1 Ageing and Ageism in India

India in recent years has been experiencing several fast-paced emerging and conflicting phenomena, such as a growing economy, a rapidly increasing overall population, declining fertility rates in southern states and a high prevalence of diabetes. The economic growth that has occurred has not resulted in an improvement in social security and health for older people. In fact, the economic growth that has fuelled urbanisation has resulted in widening inequalities (Datt et al. 2016). In addition, economic growth in India is based on urban growth, which has resulted in an influx of migrants to urban areas from rural areas. This has caused a further weakening of informal care available to older people. In particular, between 1951 and 2011, the urban population in India increased from 62 million (17% of the total population) to 377 million (31% of the population) (GOI 2011). This proportion continues to increase, due to a lack of employment opportunities in rural India. This has resulted in declining formal supports available to older people residing in rural areas, as the majority of migrants leaving their families tend to be younger.

Despite the very restricted formal state support and health care, life expectancy has increased in India, due to cheaper medicines and an expansion of the private health-care system. The percentage of elderly Indians aged 60 and above has increased from 5.3% to 5.7% between 1971 and 1981, and from 6.0% to 8.6% between 1991 and 2011 (GOI 2011), which is due to declining fertility rates as well as an increase in life expectancy. The number of older people among India's 1.2 billion residents is 103.8 million as per the 2011 census (GOI 2011), which is not dissimilar to the elderly population of the entire European Union (EU-28) aged 60 and above, which is approximately 127 million out of 500 million people (compiled from Global Age Watch Index data 2015). Moreover, one in two older Indians has at least one chronic disease such as hypertension, angina, arthritis or diabetes (Chatterji et al. 2008), and due to weaker formal health-care support, the majority of older people pay for health care out of pocket, resulting in a socio-economic gradient in health in later life.

Investigating the macroeconomics of India in relation to older adults highlights apparently contradictory results. Even though India continues to enjoy a large working age population that further fuels economic growth, the old age dependency ratio has increased from 122 in 1991 to 142 in 2011 (GOI 2011), with bigger states like Kerala, Punjab, Himachal Pradesh and Tamil Nadu topping the chart. This is a rather crude demographically driven economic indicator, that particularly undermines the economic contribution of older people in informal and agriculturally dependent economies like India. As informal economy refers to income generation from activities that are not regulated by government, the income flow and job security would be erratic especially for older workers. Besides, the contribution of older people in an informal economy would be excluded from the tax system and pension records, underestimating the contribution of older people to the economy. The majority of older people, especially from marginalised and poorer sections of Indian societies, contribute to the economy by working after retirement age, as they have little formal support available. However, a significant proportion of the work could be unpaid, or paid in the form of food and shelter. Older people, mostly women, also contribute to the economy by taking care of grandchildren, to help their children in paid employment. One would anticipate that an increase in life expectancy at birth (e.g. in India from 49.7 in 1970 to 69.6 in 2011) would have resulted in an increased direct economic contribution of older people that enjoyed better health. But this fast-paced increase in life expectancy in India also has resulted in four-generation families, and multiple-generation families, if successful, provide informal support to older people in return for childcare and an economic contribution. However, such families can result in conflicts that are either economic and/or social.

For a country aspiring to emerge as a global economic player, the question is no longer whether ageing is an issue for India, but how big the issue is. The severity of this issue is relevant, given the heterogeneity in the speed of ageing and the number of older people. Low-fertility southern states also have the highest share of older people, with Kerala topping the list of ageing states (12.6%). The majority of northern states have the lowest share of older adults, with Delhi reporting only 5.4% of older people (GOI 2011). In terms of gender composition, like European countries, India has a higher proportion of older females compared to males. The proportion of females aged 60 and above is higher than the rest of the states in all of the bigger states, except five northern and north-eastern states, namely, Assam, Bihar, Himachal Pradesh, Jammu and Kashmir and Jharkhand (GOI 2011).

20.2.2 Ageism in India

Ageing in India is a socially constructed phenomenon (Jamuna and Ramamurti 2011), and the occurrence of ageism in India is similar to global ageism to a certain extent. For instance, age-specific distinctions, stratifications, judgements and behaviours are incorporated into our pattern of thinking (Macnicol 2006). Ageism is typically expressed in the form of stereotyping individuals based on their age that could result in discrimination. This social construction of ageing, amalgamated with cultural diversity and gender norms in India, intensifies ageism. For example, older women from marginalised communities might experience ageism, as well as sexism, combined with other forms of discrimination stemming from social marginalisation due to factors such as class and ethnicity.

Ageist perspectives of theories such as the disengagement theory (Cumming and Henry 1961) have advocated that humans become inactive over time, and with increasing age older adults tend to disengage from society on their own. In the case of India, such disengagement is imposed on older people forcefully by society due to a lack of formal support, support that is even weaker for an increasing proportion of older people due to increasing urbanisation and migration. In other instances, older people are forced to engage economically despite their poor health, to 'make ends meet'. In the absence of any formal state support, many older people are not left with much of an option but to work under atrocious conditions, marked by things such as ageism and a particularly low income. Despite the social, health and economic insecurities of South Asian older people (Hassan 2007), they continue to contribute to their societies in several ways. For example, using positive frameworks such as active ageing, it is possible to take into consideration social engagement, economic activity and other positive contributions of older people. A positive framework challenges negative attitudes in emerging countries like India, by showing the heterogeneity among older people and by reinforcing their valuable contributions to society.

Recent evidence shows that employers discriminate older employees due to projected ageist characteristics such as forgetfulness, frequent sickness, weakness, wrinkles, baldness, grey hair, constipation and tendency for falls (Selveraj et al. 2011; HelpAge India 2013). Older people are also viewed as needy, worthless, lethargic and excessively spiritual or religious, making them unfit for modern life. Not surprisingly, these stereotypes and beliefs have been associated with the exclusion and marginalisation of the aged (Jamuna and Ramamurti 2011). Furthermore, such redundant subjective negative images of older people in India portray older adults in a reductionist and pessimistic way.

By constructing the active ageing index of elderly people in India, we attempt to mitigate stereotyping and highlight the contributions of older people in India. Besides, we aim to show evidence that many older people in India are ageing actively in different aspects of their lives, such as their health, education, employment, social participation and other dimensions of life, including provision of informal support. Where older people are not actively engaged, we would like to identify the policy gaps and also question the reductionist ageist perspective using the AAI. We believe that the AAI is the most comprehensive measure to date to measure positive ageing using reliable micro-data.

20.3 The Active Ageing Framework and Index

20.3.1 The WHO Active Ageing Framework and the Emergence of the AAI

The WHO active ageing framework (introduced in 2002) is based on three pillars: participation, health and security (see Fig. 20.1), and it encompasses six groups of determinants. These six determinants and the associated further aspects include:

- Health and social services (including prevention, health promotion, hospital admissions, long- and short-term care and mental health care);
- Behavioural determinants (individual behavioural factors including smoking, physical activity, dietary patterns and alcohol consumption);

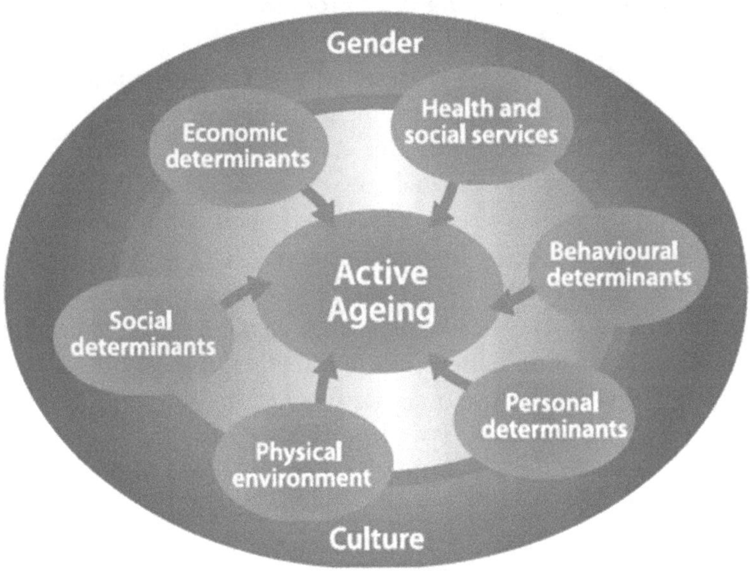

Fig. 20.1 The determinants of active ageing (WHO 2002)

- Personal determinants (individual personal factors and environmental determinants including biological and genetic factors that are non-modifiable as well as psychological factors);
- Physical environment (including housing, safety and pollution);
- Social determinants (including available social support, abuse and violence and education); and
- Economic determinants (including both macro and micro factors such as wages, work and social security).

No attempt was made to create an index using the framework until the creation of the AAI supported by the United Nations Economic Commission for Europe (UNECE) and the European Commission. In 2013, a solution to the problem of measuring active ageing has been provided by the AAI proposed by Zaidi et al. (2013) at the European Centre for Social Welfare Policy and Research, Vienna. The AAI is designed to provide reliable quantitative measures of the untapped potential of older people for active and healthy ageing across European countries. The AAI has been viewed as an emerging policy tool internationally, and has been applied successfully to quantify active life in the EU. The AAI also projects the potential of older people by measuring the participation of elderly in income and non-income generating activities, in addition to promoting active and independent later life.

The AAI is a multidimensional tool measuring four domains of life: participation in productive employment; participation in society; independent, healthy and secure living; and age-friendly environments (Zaidi et al. 2013; Zaidi 2014a). Each of these domains captures a crucial dimension of health and successful ageing. To date, the AAI has not been utilised by policy makers in developing and emerging countries including India, mostly due to a lack of quantitative information required to calculate the index. A study from Bangladesh, with a similar social and economic context to India, attempted to measure active ageing and found that urban male elderly are not only more educated but also more active in all aspects of life and have longer healthy life expectancy. Based on these findings, Tareque et al. (2012) suggest that steps should be taken to promote lifelong learning and also further education with motivation of being active in every aspect of life. The study also suggests positive community participation, providing urban recreational facilities in rural areas for the elderly and promotion of self-management of physical and economic stability among the elderly through more mass media exposure (Tareque et al. 2012). The aim of the chapter is to develop the AAI for India using micro-level data, including the recently released United Nations Population Fund (UNFPA) Ageing in India data.

20.3.2 Domains of Active Ageing Index in the Indian Context

Several studies have highlighted that gainful employment, social participation, independent living and health status are crucial aspects of active ageing. However, the workforce participation of older people continues to be a neglected area in India, with very limited data on employment of older Indians. One of the few studies available, by Selveraj et al. (2011), stated that the overall workforce participation rate declines significantly from age 58 and above, as 58 or 60 years of age is the stipulated retirement age for most Indian states. However, this decline depends on the size of the informal sector, in which the concept of retirement is irrelevant. In addition, the postretirement workforce participation rate significantly varies between the more populated age groups of 60–64, followed by 65–69 and finally 70+. A significant finding is that a substantial proportion of the oldest old (defined as people aged 70 and above in India) are economically active and mostly employed in the informal sector, which is characterised by job insecurity, insufficient wages, passive physical stress and discriminatory working atmospheres (Selveraj et al. 2011). Based on this evidence, we feel that employment in later life is not viewed positively due to the country's early retirement policies. This is a concern in countries like India, where there is inadequate formal economic support and a weak public pension system. Older people might be forced to work, irrespective of their health status, in an ageist informal working atmosphere and negative policy environment.

20.3.3 Critical Evaluation of the Construction of the Active Ageing Index

Development of a valid and accurate AAI is a challenge, as several researchers have shown that lay definitions of active ageing, successful ageing and quality of life have considerable overlap (Bowling 2009). Furthermore, any slight modification in the dimensions of an AAI could result in significant changes in outcomes. Although it is well established that physical health and functioning, social engagement and mental health should be integrated into active ageing, most research quantifying the active ageing framework has focused on one indicator only.

For example, several researchers on active ageing tend to focus only on the economic activities of older people, even for countries with excellent formal support. Such policies, by default, provide limited focus on enhancement of quality of life (Bowling 2009). In contrast to the previously constructed active ageing measures, the AAI happens to be the only index that captures the three pillars of active ageing framework introduced by WHO (refer to the Fig. 20.1). The AAI, like other contemporary research, sheds positive light on the employment of older people as supported by Boudiny (2012). In addition, unlike other unidimensional research focusing on health or economic domains, the AAI is a holistic measure that clearly distinguishes between economic and non-economic concepts. Such inclusion of non-economic indicators in the construction of the AAI is extremely important from both a policy and layperson's perspective.

20.4 Data Source and Methodology

As mentioned, the AAI captures the four domains of active ageing: participation in productive employment; participation in society; independent, healthy and secure living; and a friendly environment for active ageing. The basic criteria for choosing these domains were to reflect actual experiences of active ageing and the capacity for it. The first three domains reflect the experiences of older people, while the fourth domain captures the enabling macroenvironment.

Our analysis could not follow the AAI UNECE methodology completely, due to a lack of data for a couple of sub-domains (refer to the Table 20.1). In these cases, proxies were used which resulted in a slightly modified version of the AAI. In the domain of participation in society, due to a lack of data on the elderly taking care of adults, that sub-domain was dropped and the remaining three sub-domains were given equal weight. Also, due to an unreliable source of data on income in India, relative median monthly expenditure is taken as a proxy. Moreover, the sub-domain of physical exercises included practising yoga, as yoga is commonly practised by Indian older people more so than jogging, swimming or visiting a gym. In addition, we included spirituality and religion, as we felt that in India they play an important role in active ageing.

We also had to exclude sub-domains like life-long learning, the use of ICT and share of healthy life expectancy due to a lack of data, and weighting has been equally distributed among the other subcomponents of each domain. Most of the analysis is restricted to those 60 and older, taking into consideration lower life expectancy in India compared to the EU. Our analysis did not include the oldest old due to a lack of a reliable source of data. This raised a few concerns on the applicability of such a complex index in developing countries, where data on older people is limited. However, we felt that we should construct the AAI using every possible reliable source (and proxy where required), rather than "shelving" the index calculation. By highlighting the data issues, we can make a case for additional surveys on Indian older people that can help us to create a positive policy environment.

We computed the AAI for India and seven major states that have a higher proportion of older people. Most of the analysis included UNFPA Ageing data (UNFPA 2012), which was collected to develop a knowledge base with regard to demographic, social and economic conditions, health needs and living arrangements and entitlements. For sampling purposes, data is collected from households in states with a higher proportion of the elderly, that is in Kerala, Tamil Nadu, Maharashtra, Himachal Pradesh, Punjab, Orissa and West Bengal. The sampling frame of the

Table 20.1 Illustration of the AAI calculation for India

Employment	Participation in society	Independent, healthy and secure living	Capacity and enabling environment for active ageing
Employment rate 55–59*	Voluntary activities*	Physical exercise **(Includes mainly yoga)***	Remaining life expectancy at age 55 Sample Registration System#
Employment rate 60–64*	Care to children and grandchildren*	Access to health and dental services*	Share of healthy life expectancy at age 55 excluded due to lack of data
Employment rate 65–69*	Care to older adults excluded due to lack of data*	Independent living*	Mental well-being*
Employment rate 70–74*	Political participation*	Financial security (three indicators) **Relative median monthly expenditure rather than median income+**	Use of ICT excluded due to lack of data
		No poverty risk+	Social connectedness*
		No material deprivation (at age 60)*	Educational attainment*
		Physical safety excluded due to lack of data	
		Lifelong learning excluded due to lack of data	
<u>Weights</u>	<u>Weights</u>	<u>Weights</u>	<u>Weights</u>
Employment rate 55–59 (25)	Voluntary activities (35)	Physical exercise (10)	Remaining life expectancy at age 55 (41)
Employment rate 60–64 (25)	Care to children and grandchildren (35)	Access to health services (30)	Mental well-being (24)
Employment rate 65–69 (25)	Political participation (30)	Independent living (30)	Social connectedness (21)
Employment rate 70–74(25)		Relative median monthly expenditure (10)	Educational attainment (14)
		No poverty risk (10)	
		No material deprivation (10)	

Source: Authors
* data is taken from the UNFPA 2011 data
+ represents National Sample Survey (NSS) 2010 data
# represents Registrar General of India (RGI) 2011 data

survey ensures that the data is representative at the state level. In addition to the UNFPA data, Sample Registration System data from Registrar General of India 2011 and National Sample Survey 66th round data (2010) were used in this study to calculate domains of capacity building, health and secure living. Using the individual level data, overall and domain-specific AAI was constructed by gender and state.

20.5 Results and Discussion

20.5.1 Overall and Domain-Specific AAI by State and Gender

Table 20.2 provides the overall AAI by state and sex. The results show regional and gender inequality. Kerala (a southern Indian state), Maharashtra (a western Indian state), and Himachal Pradesh (a north Indian state) performed well. Tamil Nadu scored the lowest AAI score and also had the lowest gender gap. High performing states had the higher gender gap. To study the contribution of the domains to the overall index, domain-specific scores by state and sex were calculated (Tables 20.3, 20.4, 20.5 and 20.6). These domain-specific analyses show that certain states in India performed well on the participation in society domain, but performed poorly on the domains of labour force participation and independent living. This is due to a lower proportion of older people with financial

Table 20.2 Estimate of Active Ageing Index by states

Active Ageing Index (overall)	Total		Male		Female	
States	Value	Rank	Value	Rank	Value	Rank
Himachal Pradesh	36.40	1	40.12	1	31.69	1
Maharashtra	35.60	2	38.65	2	30.71	2
Kerala	34.72	3	37.51	3	28.80	3
Punjab	33.07	4	34.89	4	27.19	4
Orissa	28.86	5	31.53	5	25.16	5
West Bengal	27.44	6	30.77	6	23.30	6
Tamil Nadu	25.35	7	26.10	7	22.78	7
India	31.60		32.69		28.04	

Source: Author's calculations using UNFPA 2012, NSS 2010 and SRS 2011

Table 20.3 Estimate of Workforce Participation Index by states

Workforce participation	Total		Male		Female	
States	Value	Rank	Value	Rank	Value	Rank
Himachal Pradesh	4.78	7	8.58	5	0.86	7
Punjab	5.31	6	9.71	4	1.24	6
West Bengal	5.94	4	10.12	3	2.30	3
Orissa	6.47	2	10.71	2	2.30	3
Maharashtra	8.95	1	12.04	1	6.17	1
Kerala	5.98	3	6.00	7	1.88	5
Tamil Nadu	5.40	5	6.61	6	3.78	2
India	5.75		4.89		2.65	

Source: Author's calculations using UNFPA 2012

Table 20.4 Estimate of Social Participation Index by states

Participation in society	Total		Male		Female	
States	Value	Rank	Value	Rank	Value	Rank
Himachal Pradesh	66.07	1	68.88	1	63.17	1
Punjab	56.36	4	56.71	4	56.03	4
West Bengal	49.77	6	53.79	5	46.22	6
Orissa	51.00	5	52.67	7	49.32	5
Maharashtra	62.44	2	66.84	2	58.47	2
Kerala	59.69	3	64.44	3	56.31	3
Tamil Nadu	38.48	7	37.92	6	38.96	7
India	54.89		57.37		52.65	

Source: Author's calculations from UNFPA 2012

Table 20.5 Estimate of Independent Living Index by states

Independent living	Total		Male		Female	
States	Value	Rank	Value	Rank	Value	Rank
Himachal Pradesh	46.79	3	47.70	1	27.62	3
Punjab	45.94	4	43.59	5	30.10	2
West Bengal	41.85	7	40.15	7	27.26	4
Orissa	44.41	6	43.38	6	24.59	7
Maharashtra	46.97	2	45.95	3	26.52	6
Kerala	44.78	5	44.29	4	27.21	5
Tamil Nadu	47.28	1	46.54	2	30.67	1
India	45.69		44.76		28.24	

Source: Author's calculations from UNFPA 2012 and NSS 2010

Table 20.6 Estimate of Capacity Index by states

Capacity	Total		Male		Female	
States	Value	Rank	Value	Rank	Value	Rank
Himachal Pradesh	34.62	2	41.18	2	32.59	1
Punjab	34.44	3	36.43	3	20.68	6
West Bengal	18.77	7	21.93	7	17.97	7
Orissa	21.56	6	25.06	6	23.19	5
Maharashtra	29.55	4	32.22	4	27.16	3
Kerala	36.31	1	42.15	1	28.57	2
Tamil Nadu	26.33	5	29.31	5	23.76	4
India	29.03		32.10		29.31	

Source: Author's calculations from UNFPA 2012 and SRS 2011

security and employment. While Maharashtra state, that includes the financial capital Mumbai, ranks highest in the labour force participation in later life (81%), Kerala ranked the lowest for male labour force participation, despite being the state with highest education and lowest inequality. In the participation domain of the AAI, all states performed very well, with Himachal Pradesh ranking the highest and Tamil Nadu ranking the lowest. These regional differences could be attributed to migration patterns and the specific historical development of states. Though Kerala, West Bengal and Orissa have high migration rates in younger ages, their socio-economic conditions and patterns of migration differ. Himachal Pradesh, Kerala and Maharashtra ranked above the national overall AAI score. Tamil Nadu, a southern Indian state with high education and moderately high GDP, has performed poorly, which was unexpected.

The gender-specific analysis did not align with typical gender inequalities shown using indicators such as overall sex ratio and financial stability in later life. Despite poor performance of females in employment, independent living and capacity for active ageing, the overall AAI for males and females resulted in a similar score, due to higher social participation rates among older females in India. This complex mismatch in dimensions and results has also been observed in earlier studies that have used other measures of active ageing. Paul et al. (2012) highlighted the complexity in the concept of active ageing due to its multidimensional nature. Since the AAI is a more composite version of the WHO concept, it is observed that a more dominant role is being played by functional health and crude economic participation compared to other quality-based dimensions. Based on these results, we believe that the AAI is an effective tool to capture both inter-state and gender differences in India. However, we strongly recommend to look at the sub-domains to understand their contribution to the overall AAI.

20.5.2 Comparison of the AAI Indian Score with European Scores

After the construction of the AAI for Indian states, we compared the Indian AAI with the EU indices. Table 20.7 shows the EU countries by their AAI rank in ascending order (Zaidi 2014b). Such comparisons not only act as a validation tool but also provide in-depth information on comparability of the AAI sub-domains. The overall AAI for India is closer to the values of Hungary, Greece and Slovakia. The western state of Maharashtra has scored similar to Bulgaria, Latvia and Greece, while the northern states of Himachal Pradesh and Punjab are performing closer to the German level. The southern state of Kerala has a similar performance to Cyprus. However, the AAI values for the east Indian state of West Bengal and the southern state of Tamil Nadu are closer to the AAI of Poland. This raises a conceptual question: if older people in Cyprus and Kerala have similar scores, can we interpret that participation and contribution are similar? The domain-specific analysis shows that the overall scores of Cyprus and Kerala vary a lot. We observed that results differed drastically when we compare domain-specific scores of Indian states with domain-specific scores of the EU.

We feel that these index-specific similarities and domain-specific differences stem from sociocultural differences, lack of formal support in India and the nature of variables used in calculating the AAI, where the employment domain includes employment rate in various age groups. For example, the employment domain score ranges between 17.8 and 41.0 for EU countries despite including the oldest old, whereas for Indian states the score ranges between 4.8 and 5.4. Besides, India performed poorly in the domain of independent living and capacity for active ageing compared to the EU. However, in the social participation domain of the AAI, Indian states scored highly (between 39 and 66)

Table 20.7 Rank of EU countries as per AAI scores

Country	Value	Rank
Sweden	44.01	1
Denmark	40.18	2
Ireland	39.37	3
UK	39.26	4
Netherlands	38.93	5
Finland	38.84	6
Cyprus	36.32	7
Luxembourg	35.07	8
Germany	34.96	9
Austria	34.90	10
Czech Republic	34.29	11
France	34.20	12
Portugal	34.18	13
Belgium	33.48	14
Italy	33.28	15
Estonia	33.13	16
Spain	32.50	17
Lithuania	31.57	18
Malta	30.98	19
Romania	30.91	20
Slovenia	30.61	21
Latvia	30.19	22
Bulgaria	29.96	23
Greece	29.34	24
Hungary	28.23	25
Slovakia	27.73	26
Poland	27.32	27

Source: Zaidi 2014b

compared to the EU (between 12 and 25). Despite this domain-specific divergence between the EU countries and Indian states, the overall AAI score has converged.

Therefore, we recommend that a domain-specific investigation has to be carried out while engaging in a cross-cultural comparison. Based on the overall and disaggregated analysis, we believe that the AAI is indeed a dynamic index that identifies the regional variation within a country to some extent. We caution that the application of this index to a complex and diverse country like India has to take into consideration the context of the county and the formal support available. As overall index score can suppress the expression of domain-specific differences that are very relevant for policy purposes, we recommend that policy makers have to focus on domain-specific indicators, rather than basing their policy recommendations on the overall index.

20.5.3 Applicability of the AAI in Indian Setting

We feel that some of the sub-domain indicators might not be applicable to India, as its ageing experience is structurally different from those of Western countries. A very poignant argument in this context is that the sex ratio in older age is more in favour of males compared to even many developing nations of Eastern Europe and Central Asia. Furthermore, a huge statewise variation exists. As per the latest 2012 report of the UNFPA survey, there are 1021 females per 1000 males, but there are bigger states like Punjab where it is as low as 910 females per 1000 males. While as per the 2011 census, only the state of Kerala shows a favourable sex ratio for females (UNFPA 2012). Moreover, the domain participation—measured as voluntary activities by older people and care provided to grandchildren by older people—differs between Indian states and EU states. Thus, additional analyses are needed to study the labour market of older people, intergenerational solidarity and health status of older people in India, three main factors relevant for the AAI.

20.5.3.1 Employment in Later Life in India and Its Contribution to the AAI

When calculating the AAI, we have taken into consideration only those that worked more than three months in a year and were paid. This has resulted in poor performance of the economic domain of older people. However, when we look at all forms of employment (including seasonal and short-term employment) and all kinds of payment (including cash, cash in kind or no payment), the majority of older people work in India. Table 20.8 reveals interstate differences, with working proportion ranging between 81% in Maharashtra and 53% in West Bengal. As mentioned in the previous sections, the vast majority of older people in India receive no pension, and formal pensions when provided are meagre. The UNFPA

Table 20.8 Distribution of elderly by type of work for age group and states

	Type of current work/main occupation				
States	Not working	Public sector	Private sector	Informal	Total
Himachal Pradesh					
60–64	37.8	30.6	21.9	9.8	471
65–69	44.2	27.4	17.4	11.0	373
70–74	47.7	23.3	18.4	10.5	266
80+	46.2	17.7	23.1	12.9	372
Total	43.3	25.2	20.5	11.0	1482
Punjab					
60–64	46.9	11.2	15.6	26.3	437
65–69	47.5	11.7	14.0	26.9	394
70–74	44.0	12.5	17.9	25.7	257
80+	48.2	9.9	18.4	23.4	282
Total	46.8	11.3	16.1	25.8	1370
West Bengal					
60–64	40.0	11.7	17.2	31.1	453
65–69	46.0	10.0	18.2	25.8	341
70–74	44.7	9.8	17.5	28.1	235
80+	48.8	8.5	20.7	22.0	246
Total	44.2	10.3	18.2	27.4	1275
Orissa					
60–64	45.3	7.8	19.1	27.9	477
65–69	41.0	10.0	17.7	31.3	451
70–74	43.7	9.5	18.0	28.9	284
80+	41.3	9.3	17.8	31.6	269
Total	42.9	9.1	18.2	29.8	1481
Maharashtra					
60–64	21.4	5.6	31.6	41.4	519
65–69	17.9	5.8	38.7	37.6	452
70–74	15.7	5.0	38.4	40.9	242
80+	17.6	7.2	34.7	40.5	222
Total	18.8	5.8	35.5	40.0	1435
Kerala					
60–64	32.6	17.1	26.0	24.3	469
65–69	34.0	15.4	17.7	32.9	344
70–74	31.3	16.4	23.6	28.9	208
80+	35.8	12.5	21.2	30.5	344
Total	33.6	15.4	22.3	28.7	1365
Tamil Nadu					
60–64	31.7	2.1	23.1	43.1	707
65–69	21.1	4.8	16.2	57.8	351
70–74	19.6	5.4	13.2	61.8	204
80+	21.4	6.6	12.6	59.3	182
Total	26.1	3.8	18.7	51.4	1444

Source: Author's calculations from UNFPA 2012

report (2012) showed that 72% of male and 94% of female elderly did not receive any retirement or pension benefits in India. Old age and widow pensions were received only by 13.7% of males and 22.4% of female pensioners. Due to unavoidable dependency on agriculture and the informal sector, vulnerable older people are likely to have higher participation rates in the workforce. Putting it differently, older people in India, in the absence of formal and informal support, have to rely on work, with or without good health. Older men are more likely to work outside the home, whereas older women are more likely to play an important role in informal care.

South and west Indian states such as Maharashtra, Tamil Nadu and Kerala, that have higher literacy, employment rates and economic growth, exhibit a higher proportion of older working people, and this proportion decreases with increasing age in all states. This is a well-understood phenomenon, due to the progression of fragility and disability with age. The majority of older people in Tamil Nadu (51%) and Maharashtra (40%) rely on the informal sector, whereas a higher proportion of older people in Kerala rely on the public sector (15%). This may be mostly attributed to higher literacy among the elderly in the state of Kerala and a positive social policy environment. The results clearly portray the variations and the complexity of work in later life, reflecting the lack of formal and informal support. Moreover, inclusion of all forms of employment would have improved the AAI score. Further qualitative and quantitative data are required to better study the role of work in active ageing in India.

20.5.3.2 Intergenerational Support in India

Table 20.9 focuses on intergenerational living arrangements in India, which is actually a key indicator of solidarity among generations. Contrary to the work domain in India, the intergenerational support domain received very high scores for most of the Indian states. This is due to the higher weights provided to the caring of grandchildren in the participation domain of active ageing. The joint family system is a common practice in India, that encourages intergenerational exchange in care and support. The joint family model, defined as a group of people who generally live

Table 20.9 Distribution of households by intergenerational living arrangement and wealth status by age group

	Wealth status of elderly and children living together					
	Elderly households without children			Elderly households with children		
Age groups	Poor	Non-poor	Total	Poor	Non-poor	Total
60–64	56.73	43.27	1992	67.52	32.48	1804
65–69	51.2	48.82	1356	65.2	34.76	1565
70–79	52.5	47.55	1327	66.4	33.65	1453
80+	58.4	41.61	1327	68.9	31	498
Total	54.4	45.61	5247	66.7	33.35	5320

Source: Author's calculations from UNFPA 2012

under one roof, share food cooked at one hearth, hold property in common, participate in common family worship and are related to each other, is portrayed as a golden model that meets care requirements of older people in countries where formal support is limited. Mandelbaum (1957) argued that the joint family is more of a rural phenomenon than urban, and with increasing urbanisations these concerns seem to grow. The recently conducted UNFPA study supports this evidence, by showing that almost 70% of elders both in rural and urban areas reside with their spouse and children (UNFPA 2012). In rural areas, 41% of older couples lived in joint families, compared to only 39% in urban areas. However, more urban solo older people are likely to live with their children and grandchildren (33%) compared to rural areas (30%). Due to the joint family system and living together, the intergenerational domain of the AAI has scored higher, which resulted in a EU-level AAI score.

Given the importance of the intergenerational support domain, further analysis was carried out to critically evaluate the engagement aspect. The proportion of households where solo or couple older people reside with at least one young person (an individual below 15 years) was calculated by their poverty status (Table 20.9). Poor people are defined as the bottom two quintiles of wealth, whereas the top three wealth quintiles are classified as non-poor. This analysis is used as a proxy to show the quality of intergenerational solidarity, and not just mere prevalence in Indian context. A critical outcome of the analysis is that higher proportions of households where elderly and children are residing together are poor (i.e. around 67%), compared to households where they are not residing together (i.e. around 54%). This raises questions on how the domain of intergenerational solidarity should be captured especially in the context of India. For instance, in several cases, provision of care to grandchildren might be obligatory for poor older people due to a lack of formal support, and hence it cannot be interpreted as a contributory factor to active ageing. Moreover, it is likely that such care could increase health vulnerability among older people, if they are the sole care providers.

20.5.3.3 Unmet Need for Health Care in Later Life

Additional analysis on unmet need for health care (Table 20.10) shows that almost 17% of older people report unmet health care. In the UNFPA survey, the reason for not getting treatment for 20 various disease conditions was questioned. Unmet need includes responses related to affordability, accessibility and availability of health care. Further analysis by sex and marital status shows that a significant proportion of older females and a high proportion of currently single (i.e. widowed, never married or separated) elderly suffer from unmet health-care needs. Interestingly, socio-economically backward state like Orissa has lower unmet need among female elderly or currently single elderly (majorly female), as compared to a developed state like Kerala or Tamil Nadu. This inverse association stems from the

Table 20.10 Statewise distribution of elderly's unmet need of health care

	Age groups				Sex		Marital status		
States	60–64	65–69	70–79	80+	Male	Female	Currently married	Currently single	Total
Himachal Pradesh	23.3	24.5	32	20.2	44.7	55.3	54.6	45.5	253
Punjab	25.4	22.8	35.1	16.7	41.0	59.1	62.1	37.9	359
West Bengal	26.2	23.8	35.3	14.7	47.2	52.8	51.6	48.4	252
Orissa	25.1	27.3	33.5	14.1	51.5	48.5	66.5	33.5	227
Maharashtra	24.3	25.4	29.5	20.8	43.4	56.7	48.6	51.5	173
Kerala	24.4	19.9	36.6	19.1	41.5	58.5	50	50	246
Tamil Nadu	31	27.8	26.7	14.4	43.3	56.7	43.3	56.7	187

Source: Author's calculations from UNFPA 2012

lack of awareness about health and cultural setup in elderly women. It is likely that women in poorer states might be unaware of health-care needs and might report low levels of unmet need for health care. This enormous unmet need for health care clearly shows that the health domain of the AAI in countries like India requires further investigation.

20.6 Summary and Conclusion

The AAI quantifies the contributions of older people in different facets of their lives such as their participation in social, economic and political domains of society. One of the questions the chapter asked was the applicability of this widely popular European policy tool in the context of developing countries. The AAI, though accepted in Europe, has not yet managed to move beyond applied policy circles and would benefit from mainstream policy recognition. As Ney (2005) argues, the majority of policy discussions related to the inclusion of older people is merely rhetorical, and requires more applied perspective. Similarly, the active ageing agenda could become more rhetorical and could be ignored by mainstream policy makers in India. It is also likely that the between-country differences in the endorsement of the AAI could result in further uncertainty in implementing this index by non-EU nations. In India, where policies that support the health and well-being of older people are virtually non-existent, any endorsement of positive approaches to ageing could be simply rhetorical. Despite these challenges, emerging economies and developing countries might benefit from positive indices such as the AAI, that are not only multidimensional but also look at ageing in a positive way. Hence, the contribution of the AAI would make a significant impact on policy making in India by further enhancing the positive perspective.

Based on our results, we conclude that the AAI will help understand the contribution of older adults in India, by including several relevant dimensions including economic and social contribution. We hope this study will make a strong case for encouraging more positive approaches of ageing research in India. Our study shows state- and gender-level economic and social disparities in participation in India. We believe that one of the strengths is the critical evaluation of employment and social engagement domains. We feel that in developing countries like India, employment in later life could not always be interpreted as an active engagement, due to lack of formal support and prevailing ageism in the labour market. We also stress that the social engagement is not necessarily positive, as older people might be forced to participate in informal care.

The analysis also shows that it is important to look at sub-domain scores of the AAI, before making policy recommendations. This stems from the fact that, despite poor scoring in employment, independent living and capacity for active ageing, India has similar scores to the EU, due to higher social participation rate among the elderly. A similar bias was also noted when performing gender disaggregated analysis. We recommend that in-depth analysis has to be carried out to draw policy conclusions based on the AAI in developing countries including India. We conclude that older people play a vital role in India evidenced by the AAI application to this country. Older Indians contribute to economic, political and social domains, although they are ignored by the mainstream policies. By further filling the policy gaps, it is possible to further enhance the active engagement of older Indians.

References

Baltes, P. B., & Baltes, M. M. (1990). Psychological perspectives on successful aging: The model of selective optimization with compensation. In P. B. Baltes & M. M. Baltes (Eds.), *Successful aging: Perspectives from the behavioural sciences* (pp. 1–35). Cambridge: Cambridge University Press.

Beard, J., Biggs, S., Bloom, D., Fried, L., Hogan, P., Kalache, A., & Olshansky, J. (2012). *Global population ageing: Peril or promise?* Working Paper Series, No. 89. Retrieved from http://www.hsph.harvard.edu/pgda/working.htm

Boudiny, K. (2012). 'Active ageing': From empty rhetoric to effective policy tool. *Ageing & Society, 33*, 1077–1098.

Bowling, A. (2009). Perceptions of active ageing in Britain: Divergences between minority ethnic and whole population samples. *Age and Ageing, 38*, 703–710. https://doi.org/10.1093/ageing/afp175.

Butler, R., & Gleason, H. P. (1985). *Enhancing vitality in later life*. New York: Springer.

Chatterji, S., Kowal, P., Mathers, C., Naidoon, N., Smith, J. P., & Suzman, R. (2008). The health of aging populations in China and India. *Health Affairs, 27*(4), 1052–1063.

Cumming, E., & Henry, W. E. (1961). *Growing old* (p. 227). New York: Basic.

Datt, G., Ravallion, M., & Murgai, R. (2016). *Growth, urbanization, and poverty reduction in India.* Policy Research Working Paper, No. WPS 7568. Washington, DC: World Bank Group. Retrieved from http://documents.worldbank.org/curated/en/2016/02/25932338/growth-urbanization-poverty-reduction-india

Fernández-Ballesteros, R., Robine, J. M., Walker, A., & Kalache, A. (2013). Active aging: A global goal. *Current gerontology and geriatrics research.* Hindawi Publishing Corporation, ID 298012, 1–4.

Global Age Watch Index. (2015). Retrieved from http://www.helpage.org/global-agewatch/

GOI. (2011). *Census India, 2011.* Retrieved from http://censusindia.gov.in/

Hassan, M. (2007). Country statement of Bangladesh on 'High-level meeting on the regional review of the Madrid international plan of action on aging (MIPAA)', Macao, China. Retrieved from http://undesadspd.org.dnnmax.com/LinkClick.aspx?fileticket=likIb5EfS7w%3D&tabid=333

HelpAge India. (2013). *A report on Elder Abuse in India.* Retrieved from https://www.helpageindia.org/images/pdf/ElderAbuseIndia13.pdf

Jamuna, D., & Ramamurti, P. V. (2011). Perceptions of ageism across the generations. In L. McDonald & K. L. Sharma (Eds.), *Ageism and Elder Abuse.* India: Rawat Publication.

Macnicol, J. (2006). Analysing ageism and age discrimination. In *Age discrimination: Historical and contemporary analysis.* Cambridge: Cambridge University Press.

Mandelbaum. (1957). *System of Caste in India, (The), Part I. Social order. System of Caste in India, (The), Part I. Social order.* Berkeley: University of California.

Ney, S. (2005, Fall). Active aging policy in Europe: Between path dependency and path departure. *Ageing International, 30*(4), 325–342.

Paul, C., Ribeiro, O., & Teixeira, L. (2012). Active ageing: An empirical approach to the WHO model. *Current gerontology and geriatrics research.* Hindawi Publishing Corporation.

Rajan, S. I. (2007). *Social security for the elderly: Experiences from South Asia.* India: Routledge.

Rajan, S. I., Mishra, U. S., & Sharma, P. S. (1999). *India's elderly: Burden or challenge?* New Delhi: Sage.

Rowe, J. W., & Kahn, R. L. (1987). Human aging: Usual and successful. *Science, 237*(4811), 143–149.

Rowe, J. W., & Kahn, R. L. (1997). Successful ageing. *Gerontologist, 37*, 433–440.

Selvaraj, S., Karan, A., & Madeshwaran, S. (2011). *Elderly workforce participation, wage differentials and contribution to household income*. Building knowledge base for ageing in India series-I Working Paper-4. UNFPA, New Delhi.

Tareque, M. I., Hoque, N., Islam, T. M., Kawahara, K., & Sugawa, M. (2012). Active aging index and healthy life expectancy in Bangladesh. *Population Association of America Conference 2012*, San Francisco.

UNECE. (2012). *Policy brief of ageing, No. 13*. Retrieved from http://www.unece.org/fileadmin/DAM/pau/age/Policy_briefs/ECE-WG.1.17.pdf

UNFPA. (2012). *Report on status of elderly in selected states of India, 2011*. New Delhi.

Walker, A. (2002). A strategy for active ageing. *International Social Security Review, 55*(1), 121–140.

WHO. (1990). *Healthy ageing, Technical Report*. Geneva: World Health Organization.

WHO. (2002). *Active aging. A policy framework*. Geneva: World Health Organization.

Zaidi, A. (2014a). Changing the way we age: Evidence on active ageing from Europe. *Professional Briefing, Social Science Evidence and the Policy Process: International Insights, 4*(2), 4–8.

Zaidi, A. (2014b). *Active Ageing Index for 28 European Union Countries*. UNECE. Retrieved from www.esn-eu.org/news/319/index.html

Zaidi, A., Gasior, K., Hofmarcher, M. M., Lelkes, O., Marin, B., Rodrigues, R., Schmidt, A., Vanhuysse, P., & Zolyomi, E. (2013). *Active Ageing Index 2012. Concept, methodology, and final results*. Research memorandum/methodology report. European Centre Vienna, March 2013. Retrieved from http://www.euro.centre.org/data/aai/1253897823_70974.pdf

Aravinda Meera Guntupalli works as a Senior Lecturer in Public Health at The Open University, UK. Her research since 2001 has focused on health inequalities across the life course. Her recent work is concerned with life course epidemiology, gender differentials in well-being, disability in later life and non-monetary indicators of poverty across the life course. She is currently chairing the Centre for Ageing and Biographical Studies research group at The Open University—a centre that was established in 1995.

Suchandrima Chakraborty is currently a Doctoral Student at the International Institute for Population Sciences, Mumbai. She has previously worked as Research Officer at Tata Institute of Social Sciences, Mumbai, in a joint project with London School of Economics. She was also a Visiting Research Scholar in the Department of Geo-demography at Charles University of Prague. Her interest areas are ageing, gender issues, mental health, economic deprivation and social inequality.

Index[1]

A

[1] Note: Page numbers followed by 'n' refers to notes.

A. Zaidi (eds.), *Building Evidence for Active Ageing Policies*,
https://doi.org/10.1007/978-981-10-6017-5

F

G

H

I

K

FSC
www.fsc.org
MIX
Papier | Fördert gute Waldnutzung
FSC® C083411

Kolibri 360 GmbH
Ferdinand-Jühlke-Straße 7
99095 Erfurt, Deutschland
produktsicherheit@kolibri360.de